Bykov, Tsybenova, Yablonsky
Chemical Complexity via Simple Models
De Gruyter Graduate

Also of Interest

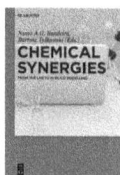

Chemical Synergies.
From the Lab to In Silico Modeling
Bandeira, Tylkowski (Eds.), 2018
ISBN 978-3-11-048135-8, e-ISBN 978-3-11-048206-5

Computational Sciences.
Ramasami (Ed.), 2017
ISBN 978-3-11-046536-5, e-ISBN 978-3-11-046721-5

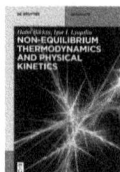

Non-equilibrium Thermodynamics and Physical Kinetics.
Bikkin, Lyapilin, 2015
ISBN 978-3-11-033769-3, e-ISBN 978-3-11-033835-5

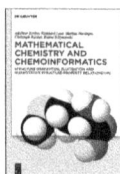

Mathematical Chemistry and Chemoinformatics.
Structure Generation, Elucidation and Quantitative Structure-Property
Relationships
Kerber, Laue, Meringer, Rücker, Schymanski, 2013
ISBN 978-3-11-030007-9, e-ISBN 978-3-11-025407-5

Valeriy I. Bykov, Svetlana B. Tsybenova,
Gregory Yablonsky

Chemical Complexity via Simple Models

MODELICS

DE GRUYTER

Authors

Prof. Valeriy I. Bykov
Russian Academy of Sciences
Emanuel Institute of Biochemical Physics
Kosygin Street 4/11
119334 Moscow
Russian Federation

Dr. Svetlana B. Tsybenova
Russian Academy of Sciences
Emanuel Institute of Biochemical Physics
Kosygin Street 4/11
119334 Moscow
Russian Federation

Prof. Gregory Yablonsky
Saint Louis University
Department of Chemistry
3450 Lindell Blvd
St. Louis MO 63103
United States of America

ISBN 978-3-11-046491-7
e-ISBN (PDF) 978-3-11-046494-8
e-ISBN (EPUB) 978-3-11-046514-3

Library of Congress Cataloging-in-Publication Data
A CIP catalog record for this book has been applied for at the Library of Congress.

Bibliographic information published by the Deutsche Nationalbibliothek
The Deutsche Nationalbibliothek lists this publication in the Deutsche Nationalbibliografie;
detailed bibliographic data are available on the Internet at http://dnb.dnb.de.

© 2018 Walter de Gruyter GmbH, Berlin/Boston
Cover image: Pobytov/DigitalVision Vectors/gettyimages
Typesetting: le-tex publishing services GmbH, Leipzig
Printing and binding: CPI books GmbH, Leck
♾ Printed on acid-free paper
Printed in Germany

www.degruyter.com

Preface

The idea of writing this book was born in 2015 during the International Conference "Mathematics in (Bio)Chemical Kinetics and Engineering" (MACKiE-2015), held in Ghent (Belgium). Professors Valeriy Bykov and Grigoriy (Gregory) Yablonsky were colleagues who worked closely at the Boreskov Institute of Catalysis (Novosibirsk, Russia) for more than a decade in the 1970s to 1980s. They established the Siberian chemico-mathematical team – together with Alexander Gorban and Vladimir Elokhin. They co-authored many books and articles related to the area of the mathematical modeling of chemical processes. Dr Svetlana Tsybenova joined this activity later in the 1990s, enhancing its computational and applied aspects.

After graduating from Novosibirsk State University in 1968, Valeriy Bykov started his scientific career in the Department of Mathematical Modeling at the Boreskov Institute of Catalysis. Professor Mikhail Slin'ko and Dr Albert Fedotov were his scientific supervisors. In 1985, Valeriy Bykov received his degree of Doctor of Physics and Mathematics (Physical Chemistry) from the Institute of Chemical Physics in Chernogolovka.

Gregory Yablonsky also worked in the same Department of Mathematical Modeling at the Boreskov Institute of Catalysis (Novosibirsk), first as a post-graduate student and then as a researcher. Professor Mikhail Slin'ko was also his supervisor. In 1989, Gregory Yablonsky received his degree of Doctor of Science (Physical Chemistry) from the Boreskov Institute.

Svetlana Tsybenova graduated from Krasnoyarsk State Technical University in 1996 and received her PhD in Technical Sciences from the same university in 1999. In 2011, she received the degree of Doctor of Physics and Mathematics (Physical Chemistry) from Bashkir State University, Ufa; her being adviser was Professor Semyon Spivak.

In 1978, there was a remarkable moment in this story when a scientific delegation from the USA, the three prominent professors Rutherford Aris, Dan Luss, and Harmon Ray, visited the Boreskov Institute in Novosibirsk. This was a starting point for Soviet–American cooperation in mathematical chemistry. Unfortunately, this cooperation met many political obstacles. Nonetheless, it became a significant stimulus for a fruitful exchange of information and ideas.

Over the last 50 years, the main directions and approaches have been determined in mathematical chemistry, both theoretical and applied. Discoveries of new experimental facts, i.e., the rate of hysteresis, chemical oscillations, chaos, etc, created new challenges in decoding the complexity of chemical reactions. Batteries of mathematical models distinguished by the level of complexity and assumed factors have been developed for imitating complex chemical behavior.

"Battery of models", "zoo of models", or "market of models" – different metaphors can be used. Nevertheless, the real alternative in contemporary modeling is between the model taken from the "market" and that produced by the individual "tailor".

https://doi.org/10.1515/9783110464948-201

Certainly, a suit from the tailor is more elegant; however, it is much cheaper and faster to buy a suit in a supermarket and adapt it if necessary.

An optimal strategy of modeling can be formulated as follows:

1. to develop typical ("simple") models for describing the phenomena of our interest;
2. to adapt them to concrete phenomena or processes.

A special question arose: what is the *simplest* model to describe newly discovered critical phenomena? This book is devoted to *basic models*, which can be used as building blocks for constructing the mathematical models of complex chemical processes. We call the methodology of selecting and analyzing these models *"modelics"*. Our book is focused on *simple nonlinear models*.

Generally, the concepts of "simplicity" and the "simple model" are complex. Einstein's advice was: "Make everything as simple as possible but not simple." On the other hand, Leonardo da Vinci said: "Simplicity is the ultimate sophistication." So, when working with simplicity, we move through the "gray zone" between science, art, and philosophy, and the inscription on the gates is: "Less is more!"

The authors express their gratitude to the colleagues who provided them with help at various times and in different situations: Professors Sergey Varfolomeev, Bair Bal'zhinimaev, Alexander Gorban', Semyon Spivak, Aizek Volpert, Konstantin Shkadinskii, Sergey Reshetnikov, Georgij Malinetskii, and Zulhair Mansurov.

Finally, we would like to thank our beloved ones for their support and understanding.

Valeriy Bykov, Moscow, 2017
Svetlana Tsybenova, Moscow, 2017
Gregory Yablonsky, St. Louis, 2017

Contents

Part III: **Modelics everywhere**

Part I: **General part**

1 Introduction. How to describe complex processes using simple models: Modelics

In contemporary scientific folklore, there is a legend which is transferred from one book to another one. At the end of the 1950s there was a meeting organized on an important and highly secret project, say on the possible constructing of military bases on the Moon. One intellectual from the Rand Corporation ("highbrow" or even "egghead") presented an invited talk involving many formulas and numbers. A general from the Pentagon interrupted him: "Excuse me, sir! What is a source of your information?" – "What do you mean, sir?" "How do you know all these estimates, equations, numbers etc." – The scientist replied immediately: "Sir, we have a model!", and the general was completely satisfied by this statement.

1.1 Model...modeling...

Today, on August 30, 2017, Google presented these numbers

Model	5,600,000,000
Modeling	553,000,000
Energy	1,690,000,000
Force	1,600,000,000
Physical model	650,000,000
Physical energy	819,000,000
Physical force	41,700,000
Mathematical modeling	4,870,000
Mathematical model	10,100,000

Clearly "*modeling*" and "*model*" are terms extremely popular in science and engineering.

It is easy to define modeling as the study of processes using models. However, what is a *model* in "hard sciences"?

There are many definitions of this term which pretend to be rigorous to some extent. We prefer the following one:

Object **M** is a model of object **A** with respect to a certain group of characteristics (properties), if **M** is constructed (or chosen) to simulate **A** in accordance with these characteristics.

In physics, chemistry, and biology ("hard sciences") and psychology and sociology ("soft sciences"), mathematical models are symbolic descriptions which represent the different dependencies of process characteristics or/and material properties in terms of controlled parameters (temperature, pressure, composition, electrical con-

https://doi.org/10.1515/9783110464948-001

ductivity, etc.). A mathematical model can be a number, a geometrical image, a function, a set of equations, etc.

What should we know before "constructing" any model? First, its basic elements, secondly, its main principles and laws, and thirdly, the algorithm for the model construction.

"Complex" and "simple" are key words in the development of models. The word "complex" comes from the Latin *complexus*, past participle of *complecti* (to entwine, encircle, compass, infold), from *com* (together) and *plectere* (to weave, braid). This concept reflects the multilevel and multicomponent structure of the world. The concept of "simplicity" is deep as well. There are different meanings of this term, positive and negative. St. Augustine said "Ignorance and stupidity are given the names of simplicity and innocence". Etymologically, "simple" originated in the medical science of the Dark Ages as related to a medicine made from one constituent, especially from one plant.

In contemporary science, "simplicity" is about the ability to understand or explain in an easy way with a minimum of assumed concepts, and, finally, about elegancy and parsimony. When we are talking about simplicity, we always remember "Occam's razor", the principle of simplicity: "the simpler explanation is usually better" (William of Ockham was a Franciscan friar, a philosopher of the 14th century). It is not true in general, however it is a good starting point of reasoning.

In real science and modeling, the "Holy Grail" is the model which represents an efficient compromise between "complexity" and "simplicity": Simple, but not too simple; complex, but not too complex. This compromise is determined by two primary characteristics of the model, i.e., its goal and its number of assumed variables.

1.2 Top-down and bottom-up

In modeling, two different strategies can be distinguished, *top-down* and *bottom-up*.

The top-down approach became possible and popular since the start of the computer era. In this approach a "large" complex model constructed, say via combinatorial methods, is decomposed into "small" simple submodels in accordance with some hierarchy. In chemistry and chemical engineering, the kinetic model is a foundation of the mathematical modelling of chemical reactions, reactors, and processes [1–3]. There exists a hierarchy of models of complex catalytic process: kinetic model, catalyst pellet model, catalyst bed model, contact reactor model, aggregate model, and, finally, model of the chemical plant. In this hierarchy of models, the kinetic model is the first level. None of the calculations that are of interest for chemical technology are carried out without kinetic models. For dynamic (kinetic) models of physicochemical processes, the basic elements are chemical substances and elementary acts; the main laws are the mass-action-law and surface-action-law; one of the algorithms for model construction is the quasi-steady-state method. Later we'll explain it in more detail.

The bottom-up approach is opposite to the *top-down* one. It starts from a "seed" simple submodel, combining it with another simple model "model-by-model"), and, finally, developing the model of a complex multilevel material or process.

Summing up, in the top-down modeling, simple models are obtained by the decomposition of a complex model. Such models approximate complex behavior in a certain parametric or temporal domain. The bottom-up methods take simple submodels as the initial ones. In any case, in both approaches, *top-down* or *bottom-up*, the simple models are unavoidable elements of modeling. Moreover, it is obvious that the simple models are more reliable for wide application in the "modeling industry" than the complex ones because of the amount and quality of information. In some situations, simple models exhibit very complex properties. However, knowing the properties of simple models we will be able to achieve an understanding which is a final goal of modeling, not just a calculated number.

Simple nonlinear models are the center of our interest. These models must reflect the main features of the chemical system studied, with the goal "not numbers, but understanding". For example, if the rate hysteresis is experimentally observed, the corresponding simplest kinetic model must be nonlinear and have a dimension of 2 including the special nonlinear term of "xy" type.

If the chemical self-oscillations of the chemical rate are found, the simplest mathematical model must be nonlinear as well and have a dimension of 3. If complex spatiotemporal structures occur, it is necessary to use a "reaction–diffusion" model with mass-action-law nonlinearity.

Our experience shows that with a well-developed system of basic models, it is much easier to construct and understand the specific mathematical model of the real process.

That is why this book is focused on the analysis of typical simple models using a special term, *modelics*, for modeling via simple models.

Bibliography

[1] Aris R. Introduction to the analysis of chemical reactors. Englewood Cliffs, NJ, USA, Prentice-Hall, 1965.
[2] Yablonskii GS, Bykov VI, Gorban AN, Elokhin VI. Kinetic models of catalytic reactions. Amsterdam, Oxford, New York, Tokyo, Elsevier, 1991.
[3] Marin GB, Yablonsky GS. Kinetics of chemical reactions. Decoding complexity. NJ, USA, Wiley-VCH, 2011.

2 Categorization of models

2.1 Physical framework of model design

Models of closed, open and semiopen systems.
Local and global models.
Time in modeling. Steady-state and non-steady-state models

Models differ by the factors and processes which they reflect. In fact, the goal of a model is a description of the specific processes based on reasonable assumptions and considering certain factors.

In model design, the key words are: *open* and *closed*; *local* and *global*; *steady state* and *non-steady-state*. As known from thermodynamics, systems can be classified as either open or closed, depending on whether there is exchange of matter with the surroundings. Closed systems can exchange energy with the surroundings, but they cannot exchange matter, while open systems can exchange either matter and energy or only matter. Semiopen (or semiclosed) systems also exist, in which only some type of material is exchanged with the surroundings. In chemical kinetics and engineering, the closed reactor is better known as the batch reactor and the open reactor as the continuous-flow reactor. In the pulse reactor, a small quantity of a chemical substance is injected into the reactor.

The general equation, which reflects the material balance for any component in any system, open or closed, can be represented qualitatively as follows:

$$\begin{matrix} \text{temporal change of} \\ \text{amount of component} \end{matrix} = \begin{matrix} \text{transport} \\ \text{change} \end{matrix} + \begin{matrix} \text{change due to} \\ \text{reaction} \end{matrix} \qquad (2.1)$$

in which the temporal change of the amount of component, often termed *accumulation*, is its change with respect to time at a fixed position, the transport change is the change caused by motion of the component and the reaction change is the change caused by chemical reaction. It is the model of a non-steady-state process (*non-steady-state model*), $dc_i/dt \neq 0$, where c_i is the concentration of i-th component, t is time.

If the temporal change is assumed to be zero, $dc_i/dt = 0$, the differential (2.1) is transformed to an algebraic equation. It becomes *the steady-state model*, i.e., the model of the steady-state process.

Rigorously speaking, (2.1) is the so-called *continuity equation* (see the classical monograph by Bird, Stewart, and Lightfoot [1]) with two terms, which are the "transport term" and "chemical term", respectively. All isothermal models represent different cases of this continuity equation. The chemical term is *local*, reflecting the changes at the given place of space. The transport term is *global*, corresponding to the exchange between different places of space. Equation (2.1) can be used for the classification and

https://doi.org/10.1515/9783110464948-002

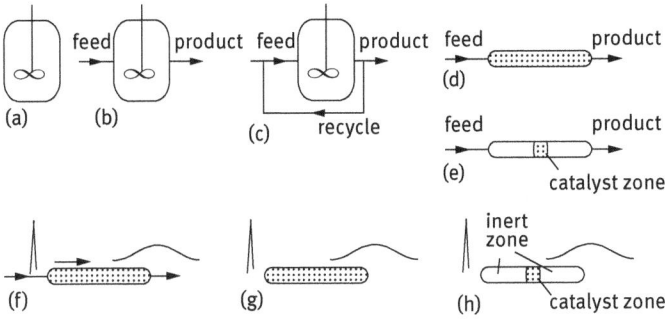

Fig. 2.1: Reactors for kinetic experiments. (a) batch reactor; (b) continuous stirred-tank reactor; (c) continuous-flow reactor with recirculation; (d) plug-flow reactor; (e) differential plug-flow reactor; (f) convectional pulse reactor; (g) diffusional pulse reactor or TAP reactor; (h) thin-zone TAP reactor

qualitative description of different types of systems and reactors [2–5]. All isothermal models represent different cases of this continuity equation.

Figure 2.1 shows schematic representations of several reactor types which are widely used for purposes of chemical engineering. These reactors are supplied by typical simple models with corresponding names, and these models are models of applied kinetics

2.1.1 Models of transport

Typically, transport processes are quite complicated, including at least two types of processes: convection and diffusion.

For convection, the molar flow rate F_i (mol $\cdot$ s^{-1}) of a component i is determined as the product of the total volumetric flow rate q_V (m$^3 \cdot$ s^{-1}) and the concentration of the component c_i (mol $\cdot$ m^{-3}):

$$F_i = q_V c_i \,. \tag{2.2}$$

For diffusion, in the simplest case the molar flow rate of a component is determined in accordance with Fick's first law:

$$F_i = -D_i A \frac{dc_i}{dz} \,, \tag{2.3}$$

where D_i is a diffusion coefficient (m$^2 \cdot$ s^{-1}), A is the cross-sectional area of the reactor available for fluid flow (m^2), and z is the axial reactor coordinate (m).

Pure convection or pure diffusion are examples of well-defined regimes. These hydrodynamic regimes with their corresponding mathematical descriptions are used as "measuring sticks" for extracting the intrinsic kinetic dependencies of chemically active materials, adsorbents, catalysts, membranes, etc.

In the model describing a batch reactor, the transport change term is completely absent. In perfectly mixed convectional systems and reactors, the "transport change" can be represented as the difference of convectional molar flow rates, $q_{V0}c_{i0} - q_V c_i$, where q_{V0} and q_V are the inlet and outlet volumetric flow rates and c_{i0} and c_i are the inlet and outlet concentrations, or $q_V(c_{i0} - c_i)$ if $q_V = q_{V0}$.

In purely diffusional systems and reactors, the "transport change" in the simplest case can be represented as the difference between diffusional flow rates in and out, F_{i0} and F_i. Both flow rates are written in accordance with Fick's first law:

$$F_{i0} = -D_i A \frac{dc_i}{dz}\Big|_z , \qquad F_i = -D_i A \frac{dc_i}{dz}\Big|_{z+\Delta z} . \tag{2.4}$$

Then,

$$F_{i0} - F_i = \left(-D_i A \frac{dc_i}{dz}\Big|_z\right) - \left(-D_i A \frac{dc_i}{dz}\Big|_{z+\Delta z}\right) = D_i A \frac{d^2 c_i}{dz^2}\Delta z . \tag{2.5}$$

2.1.2 The batch reactor

In an ideal batch reactor, i.e., a non-steady-state closed reactor with perfect mixing, (2.1) becomes

$$\begin{array}{ccc} \text{temporal change of} & = & \text{change due to} \\ \text{amount of component} & & \text{reaction} \end{array} \tag{2.6}$$

The simplest mathematical model for the temporal change of any component in a batch reactor of constant reaction volume is

$$\frac{dc_i}{dt} = R_i = v_i r , \tag{2.7}$$

where R_i is the net rate of production of the component per unit reaction volume (mol m^{-3}s^{-1}), v_i is the so-called stoichiometric coefficient, and r is the reaction rate (mol m^{-3}s^{-1}).

For a reversible reaction, the reaction rate is a combination of the rates of the forward and reverse reactions:

$$r = r^+ - r^- . \tag{2.8}$$

The state in which $dc/dt = 0$ is called equilibrium. In this state $r = r^+ - r^-$ and $r^+ = r^-$.

2.1.3 The continuous stirred-tank reactor

A continuous stirred-tank reactor (CSTR) is an open reactor with perfect mixing (gradientless reactor) and only convective flow. This mixing can be achieved not only by internal but also by external recirculation. The material balance for any component in a non-steady-state CSTR can be written as

$$\frac{dc_i}{dt} = R_i + \frac{q_{V0}c_{i0} - q_V c_i}{V} \tag{2.9}$$

with V the reaction volume (m^3). At steady state, the net rate of production of component i can be determined from

$$R_i = -\frac{q_{v0}c_{i0} - q_v c_i}{V} .$$ (2.10)

If $q_v = q_{v0}$, (2.10) can be expressed as

$$R_i = -\frac{q_{v0}(c_{i0} - c_i)}{V} = -\frac{c_{i0} - c_i}{\tau} ,$$ (2.11)

where $\tau = V/q_{v0}$ is the space time (s). It is denoted as space time because its definition involves a spatial variable, V, which distinguishes it from the "astronomic" time. It corresponds to the average residence time in an isothermal CSTR.

2.1.4 The plug-flow reactor

In an ideal plug-flow reactor (PFR), it is assumed that perfect uniformity is achieved in the direction perpendicular to that of the flow, i.e., in the radial direction. Axial diffusion effects are also neglected. The composition of the fluid phase varies along the reactor, so the material balance for any component must be made for a differential element:

$$dV\frac{dc_i}{dt} = R_i dV - q_v dc_i .$$ (2.12)

In a more rigorous form, (2.12) can be written as a partial differential equation:

$$dV\frac{\partial c_i}{\partial t} = R_i dV - q_v \frac{\partial c_i}{\partial z} dz .$$ (2.13)

Using $q_v = uA$ and $dV = Adz$, where u is the superficial fluid velocity (m s^{-1}), (2.13) can be written as

$$\frac{\partial c_i}{\partial t} = R_i - u\frac{\partial c_i}{\partial z} .$$ (2.14)

or

$$\frac{\partial c_i}{\partial t} + \frac{\partial c_i}{\partial \tau} = R_i$$ (2.15)

with $\tau = z/u$.

For the steady-state case, $\partial c_i/\partial t = 0$ and the model equation for an ideal PFR can be expressed by the ordinary differential equation

$$\frac{dc_i}{d\tau} = R_i ,$$ (2.16)

which remarkably is identical to the expression for a batch reactor, (2.7). The only difference is the meaning of the term time used. In the model for the batch reactor, the time is the time of the experimental observation or "astronomic time", whereas the time in the model for the plug-flow reactor is the space time, τ.

2.1.5 The pulse reactor

The pulse reactor is, by definition, the non-steady-state system. In a pulse reactor, which typically contains a fixed active zone, e.g., catalytic material, a small amount of a component is injected into the reactor during a small interval. In a conventional pulse reactor, the component is pulsed into an inert steady carrier-gas stream. The relaxation of the outlet composition following the perturbation by this pulse provides information about the mechanism of complex chemical process.

In the TAP reactor, created by John Gleaves in the 1980s [5], no carrier gas stream is used and the component is pulsed directly into the reactor. Transport only occurs by Knudsen diffusion, in which gas molecules collide with the wall, not with other molecules. In a thin-zone temporal-analysis-of-products reactor (TZTR) [6], the active material (catalyst) is located only within a narrow zone. The net rate of production in the catalyst zone of the TZTR is the difference between two diffusional flow rates at the boundaries of the thin active zone divided by the mass of catalyst in the reactor:

$$R_{W,i} = \frac{F_i(t) - F_{i0}(t)}{W_{cat}} . \tag{2.17}$$

This is analogous to the case of the steady-state CSTR, in which the reaction rate is given by the difference between convectional flow rates.

2.2 How to simplify complex models? Principles of simplification

This section is devoted to approaches of simplification of chemico-mathematical models. Many of them have been categorized in the recent monograph by Constales et al. [7] Most of this activity is performed for *models of detailed kinetics (microkinetics)*, i.e., models based on a detailed chemical mechanism.

In science and engineering, simplification is not only a method for the easy and efficient analysis of processes, but it also is a necessary step in understanding their behavior. In many cases, "to understand" means "to simplify". Now the main question is: "Which separate process or set of processes are responsible for the observed characteristics?" Frequently, simplification is defined as a reduction of the "original" set of system factors (processes, variables, parameters) to the "essential" set for revealing the behavior of the system, observed through real or virtual (computer) experiments. Every simplification must be correct. In physical science and chemical engineering, the answer to this question very much depends on the details of the reaction mechanism and on the temporal domain that we are interested in.

As a basis of simplification, many physicochemical and mathematical principles/methods/approaches or their efficient combination are used, such as fundamental laws of mass conservation and energy conservation, the dissipation principle, the principle of the detailed equilibrium, etc. Based on these concepts, many advanced methods of simplification of complex chemical models have been developed [3, 4].

In the mathematical sense, simplification can be defined as "model reduction", that is, the rigorous or approximate representation of complex models by simpler ones. For example, in a certain domain of parameters or times, a model of partial differential equations ("diffusion-reaction" model) is approximated by a model of differential equations, or a model of differential equations is approximated by a model of algebraic equations, etc. See also [8–11].

2.2.1 Physicochemical assumptions of simplification of chemico-mathematical models

Typically, assumptions are made on substances, on reactions and their parameters, on transport-reaction characteristics and experimental procedures as well. Presenting these assumptions, we follow the monograph by Constales et al. [7].

2.2.1.1 Assumptions on substances

1. Abundance of some substances in comparison with others, so their amount/concentration can be assumed to be constant during the process, either steady state or non-steady state. For example, in aqueous-phase reactions, the water concentration is often taken as a parameter in kinetic reaction models.
2. Insignificant change of some substance amount/concentration in comparison with its initial amount/concentration during a non-steady-state process. For example, in pulse-response experiments under high vacuum conditions in a temporal-analysis-of-products (TAP) reactor the total number of catalytic active sites is much larger than the amount of gas molecules injected in one pulse. Therefore, the concentration of active catalyst sites may be assumed to remain approximately equal during a pulse-response experiment.
3. Dramatic increase of the concentration/temperature at the very beginning of a process in a batch reactor or at the inlet of a continuous-flow reactor, typically is presented by a delta function or step function.
4. Complete conversion of some substances in time during the process or at the very end (the final section) of the chemical reactor.
5. Gaussian distribution of the chemical composition regarding some physicochemical properties, e.g., the molecular weight of polymers.
6. Assumptions on intermediates of complex chemical reactions:
    (a) Abundance of some intermediates. Frequently, the concentrations of many intermediates are very small compared to the concentrations of others. At the limit, only one intermediate dominates. For heterogeneous catalysis, the term "most abundant reaction intermediate" (MARI is used. This term introduced by Boudart means the only important surface intermediate on the catalyst surface under reaction conditions.

(b) Quasi-steady state (QSS) for some intermediates. Some intermediates can be in a quasi-steady state, or pseudo-steady state (PSS). In the literature, the meanings of the terms "pseudo" and "quasi" are a bit different. "Pseudo" is from Greek, meaning "false" or "not real(ly)" and is typically used for situations where deception is deliberate. "Quasi" is from Latin, meaning "almost", "as if" or "as it were". It is often used to describe something that for the most part behaves like something else, but not completely. Hence, we prefer to use "quasi" to describe this type of (non)steady state.

A quasi-steady-state assumption relates to reaction intermediates whose rate of change follows the time evolution of the concentrations of other species. Per the quasi-steady-state assumption the rates of production and consumption of intermediates are approximately equal, so their net rate of production is approximately equal to zero.

Two typical uses of the quasi-steady-state assumption are in:

1. Gas-phase chain reactions (e.g., oxidation reactions) are propagated by free radicals, that is, species having an unpaired electron (H, O, OH, etc.). The kinetic parameters of reactions in which these short-lived, highly reactive free radicals participate are much larger than the kinetic parameters of reactions involving other species. Their concentration in the quasi-steady state is necessarily small.

2. Gas-solid catalytic reactions occur through catalytic surface intermediates. These are not necessarily short-lived, but their concentrations are much smaller than the concentrations of reactants and products of the overall reaction. Therefore, the kinetic dependencies of the surface intermediates are governed by the concentrations of the gaseous species. A similar reasoning holds for enzyme-catalyzed biochemical reactions, in which the number of active enzyme sites is small compared to the number of substrate and product molecules.

2.2.1.2 Assumptions on (processes) reactions and their parameters

Assumption on irreversibility of processes (reaction steps):

1. All Processes (reaction steps) are irreversible, i.e., strong irreversibility or
2. Some processes (reaction steps) are irreversible, i.e., weak reversibility.

Rigorously speaking, all reaction steps are reversible. If the rate of the forward reaction is much larger than that of the reverse reaction, we consider the reaction step to be irreversible. If in a sequence of steps, say in a heterogeneous catalytic cycle, at least one step is irreversible, the overall reaction can be irreversible.

Assumption on "rate-limiting or rate-determining step":

In a sequence of reaction steps there usually are "fast" steps and "slow" steps. The kinetic parameters of the slow steps are much smaller than those of the fast steps, reversible or irreversible, and kinetic dependencies are governed by these small parameters. If there is only a single slow step, this is called the rate-limiting or rate-determining step. However, this is not a rigorous definition of the "rate-limiting-step" concept, which remains a subject of permanent fierce discussions [8, 10, 12]. In their paper [12], Kozuch, and Martin express a provocative opinion on this subject.

Assumption of "quasi equilibrium" or "fast equilibrium":
If in a sequence of steps both the forward and reverse reactions of some reversible steps are much faster than other reaction steps, the assumption can be made that the forward and reverse reactions of such fast steps occur at approximately equal rates, i.e., are at equilibrium. Typically, this assumption is justified by the fact that the kinetic parameters of these fast steps are much larger than the kinetic parameters of the other, slow steps. For many chemical systems, the assumption of quasi equilibrium is complimentary to the assumption of a rate-limiting step; if one step is considered to be rate limiting, other, reversible, steps can be assumed to be at equilibrium.

Assumption of equality or similarity of chemical activity:
Based on a preliminary analysis some groups of species with identical or similar chemical functions or activities can be distinguished, e.g., a family of hydrocarbons of similar activity can be represented by just one hydrocarbon. This is the so-called lumping procedure.

Additional assumptions on parameters:
1. Assumption of equality of parameters of some steps, e.g., kinetic parameters of some adsorption steps or even coefficients of all irreversible reactions are equal.
2. Assumption of "fast step", that is, the kinetic parameter of a certain step is assumed to be much larger than the kinetic parameters of other steps.
3. Assumption regarding the hierarchy of kinetic parameters, e.g., in catalytic reactions adsorption coefficients are usually much larger than the kinetic parameters of reactions between different surface intermediates.

Principle of critical simplification:
In accordance with this principle (Yablonsky et al., [13]), the behavior near critical points, for instance ignition or extinction points in catalytic combustion reactions, is governed by the kinetic parameters of only one reaction – adsorption for ignition and desorption for extinction – which is not necessarily the rate-limiting one.

2.2.1.3 Assumptions on transport-reaction characteristics
1. Assumption of continuity of flow. When a fluid is in motion, it must move in such a way that mass is conserved.
2. Assumption of uniformity of chemical composition, and/or temperature, and/or gas pressure in a chemical reactor.

3. Assumption of transport limitation, i.e., an assumption under which a model only comprising transport can be used (fast reaction and slow transport, in particular diffusion limitation).
4. Assumption of kinetic limitation, i.e., an assumption under which a model only including reaction can be used (fast advection or fast diffusion and slow reaction, kinetic limitation).

2.2.1.4 Assumptions on experimental procedures

1. Assumption of insignificant change of the system characteristics during an experiment involving a small perturbation of the system, i.e., a state-defining experiment.
2. Assumption of controlled change of the system characteristics during an experiment, i.e., a state-altering experiment.
3. Assumption of instantaneous change, i.e., instantaneous injection of a reactant into a chemical reactor.
4. Assumption of linear change of the controlled parameter, e.g., a linear temperature increase during thermodesorption.

2.2.1.5 Combining assumptions

It should be noted that some physicochemical assumptions are overlapping and some are complimentary. For example, if some steps are fast, we automatically assume that other steps are slow. In the simplest case – the two-step mechanism – the assumption of a fast first step is identical to the assumption of a rate-limiting second step. Assumptions on the abundance of species and rate-limiting steps can be made both for reversible and irreversible reaction steps. In contrast, the quasi-equilibrium assumption cannot be applied to a set of reaction steps that are all reversible. Sometimes many assumptions, not just one or two, are used for the development of a model. An example is the Michaelis–Menten model that is well known in biocatalysis. In this model,

1. The total amount of active enzyme sites is much smaller than the amounts of liquid-phase substrate (S) and product (P). Because of that quasi-steady-state behavior of the enzyme species (free enzyme E and substrate-bound enzyme ES) is observed.
2. The first step (E + S = ES) is assumed to be reversible, while the second step (ES $\rightarrow$ P + E) is assumed to be irreversible.
3. The kinetic parameters of the first step are assumed to be much larger than those of the second step, i.e., the first step is fast and the second step is slow.

Therefore, there are two simultaneous assumptions, i.e., the assumption of quasi-equilibrium of the first step and the assumption that the second step is rate-limiting.

2.3 Mathematical concepts of simplification in chemical kinetics

A primary analysis of different types of mathematical simplifications was done in the monograph by Constales et al. [7] In modeling, it is not enough to represent assumptions or simplifications expressed in a verbal way. Physicochemical assumptions have to be translated into the language of mathematics. In 1963, Kruskal [14] introduced a special term for this activity, "asymptotology". See Gorban et al. [8] for a detailed analysis. Mathematical models should be developed based on assumptions with a clear physicochemical basis. Every physicochemical assumption has a domain of its correct application, and this domain must be validated. Typically, this is done using the "full" model that includes the "partial" model, the validity of which is tested. The partial model is generated asymptotically from the full model and the correctness of this asymptotic procedure must be proven [10]. As stressed by Gorban et al. [8] "…often we do not know the rate constants for complex networks, and kinetics that is ruled by orderings rather than by exact values of rate constants may be very useful in practically frequent situations when the values of the various reaction constants are unknown or poorly known".

A mathematical analysis founded on the basic laws of physics, e.g., laws of thermodynamics, may provide us with an understanding of "tricks" of which the physicochemical meaning was previously unclear or even with a formulation of new fundamental concepts. The lumping procedure, a commonly-used approach to reduce the number of chemical species and reactions to be handled by grouping together species having similar chemical functions or activities into one pseudo-component or lump, was theoretically grounded and realized by Wei and Prater [15] and Wei and Kuo [16]. Complex chemical behavior that was discovered in chemical systems in the 1950s–1970s, such as bistability, oscillations, chaotic behavior, etc. has been understood only by transferring and adapting the concepts of the mathematical dynamic theory (stability, bifurcation, catastrophes, chaos, etc.). Maas and Pope [11] efficiently used the mathematical technique of manifolds for understanding combustion processes. At the same time, many mathematical tools applied to chemical problems still remain "purely mathematical", not having a special chemical content, e.g., many methods of statistical analysis, sensitivity analysis, etc.

2.3.1 Mathematical status of the quasi-steady-state (QSS) approximation

In chemico-mathematical modeling, revealing the rigorous mathematical status of the quasi-steady-state approximation was one of the most challenging problems. This assumption was introduced into chemistry at the very beginning of the twentieth century. However, it was clarified only about fifty years later via the mathematical theory of singular perturbations, and even now this knowledge is not sufficiently widespread within the chemical and chemical engineering community. One can say that this approximation is the most applied and the least understood. It can be called *"the most complicated simplification"*.

First, the quasi-steady state is not a steady state; it is a special type of non-steady state. The popular version of the quasi-steady-state approximation can be formulated as follows. During a chemical process, the concentrations of both species present in large amounts, usually the controllable and observed species, and species present in small amounts (intermediates such as radicals and surface intermediates), usually the uncontrollable and unobserved species, change in time. In the quasi-steady-state approximation, the concentrations of the intermediates become functions of the concentrations of the observed abundant species; they "adapt" to the concentrations of the observed species as if they were steady-state concentrations.

Within the traditional mathematical QSS-procedure, three steps can be distinguished:

1. Write the non-steady-state model, that is, a set of ordinary differential equations for both the observed species and the unobserved intermediates;
2. Then replace the differential equations for the intermediates with the corresponding algebraic equations by setting their rates of production equal to their rates of consumption, so that the net rate of production is zero, which in the case of catalytic surface intermediates translates into putting:

$$\frac{d\theta_j}{dt} = 0 \, ,$$

   where θ_j is the normalized concentration of surface intermediate j, and then solving these equations, such that the concentrations of intermediates are expressed as a function of the concentrations of the observed species and temperature. In fact, solving this set of equations is quite easy for linear models, but for nonlinear models this may not be so simple;
3. Finally, expressions for the reaction rates of the observed species can be constructed terms of the reactant and product concentrations of the overall reaction only.

The rigorous mathematical theory of quasi-steady-state approximation is the following. A complex reaction mechanism consisting of a combination of subsystems related to the observed variables x and unobserved variables y can be described by the general model:

$$\frac{dx}{dt} = f(x, \, y) \, ,$$

$$\frac{dy}{dt} = g(x, \, y) \, .$$

The subsystems are called subsystems of "slow" and "fast" motion, respectively. The mathematical validity of the quasi-steady-state approximation can be illustrated by scaling the original set of equations and writing it in dimensionless form as

$$\frac{d\bar{x}}{dt} = f(\bar{x}, \bar{y}),$$

$$\varepsilon \frac{d\bar{y}}{dt} = g(\bar{x}, \bar{y}),$$

in which ε is the so-called "small parameter" ($\varepsilon \ll 1$). At the limit $\varepsilon \to 0$, this system transforms into the so-called "degenerated" set of equations

$$\frac{d\bar{x}}{dt} = f(\bar{x}, \bar{y}),$$

$$0 = g(\bar{x}, \bar{y}).$$

Different actual systems generate the small parameter ε in different ways, and Yablonskii et al. [3] have indicated different scenarios for reaching quasi-steady-state regimes. For example, in homogeneous chain reactions, the small parameter is a ratio of rate coefficients. It arises because the reactions in which unstable and thus short-lived free radicals participate are much faster than the other reactions.

In heterogeneous gas-solid catalytic systems, the small parameter is the ratio of the total amount of surface intermediates $n_{t,int}$ to the total amount of reacting gas molecules $n_{t,g}$ which are present in the reactor. Summing up the theoretical analysis of the quasi-steady-state problem, we can distinguish two types of behavior:
1. a quasi-steady state caused by a difference in kinetic parameters (rate-parametric QSS);
2. a quasi-steady state caused by a difference in mass balances of species (mass-balance QSS).

2.3.2 Limits of simplification: optimal model

Every simplification has a limit. Please remember: Simple, but not too simple; complex, but not too complex. Obviously, the level of minimal complexity depends on the amount of available information. In the literature, the corresponding model is termed a "minimal", or "optimal", or "rational", or "skeleton" model. See the concept of the minimal mechanism described by Marin and Yablonsky [4]. The question answered was: "What is the minimum number of steps of the detailed mechanism?". Certainly, this number is not smaller than two because otherwise there would not be a catalytic cycle, which should include not smaller than two steps. E.g., for the Michaelis–Menten mechanism with one substrate (S) and one product (P), the minimal mechanism has two steps, not more. Then the properties of the minimal mechanism are summarized in a very simple way, depending on whether the number of reactant molecules in the overall reaction is larger or smaller than or equal to the number of product molecules. If the number of reactant molecules is bigger than the number of product molecules, the number of steps is equal to the number of reactant molecules. Similarly, if the number of product molecules is bigger than the number of reactant molecules, the number of steps is equal to the number of product molecules. Based on the minimal mechanism, the minimal kinetic model will be constructed.

There is no a general theory how to determine the "optimal model". In every concrete case, such a model is generated via special systematic studies.

Bibliography

[1]   Bird RB, Stewart WE, Lightfoot EN. Transport Phenomena. New York, USA, John Wiley Sons, 1960.
[2]   Aris R. Introduction to the analysis of chemical reactors. Englewood Cliffs, NJ, USA, Prentice-Hall, 1965.
[3]   Yablonskii GS, Bykov VI, Gorban AN, Elokhin VI. Kinetic models of catalytic reactions. Amsterdam, Oxford, New York, Tokyo, Elsevier, 1991.
[4]   Marin GB, Yablonsky GS. Kinetics of chemical reactions. Decoding complexity. NJ, USA, Wiley-VCH, 2011.
[5]   Gleaves JT, Ebner JR, Kuechler TC. Temporal analysis of products (TAP) a unique catalyst evaluation system with submillisecond time resolution. Catal Rev-Chem Eng 1988, 30, 49–116.
[6]   Shekhtman SO, Yablonsky GS, Chen S, Gleaves JT. Thin-zone TAP-reactor – theory and application. Chem Eng Sci 1999, 54, 4371–4378.
[7]   Constales D, Yablonsky GS, Thybaut JW, D'hooge DR, Marin GB. Advanced data analysis and modeling in chemical engineering. Amsterdam, Netherlands, Elsevier, 2017.
[8]   Gorban AN, Radulescu O, Zinovyev AY. Asymptotology of chemical reaction networks. Chem Eng Sci 2010, 65(7), 2310–2334.
[9]   Gorban AN, Karlin IV. Invariant manifolds for physical and chemical kinetics, Vol. 660 of Lect Notes Phys Springer, Berlin, Heidelberg. 2005.
[10] Gorban AN, Radulescu O. Dynamic and static limitation in multiscale reaction networks, revisited. Adv Chem Eng 2008, 34, 103–176.
[11] Maas U, Pope SB. Simplifying chemical kinetics: intrinsic low-dimensional manifolds in composition space. Combust and Flame 1992, 88(3), 239–303.
[12] Kozuch S, Martin JM. The rate-determining step is dead. Long live the rate-determining state! Chem Phys Chem 2011, 12(8),1413–1421.
[13] Yablonsky GS, Mareels IM, Lazman M. The principle of critical simplification in chemical kinetics. Chem Eng Sci 2003, 58(21), 4833–4875.
[14] Kruskal JB. The number of simplices in a complex. In: Bellman R, ed, Mathematical optimization techniques. Berkeley and Los Angeles, CA, USA, 1963, 251–329.
[15] Wei J, Prater CD. The structure and analysis of complex reaction systems. Adv Catal 1962,13, 203–392.
[16] Wei J, Kuo JC. Lumping analysis in monomolecular reaction systems. Analysis of the exactly lumpable system. Ind Eng Chem Fundam 1969, 8(1), 114–137.

Part II: **Chemical modelics**

3 Basic models of chemical kinetics

This chapter is devoted to a gallery of mathematical models which describe the critical phenomena of a pure chemical, nonthermal nature. Nonlinearity of systems is caused by reaction mechanisms, which are nonlinear. In typical cases, these systems are systems of ordinary differential equations with right parts which contain nonlinearities of type $X_m Y_n$. As shown in [1, 2], the presence of interaction steps of various substances (e.g., $X + Y \rightarrow$) is a necessary condition for multiplicity of steady states. Sufficient conditions of their occurrence can be determined by different factors: the presence of competition in the different stages (their different kinetic order), the presence of so-called buffer steps, the nonideality of elementary processes, the ratio of special parameters, etc. One of the most important kinetic characteristics is the presence of autocatalytic stages of type $A + Z \rightarrow 2Z$ in a reaction mechanism. We must say that such stages are present in the abstract mechanisms of chemical reactions studied in the works of I. Prigogine and his school [3–5]. These are all sorts of "oregonator", "brusselator", etc. The formal model of O. Rossler [6] also contains members which can be interpreted as the autocatalysis.

The models with the autocatalysis are the simplest models in sense of their degree of nonlinearity and the number of phase variables. Catalytic schemes of transformations which do not contain autocatalytic stages are more realistic, but they lead to systems of larger numbers of variables. The main complicating factors as a flow of the system, the presence of two or more types of active centers (multifunctional catalysis), and chemical nonideality are studied here. Main results of the analytical study are accompanied by graphical pictures. In particular, each model is characterized by its parametric portrait, where bifurcation curves divide the all parameter region into subregions of parameters with different number and type of stability of steady states.

3.1 Equations of chemical kinetics and a scheme of parametric analysis

3.1.1 Experimental background

Recently the research area of critical phenomena in kinetics has increased considerably. The discovery and study of such effects have a long history [7]. The beginning of a modern step is considered to be the detection by B.P. Belousov and the study by A.M. Zhabotinsky of oscillations in the oxidation reaction of malonic acid by bromine (catalysis – ions of the cerium) [8, 9]. It should be noted that publications on experimental studies of oscillation modes of chemical reactions appear in the beginning of the last century [7]. However, only this reaction (now called the B.Z.-reaction) has

https://doi.org/10.1515/9783110464948-003

caused a surge of attention among chemists, physicists, biologists and mathematicians.

Homogeneous systems are more simple than heterogeneous ones. Therefore, a study of homogeneous models will facilitate the understanding of the character of chemical oscillations [10]. Overviews of the main part of the experimental works on the Belousov–Zhabotinsky reaction and its modifications are presented in the book [8]. However, a single comprehensive mathematical model is not built. There is a significant number of kinetic models describing individual properties of these systems with varying degrees of detail. It seems to us that in this case there is a certain contradiction between the "want to know all" and "select the most important factors".

Lately, oscillatory reactions occurring in the gas phase have been investigated [11, 12] intensively. One of the most studied isothermal systems is the reaction of oscillatory oxidation of carbon monoxide. Gas-phase oscillators can be divided into isothermal and thermokinetic types. The first of them are characterized by nonlinearities of a nonthermal nature, which correspond to the presence of nonlinear stages in the reaction mechanism. The nonlinear nature of thermokinetic oscillations due to nonlinear dependence (in the simplest case the Arrhenius equation) of the reaction rate on temperature.

I. Prigogine's scientific school introduced a significant contribution to the development of understanding and interpretation of complex dynamic behavior of open chemical systems far from thermodynamic equilibrium [3–5]. A brief review of relevant models is given in [6].

Critical phenomena were discovered in many heterogeneous catalytic reactions, some of which have important application value to chemical technology. G.K. Boreskov, M.G. Slinko and employees studied a hydrogen oxidation reaction on nickel, palladium and platinum. They found that different values of the stationary reaction rate correspond to the same composition of the gas phase in definite region of parameters [13–25]. Since the early 70s there has been a new wave of intensive experimental and theoretical studies of critical effects in heterogeneous catalysis. It was the first cycle of works by M.G. Slinko and his disciples [26–34].

Critical phenomena in oxidation reactions of CO over Pt/Al_2O_3 and platinum wires are described in the works of E. Wicke (see, e.g., [27]). In these works, the authors showed that the multiplicity of steady states is due to the nonlinearity of the rate of formation and spending of intermediates on the catalyst surface. An important conclusion of E. Wicke and coauthors is that the cause of a complex dynamic behavior of a reaction lies in the complex chemistry of the processes on the catalyst surface. E. Wicke's line was continued by G. Eigenberger. To describe auto-oscillations of the rate of a catalytic reaction, he used the "buffer" step. Reversible stages of the formation of nonreactive able forms of oxygen or inert substance were added to E. Wicke's scheme. The "buffer" step performs the role of "feedback", but this step has allowed G. Eigenberger to describe the oscillations of rates obtained in experiments. [35]

The existence of kinetic auto-oscillations in heterogeneous reactions has been shown in the works of M.G. Slinko, V.D. Belyaev, and others [36–38]. The main nonlinear factor was accepted as the dependence of the activation energy on the degree of coverage of the catalyst surface by one of the intermediates. The mathematical models were built using this.

In the works of V.V. Barelko and A.S. Zhukov isothermal effects were investigated in a number of heterogeneous catalytic reactions with the help of a device they created, allowing them to exclude the influence of the thermal factor [39–42].

The number of works devoted to nonstationary processes in catalysis is constantly growing. To represent time a great deal of experimental data on the phenomena of self-organization in heterogeneous catalysis has been accumulated. Critical phenomena were observed for many heterogeneous oxidation reactions of hydrogen, carbon monoxide, carbon monoxide, nitrogen, and sulfur dioxide [27]. Such metals as platinum, palladium, iridium, and nickel are used in all works as a catalyst. Authors mainly consider the multiplicity of steady states and auto-oscillations [45–68].

In our book the central theme is a theme associated with the building and analysis of so-called basic mathematical models of critical phenomena of both thermal and nonthermal nature. In a sense they are the simplest models describing the multiplicity of steady states (triggers) and auto-oscillations (oscillators). The concept of basic models, developed by us, primarily supposes an understanding of the studied phenomenon. Then, a detailed description is created. The basic models are the kinds of blocks which must be put in the foundation of each detailed model. At this step we achieve not only a qualitative but also a quantitative description of the complex nonlinear and dynamic properties of the studied processes.

3.1.2 Equations of chemical kinetics

The equations of chemical kinetics are written as follows [1]. First, the list of substances is set as

$$X_1, X_2, \ldots, X_n ,\tag{3.1}$$

and the list of reactions becomes

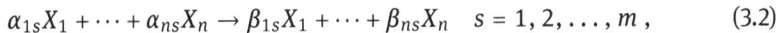

$$\alpha_{1s}X_1 + \cdots + \alpha_{ns}X_n \rightarrow \beta_{1s}X_1 + \cdots + \beta_{ns}X_n \quad s = 1, 2, \ldots, m ,\tag{3.2}$$

where α_{1s}, β_{1s} are stoichiometric coefficients.

Further, the rates of stages in scheme (3.2) are defined:

$$w_s = w_s(T, \mathbf{x}) , \quad s = 1, 2, \ldots, m ,\tag{3.3}$$

where T is temperature and $\mathbf{x} = (x_1, x_2, \ldots, x_n)$ is a vector of the concentrations of the substances (3.1).

The most simple kinetic functions (3.3) satisfy the law of mass action:

$$w_s(T, \mathbf{x}) = k_s(T) x_1^{\alpha_{1s}} \cdots x_n^{\alpha_{ns}} . \tag{3.4}$$

The temperature dependencies are defined by Arrhenius:

$$k_s(T) = k_s^0 \exp\left(-\frac{E}{RT}\right), \quad s = 1, 2, \ldots, m .$$

The law of conservation of mass in a nonstationary case at $T = $ const takes the form

$$\frac{dx_i}{dt} = \sum_{i=1}^{m} \gamma_{is} w_s, \quad i = 1, 2, \ldots n , \tag{3.5}$$

where $\gamma_{is} = \beta_{is} - \alpha_{is}$.

The equation of chemical kinetics (3.5) can be written down in compact vector form:

$$\frac{d\mathbf{x}}{dt} = \mathbf{\Gamma} \mathbf{w}(T, \mathbf{x}) , \tag{3.6}$$

where $\mathbf{w} = (w_1, w_2, \ldots, w_s)$ is a vector of the rates of the stages. The stoichiometric matrix $\mathbf{\Gamma} = ((\gamma_{is}))$ is such that the law of conservation of mass is performed for (3.5):

$$B_j : \quad \sum_{i=1}^{n} m_{ij} x_i = b_j, \quad j = 1, 2, \ldots, l ,$$

where b_j are the quantities of balances and m_{ij} are quantities proportional to the number of atoms of the j-th type contained in the i-th substance X_i.

The system (3.6) is defined in the polyhedron of the reaction in which the conditions of nonnegativity and balances are set B_j:

$$\mathbf{M} = \{x : x \geq 0, x \in B_j\} .$$

The model (3.6) represents a system in which no exchange of substances occurs with the environment. It can be a chemically reacting closed system or an open system far from equilibrium due to the persistence of some part of the reagent.

Thus, the basic model of chemical kinetics is a system of nonlinear differential equations (3.6) in the general case, in which the right-hand sides are formed according to the mechanism of transformations. General analysis of models of type (3.6) is given in [1]. Here we only note that a necessary condition for the multiplicity of steady states is the presence in the mechanism of the reaction stages of interaction of various substances. If there are no such stages, the behavior of (3.6) will be quasithermodynamic. For all initial conditions from the polyhedron reactions all solutions of the dynamic model (3.6) for $t \to \infty$ will approach a single stationary state. The equations of chemical kinetics (3.6) are a special case of dynamical systems of the general type. In the simplest case, the right-hand sides of equations (3.6) are polynomials in the phase variables $x_1, \ldots, x_n$.

However, they have sufficient commonality. Multitudes of equations of chemical kinetics are everywhere dense in the set of dynamical systems. The probability of this assertion is very high because on any finite time interval, any sufficiently smooth dynamical system can be approximated with any prescribed accuracy by a dynamical system whose right-hand sides represent finite Taylor series. After that you can use E.M. Korzukhin's theorem [8] on the possibility of approximating a dynamical system with polynomial right-hand sides by a system of equations of chemical kinetics.

3.1.3 Scheme of parametric analysis

Here, we briefly describe a scheme of parametric analysis of a dynamical system. The investigated real process usually occurs under certain external conditions, which can be characterized by some parameter values. These parameters are included in the appropriate kinetic model, which is a system of ordinary differential equations of the form

$$\dot{x}_i = f_i(x_1, \ldots, x_n, p_1, \ldots, p_m), \quad i = 1, \ldots, n, \tag{3.7}$$

where x_i are phase variables (temperature and concentration of substances) and p_j, $j = 1, \ldots, m$, are parameters which can be varied within certain limits. Parameters are usually chosen as thermal, physical and geometrical characteristics of a real process, such as temperature and concentrations of substances at the inlet to the reactor, their initial values, the volume of the reactor, and so on.

The first step in the parametric analysis of system (3.7) is the definition of its steady states. For model (3.7) the steady states are solutions of the system of equations

$$f_i(x_1, \ldots, x_n, p_1, \ldots, p_m) = 0, \quad i = 1, \ldots, n, \tag{3.8}$$

in the unknown variables $x_1, \ldots, x_n$. If system (3.8) can be solved in explicit form

$$x_i = \varphi_i(p_1, \ldots, p_m), \quad i = 1, \ldots, n,$$

we obtain the required parameter dependencies of steady states on parameters. Generally, if we cannot solve (3.8) in explicit form, the steady states and their dependencies on parameters should be found by rather hard computational procedures. However, the system of stationarity equations often can be reduced to one equation by elementary transformations

$$F(x, p_1, \ldots, p_m) = 0, \tag{3.9}$$

where x is one of the phase coordinates x_i. The nonlinear equation (3.9) may have multiple solutions, which leads to a multiplicity of steady states. Variation of p_j, $j = 1, \ldots, m$, in this case leads to the hysteresis of stationary dependencies on parameter.

The second step of the procedure of parametric analysis of system (3.7) is the investigation of the stability of steady states. It is necessary to form the Jacobian matrix with elements:

$$a_{ij} = \frac{\partial f_i}{\partial x_j}(x^*), \qquad i, j = 1, \ldots, n,$$

where the value x^* is responsible for a steady state. The stability of steady states is determined by the eigenvalues λ_i, $i = 1, \ldots, n$, of the Jacobian matrix. If all λ_i have nonzero real part, the steady state will be rude. Its stability is determined by the sign of $\operatorname{Re} \lambda_i$.

An important step of parametric analysis is the construction of the parametric dependencies of $\varphi_i(p_1, \ldots, p_m)$. In the general case this task is related to the solution of nonlinear systems with parameters. Computing and software tools for their solution are presented in [69–79]. In the study of specific systems it is sometimes possible to significantly simplify the procedure of constructing the parametric dependencies. Our experience shows that the parameters in the stationary equation (3.9) are generally such that you can write the parametric dependencies from (3.9) in explicit form.

$$p_j = \eta_j(x), \quad j = 1, \ldots, m,$$

where x is varied. Thus, it is possible to obtain the function which is the reverse of the required parametric dependence $x = \varphi(\text{integer})$, $j = 1, \ldots, m$. If, for example, the functions $\eta_j(x)$ are given in graphical form, the function $\varphi(\text{integer})$ will be obtained simply by inverting the coordinates x and p_j.

If one of the parameters, for example, p_1, is changed, there exist its special (bifurcation) values for which the number and stability of steady states is changed. Changing the second parameter p_2 leads to the result that in the (p_1, p_2) plane the bifurcation values of p_1 describe some curves which are bifurcation curves. In the simple case of dynamical systems ($n = 2$) there are two basic bifurcation curves on the plane: the curve of multiplicity of steady states L_Δ and the neutrality curve L_σ. The stability of steady states is determined by the roots of the second order characteristic equation

$$\lambda^2 - \sigma\lambda + \Delta = 0,$$

where $\sigma = a_{11} + a_{22}$ and $\Delta = a_{11}a_{22} - a_{12}a_{21}$. Let's take the two parameters p_1, p_2, and plot L_Δ, L_σ in the plane of these parameters. The boundaries of the region of the multiplicity of steady states is defined as a solution of the system

$$F(x, p_1, p_2) = 0,$$
$$\Delta(x, p_1, p_2) = 0,$$

which can be represented in the form

$$p_2 = \xi_2(x),$$
$$p_1 = \xi_1(x, \xi_2(x)),$$

where x is the changed value.

To plot the curve of neutrality L_σ, the stationarity equation must be added by the condition

$$\sigma(x,\ p_1,\ p_2) = 0 .$$

From this condition the equation for the curve of neutrality can be often written in explicit form

$$p_2 = \xi_2(x) ,$$

$$p_1 = \xi_3(x,\ \xi_2(x)) .$$

Analysis of mutual location of curves L_Δ, L_σ in the plane $(p_1,\ p_2)$ allows us to define the parametric portrait of the system. Curves L_Δ, L_σ allow for any values of the selected parameters to determine the number and stability of the steady states.

Useful information about the possible dynamic behavior of solutions of model (3.7) gives the plotting of phase portraits. The plotting of the phase portrait of system (3.7) for each given set of parameters from the selected region on the parametric portrait is performed by using numerical integration of (3.7) with different initial data. Each region of the parametric portrait matches its own type of phase portrait. The enumeration of all possible types of phase portraits is a rather time-consuming task [78]. However, in some specific cases, such a complete study is possible [79–83].

Complete representation of the dynamics of system (3.7) is given by time dependencies $x_i = x_i(t,\ p_1, \ldots, p_m)$. The solutions $x_i(t)$ are usually found by numerical integration of differential equations (3.7) for a fixed set of parameters p_j. Technical difficulties are present, and here they are connected first of all with the "rigidity" of the system of ordinary differential equations (3.7).

Consider a system of nonlinear equations with a parameter:

$$\mathbf{f}(\mathbf{x}, \alpha) = 0 , \tag{3.10}$$

where $\mathbf{x}$ is a vector of unknowns, α is a parameter, and $\mathbf{f}$ is a vector function. The system (3.10) implicitly specifies the dependence:

$$\mathbf{x} = \mathbf{x}(\alpha) . \tag{3.11}$$

Plotting this dependence is the main goal of a parametric analysis of the solutions of system (3.10). The general scheme of the method of parameter continuation [73, 78] is as follows. Using the substitution (3.11) in (3.10) and differentiating the resulting identities we have

$$\mathbf{J}\frac{d\mathbf{x}}{d\alpha} + \frac{\partial \mathbf{f}}{\partial \alpha} \equiv 0 , \tag{3.12}$$

where $\mathbf{J}$ is the Jacobian matrix of system (3.10):

$$\mathbf{J} = \frac{\partial \mathbf{f}}{\partial \mathbf{x}}(\mathbf{x}, \alpha) . \tag{3.13}$$

We consider the identity (3.12) as a system of linear equations for $d\mathbf{x}/d\alpha$. The equations of motion for the parameter can be gotten from the identity (3.12).

$$\frac{d\mathbf{x}}{d\alpha} = -\mathbf{J}^{-1}\frac{\partial \mathbf{f}}{\partial \alpha} . \tag{3.14}$$

The required parametric dependence of $\mathbf{x}(\alpha)$ is the solution of a system of ordinary differential equations (3.14) for some given initial data

$$\mathbf{x}(\alpha_0) = \mathbf{x}^0 .$$

The specificity of system (3.14) is that at the bifurcation points of solutions of (3.10) the Jacobian matrix $\mathbf{J}$ is special. Therefore, for the numerical integration of (3.14) we proceed to the parameterization by the arc length of the curve $\mathbf{x}(\alpha)$ in the appropriate space of dimension dim $\mathbf{x} + 1$. Note that for the integration of system (3.14) special methods should be applied, including those based on the calculation of the Jacobian matrix [78]. In this case, you must have the partial derivatives $\frac{\partial}{\partial \mathbf{x}}\left(\mathbf{J}^{-1}\frac{\partial \mathbf{f}}{\partial \mathbf{x}}\right)$, and that leads to additional technical difficulties. System (3.14) is often written on the basis of the solution of (3.12), for example by the Gauss method relative to $d\mathbf{x}/d\alpha$. When solving (3.14) a problem of getting sufficient accuracy initial data appears. All of this suggests that the numerical implementation of the method of parameter continuation is a rather time-consuming computational task. The degree of its complexity depends essentially on the dimension of system (3.10). The computational cost is significantly reduced by lowering, if possible, the number of equations in (3.10) due to the exclusion of some variables.

Consider the special case when the original system (3.10) can be reduced to one equation

$$g(x, \alpha) = 0 ,$$

where g is a scalar function of one argument x and parameter α. Similarly to (3.12) from differentiating the identity $g(x(\alpha), \alpha) \equiv 0$ we have

$$\frac{dx}{d\alpha} = -\frac{\partial g}{\partial \alpha} \Big/ \frac{\partial g}{\partial x} \tag{3.15}$$

or

$$\frac{d\alpha}{dx} = -\frac{\partial g}{\partial x} \Big/ \frac{\partial g}{\partial \alpha} . \tag{3.16}$$

The required parametric dependence $x(\alpha)$ or the inverse dependence $\alpha(x)$ is found by numerical integration of one of the equations (3.15) or (3.16). Equation (3.15) or (3.16) is integrated (at moving on a curve $x(\alpha)$) depending on a value of the right side of the equation. At the turning points $(\partial g/\partial x)$ or $(\partial g/\partial \alpha)$ is zero. Therefore, the following inequality can be accepted by the selection condition of motion on α or on x

$$\left| \frac{\partial g}{\partial \alpha} \Big/ \frac{\partial g}{\partial x} \right| < 1 . \tag{3.17}$$

In the numerical integration of the equations (3.15) or (3.16), there must be an "inversion" of the system, i.e., the transition from equation (3.15) to equation (3.16) or on the contrary depending on the implementation of (3.17). The integrator must include verification of the value of the right side of (3.15) or (3.16) and implement this inversion for a given condition. The advantage of the modified method of continuation on parameter for the scalar equation $g = 0$ is the absence of matrix computations for the inverse Jacobian matrix and the difficulties associated with bifurcations where the derivative $(\partial g/\partial x)$ or $(\partial g/\partial \alpha)$ becomes zero. The last difficulties are overcome in our approach by changing dependent and independent variables (x on α or α on x).

Thus, if the system of stationarity equations can be reduced to one equation, instead of the overall procedure of the method of continuation on a parameter, it is better to use the proposed approach based on the integration of the equations of type (3.15), (3.16). The proposed procedure of continuation on parameter cannot be applied for systems (3.10). However, if the system can be reduced to one equation due to the exclusion of unknowns, then parametric analysis of the solutions of the original model is preferable in order to carry out the proposed scheme. For those variables that are fixed in (3.10) algebraically, the exception can be carried out in letter form by the methods of computer algebra [85, 86]. The resulting expressions for the resultant of the system of nonlinear algebraic equations is quite cumbersome. For the above reasons, it can be hoped that parametric analysis of the solutions of (3.10) for one equation will be more effective.

The general scheme of parametric analysis of solutions of a system of nonlinear equations was modified for the important and rather particular case when the system can be reduced to one equation as a result of the procedure of exclusion of variables. However, in the analysis of specific nonlinear models it is not always possible to implement such an exception. Often the original system can be reduced to a system of two equations. The general scheme of the method of continuation on a parameter can be modified in this case. Its computer implementation is much simpler compared to the common approach [78].

Consider a system of two nonlinear equations with a parameter

$$f(x, y, \alpha) = 0 ,$$
$$g(x, y, \alpha) = 0 , \tag{3.18}$$

where x, y are the unknowns, α is a parameter, and f, g are nonlinear systems in the general case. The system of two equations (3.18) explicitly sets the dependencies:

$$x = x(\alpha) , \qquad y = y(\alpha) . \tag{3.19}$$

The purpose of the parametric analysis of the solutions of system (3.18) is to plot the functions (3.19). According to one approach of the implementation of the method of continuation on a parameter after substituting (3.19) in (3.18) and the differentiation of the obtained identities we have

$$\frac{\partial f}{\partial x}\frac{dx}{d\alpha} + \frac{\partial f}{\partial y}\frac{dy}{d\alpha} + \frac{\partial f}{\partial \alpha} = 0 ,$$
$$\frac{\partial g}{\partial x}\frac{dx}{d\alpha} + \frac{\partial g}{\partial y}\frac{dy}{d\alpha} + \frac{\partial g}{\partial \alpha} = 0 . \tag{3.20}$$

We introduce the following notation

$$\Delta = \begin{vmatrix} \dfrac{\partial f}{\partial x} & \dfrac{\partial f}{\partial y} \\ \dfrac{\partial g}{\partial x} & \dfrac{\partial g}{\partial y} \end{vmatrix}, \qquad \Delta_1 = \begin{vmatrix} -\dfrac{\partial f}{\partial a} & \dfrac{\partial f}{\partial y} \\ -\dfrac{\partial g}{\partial a} & \dfrac{\partial g}{\partial y} \end{vmatrix}, \qquad \Delta_2 = \begin{vmatrix} \dfrac{\partial f}{\partial x} & -\dfrac{\partial f}{\partial a} \\ \dfrac{\partial g}{\partial x} & -\dfrac{\partial g}{\partial a} \end{vmatrix}.$$

From (3.20) we get

$$\frac{dx}{da} = \frac{\Delta_1}{\Delta},$$

$$\frac{dy}{da} = \frac{\Delta_2}{\Delta}. \tag{3.21}$$

To plot the required parametric dependencies (3.19), a system of two differential equations (3.21) is solved for given initial data

$$x(a_0) = x_0, \qquad y(a_0) = y_0. \tag{3.22}$$

A feature of system (3.21) is that in the turning points of the curve (3.19) in three-dimensional space (x, y, a), one of the determinants Δ, Δ_1, Δ_2 is equal to zero. Therefore, following [78], one can introduce the so-called current parameter (one of the variables (x, y, a)) together with (3.21) and consider two systems of differential equations:

$$\frac{da}{dx} = \frac{\Delta}{\Delta_1},$$

$$\frac{dy}{dx} = \frac{\Delta_2}{\Delta_1}, \tag{3.23}$$

or

$$\frac{dx}{dy} = \frac{\Delta_1}{\Delta_2},$$

$$\frac{da}{dy} = \frac{\Delta}{\Delta_2}. \tag{3.24}$$

Thus, the motion on the curve parametric dependencies (3.19) can be implemented by integrating one of the three systems (3.21), (3.23) or (3.24). The alternation of these systems is determined by the values of three quantities: Δ, Δ_1, Δ_2. If $\Delta_1 < \Delta$ and $\Delta_2 < \Delta$, the parameter a is selected as the current parameter and integrated system (3.21). If $\Delta < \Delta_1$ and $\Delta_2 < \Delta_1$, then (3.22) is integrated on the current parameter x. If $\Delta_1 < \Delta_2$ and $\Delta < \Delta_2$, the current parameter is then y, i.e.,

$$\text{if } \Delta_1 < \Delta, \quad \Delta_2 < \Delta, \quad \text{then (3.21)};$$
$$\text{if } \Delta < \Delta_1, \quad \Delta_2 < \Delta_1, \quad \text{then (3.23)};$$
$$\text{if } \Delta_1 < \Delta_2, \quad \Delta < \Delta_2, \quad \text{then (3.24)}.$$

At the computer implementation of the proposed scheme of continuation on a parameter a motion along curve in the space (x, y, a) must be performed so that the corresponding integrator must contain a comparison of the values Δ, Δ_1, Δ_2, a choice of a current parameter and integrating one of the systems (3.21), (3.23) or (3.24). In our implementation as integrator scheme, the (m, k)-method proposed by E.A. Novikov [84] was used for solving rigid systems. The advantage of the proposed method of continuation on a parameter compared to the general case is the absence of technical difficulties associated with the passage of bifurcation points on the curve (3.19). Degeneration of the Jacobian matrix of system (3.18) occurs at the bifurcation points. In this case the motion on the current parameter occurs with a given constant step. However, as in the general case, it is difficult to set a good initial approximation (3.22) and the direction of the motion on a parameter. Initial data (3.22) usually are specified as a result of several Newton iterations. In addition, the limits of variation of parameters are set as the input data. Various branches (if they exist) of the curve of parametric dependencies (3.19) are plotted by varying the initial data in a wide range. Note that in some implementations of the general scheme of the method of continuation on a parameter of solutions of a system of n equations, the motion in the space $(x_1, \ldots, x_n, \alpha)$ is performed by integration of the system of n differential equations. The arc length of the curve $x_i(\alpha)$, $i = 1, \ldots, n$ is set as the independent variable, but not the parameter α [78]. In this case, the problem of passing turning points is removed. However, the right sides of the integrable system become more bulky. In our implementation this technical difficulty is absent. The motion along the arc length is replaced by the motion on the current parameter, with the simple logic of the procedure of comparing the current values of Δ, Δ_1, Δ_2.

Thus, for a system of two nonlinear equations with parameters the general scheme of the implementation of the method of continuation on a parameter can be considerably simplified. Motion on the curve of the parametric dependence of solutions of a system is defined by integration of a system of two differential equations of type (3.21), (3.23) or (3.24) on one of the three current variables. The transition from one variable to another is done by comparing the values Δ, Δ_1, Δ_2. For systems of more than three equations the proposed way can also be implemented. However, it becomes quite bulky. The problem of analytical calculation of n^2 elements of the Jacobian matrix of the system plus n derivatives on the parameter appears. In the general case, the numerical procedure of differentiation can hardly be recommended because in a neighborhood of the bifurcation points the sensitivity of solutions of the system to variations of the parameters increases dramatically. All of this suggests that the "preprocessing" of systems of nonlinear equations is necessary for parametric analysis. For example, the exclusion of some variables is desirable as far as possible to reduce the order of the system [85, 86]. If the original system can be reduced to one or at least two equations, the scheme of the method of continuation on a parameter, given above, can be applied for its parametric analysis.

The procedure of parametric analysis is implemented in its simplest form for a series of so-called basic models of kinetics and thermokinetics of catalytic reactions. The considered models, as a rule, have the minimum number of variables and parameters. In the general case, the procedure of a full parametric analysis is rather time-consuming. Here we implement only the initial stages of parametric analysis. However, its results already give a lot of information about the possible dynamic behavior of the considered thermokinetic models. In particular, it becomes clear that the interaction of the temperature nonlinearity with the kinetic characteristics of chemical reactions gives a great variety of dynamics of physical and chemical processes. Moreover, the procedure of parametric analysis allows us to obtain quantitative characteristics of chemical dynamics. It is important not only from a theoretical, but also from a more applied point of view. For example, in the chemical engineering of catalytic processes it is important to know the dangerous and safe boundaries of the varied parameters which can determine an emergency or technologically undesirable regimes [87, 88].

3.2 Autocatalytic models

3.2.1 Autocatalytic trigger

Consider some autocatalytic steps, which allow to construct the simplest models. They are characterized by a multiplicity of steady states (triggers) and auto-oscillations (oscillators).

We consider a reaction mechanism which contains two stages:

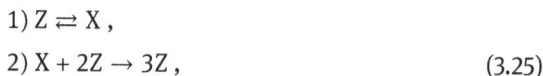

$$1)\, Z \rightleftarrows X \,,$$
$$2)\, X + 2Z \rightarrow 3Z \,, \tag{3.25}$$

where Z is a catalyst and X is an intermediate substance. Stage 2) is autocatalytic.

A formal scheme like (3.25), for example, may be interpreted as the process of adsorption-desorption of some substance A from the gas phase on the catalyst Z:

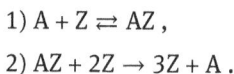

$$1)\, A + Z \rightleftarrows AZ \,,$$
$$2)\, AZ + 2Z \rightarrow 3Z + A \,.$$

The first step is a conventional monomolecular (on the intermediate substance) step adsorption-desorption of a substance A from the gas phase. The second step means the possibility of a second variant of desorption (in the presence of two neighboring vacant active places on the catalyst surface). When the gas phase ($p_A \equiv$ const) is constant, the scheme of transformations of intermediates (AZ, Z) can be written in the form (3.25). An autocatalytic step in the mechanism is nonlinear. It determines the possibility of the existence of critical phenomena. This type of autocatalytic stage is often used for the interpretation of critical phenomena of a nonthermal nature [3, 5, 9, 12]. They allow the simplest way to plot a kinetic model possessing essentially nonlinear properties.

A kinetic model for scheme (3.25) can be written in the form

$$\dot{x} = k_1 z - k_{-1} x - k_2 z^2 x \,, \tag{3.26}$$

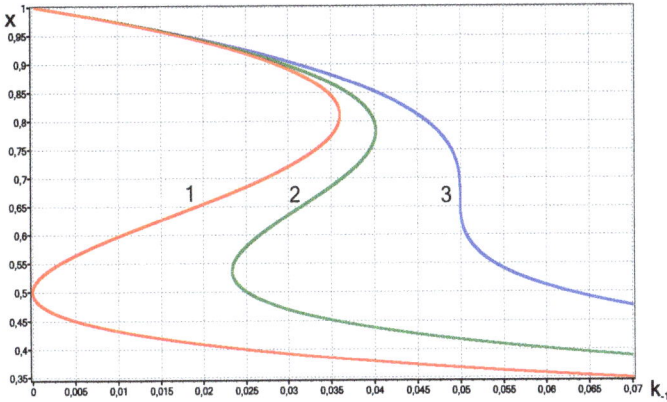

Fig. 3.1: Parametric dependencies $z(k_{-1})$ for $k_1 = 0.4$ at 1) $k_2 = 1.6$; 2) 1.5; 3) 1.35

where in accordance with the law of conservation of mass $x + z = 1$, i.e.,

$$z = 1 - x \,. \tag{3.27}$$

The rate constants of the stages k_i are the parameters of model (3.26), (3.27).

The equation of stationarity for (3.26) is a cubic equation in x:

$$k_1(1 - x) - k_{-1}x - k_2x(1 - x)^2 = 0 \,. \tag{3.28}$$

From (3.28) we obtain the following parametric dependencies:

$$k_1(x) = \frac{k_{-1}x + k_2x(1 - x)^2}{1 - x} \,,$$

$$k_2(x) = \frac{k_1(1 - x) - k_{-1}x}{x(1 - x)^2} \,,$$

$$k_{-1}(x) = \frac{(1 - x)(k_1 - k_2x(1 - x))}{x} \,. \tag{3.29}$$

The formulas (3.29) allow us to plot the parametric dependencies of the steady states of the parameters $x(k_i)$ relatively easy. One of them $(x(k_{-1}))$ is shown in Fig. 3.1.

Stability of the steady states for (3.26) is determined by the sign of the derivative

$$f'_x = -k_1 - k_{-1} - k_2z^2 + 2k_2xz \tag{3.30}$$

in some steady state x^*. If $f'_x(x^*) > 0$, then the steady state is unstable and if $f'_x(x^*) < 0$, the steady state is stable.

Curves of the multiplicity of steady states according to (3.28), (3.30) are specified by the system of equations

$$k_1(1 - x) - k_{-1}x - k_2x(1 - x)^2 = 0 \,,$$

$$k_1 + k_{-1} + k_2(1 - x)^2 - 2k_2x(1 - x) = 0 \,. \tag{3.31}$$

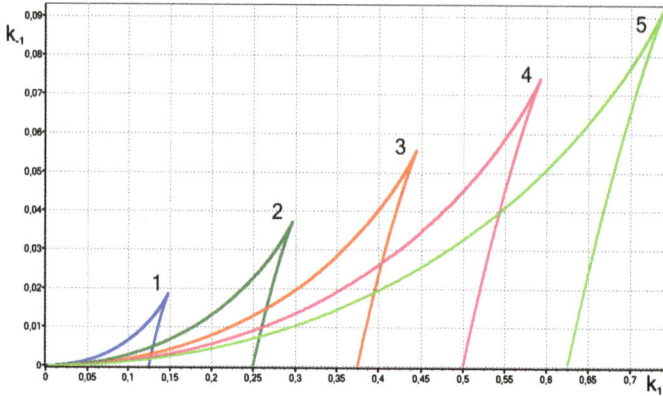

Fig. 3.2: Curves of multiplicity $L_\Delta(k_1, k_{-1})$ at 1) $k_2 = 0.5$; 2) 1; 3) 1.5; 4) 2; 5) 2.5

From (3.31) the explicit expressions for curves of multiplicity L_Δ for different combinations of parameters can be obtained. For example,

$$L_\Delta(k_1, k_2): \quad \begin{aligned} k_2(x) &= \frac{k_{-1}}{(1-x)^2(2x-1)}, \\ k_1 &= k_1(x, k_2(x)), \end{aligned}$$

$$L_\Delta(k_1, k_{-1}): \quad \begin{aligned} k_{-1}(x) &= k_2(1-x)^2(2x-1), \\ k_1 &= k_1(x, k_{-1}(x)), \end{aligned}$$

where the dependence of $k_1(x)$ is computed by the formula (3.29). The computed results are given in Fig. 3.2. Here a series of curves L_Δ under variations of the parameter k_2 is shown. The explicit form of the dependencies L_Δ allows us to investigate changes in curves of multiplicities without technical difficulties by changing the third parameter.

3.2.2 Autocatalytic oscillators

A series of autocatalytic mechanisms is considered in this subsection which differ in the form of the third step. This step is called a buffer step, because the substance Y appears only then and for the existence of auto-oscillations in the system it must be quite slow compared to the other steps. Different bifurcation curves and parametric dependencies are plotted for each of the mechanisms, and the influence of buffer step is also analyzed based on the characteristics of auto-oscillations.

Autocatalytic oscillators with buffer step Z ⇌ Y

Consider the following scheme of transformations:

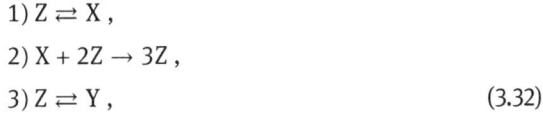

$$1)\ Z \rightleftharpoons X\,,$$
$$2)\ X + 2Z \rightarrow 3Z\,,$$
$$3)\ Z \rightleftharpoons Y\,, \tag{3.32}$$

where Z is a catalyst and X, Y are intersection substances. The step 3) is a buffer step. The substance Y is not involved in the other stages. All stages (3.32) are the simplest in the sense that the corresponding kinetic model is minimal in dimension and number of nonlinear members. It has auto-oscillations for the definite set of parameters. Therefore, mechanism (3.32) may be called the simplest autocatalytic oscillator. The kinetic model corresponding to scheme (3.32) is:

$$\dot{z} = -k_1 z + k_{-1} x + k_2 x z^2 - k_3 z + k_3 y\,,$$
$$\dot{y} = k_3 z - k_{-3} y\,, \tag{3.33}$$

where $x = 1 - y - z$ and $k_1, k_{-1}, k_2, k_3, k_{-3}$ are the parameters of model (3.33).

Steady states for (3.33) are defined from the equations

$$y = \frac{k_3}{k_{-3}} x\,,$$
$$-k_1 z + k_{-1}(1 - \alpha z) + k_2 z^2 (1 - \alpha z) = 0\,, \tag{3.34}$$

where $\alpha = 1 + k_3/k_{-3}$. Next we find the parametric dependencies from (3.34):

$$k_1(z) = \frac{(1 - \alpha z)(k_{-1} + k_2 z^2)}{z}\,,$$
$$k_{-1}(z) = \frac{k_1 z - k_2 z^2 (1 - \alpha z)}{1 - \alpha z}\,,$$
$$k_2(z) = \frac{k_1 z - k_{-1}(1 - \alpha z)}{z^2 (1 - \alpha z)}\,,$$
$$k_3(z) = \frac{k_{-3}[(1 - z)(k_{-1} + k_2 z^2) - k_1 z]}{z(k_{-1} + k_2 z^2)}\,,$$
$$k_{-3}(z) = \frac{k_3 z(k_{-1} + k_2 z^2)}{(1 - z)(k_{-1} + k_2 z^2) - k_1 z}\,. \tag{3.35}$$

As an example one of the parametric dependencies (3.35) is given in Fig. 3.3 for the set of parameters: $k_1 = 0.08$, $k_{-1} = 0.001$, $k_2 = 1$, $k_3 = 0.0032$, $k_{-3} = 0.002$.

The figure shows that the region of multiplicity of steady states expands when a value of the varied parameter k_2 is increased.

The elements of the Jacobian matrix for the right part of (3.33) are

$$a_{11} = -k_1 - k_{-1} - k_2 z^2 + 2k_2 xz - k_3\,,$$
$$a_{12} = -k_{-1} - k_2 z^2 + k_{-3}\,,$$
$$a_{21} = k_3\,,$$
$$a_{22} = -k_{-3}\,.$$

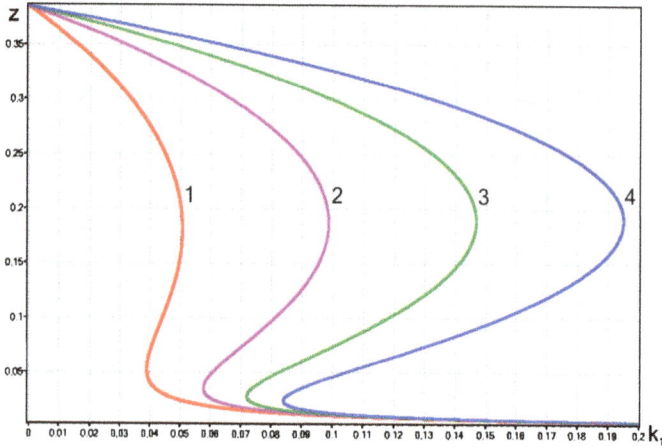

Fig. 3.3: Parametric dependencies $z(k_1)$ at 1) $k_2 = 0.5$; 2) 1; 3) 1.5; 4) 2

Stability of steady states is determined by the roots of the characteristic equation

$$\lambda^2 - \sigma\lambda + \Delta = 0 ,$$

where $\sigma = a_{11} + a_{22} = -k_1 - k_{-1} - k_2 z^2 + 2k_2 xz - k_3 - k_{-3}$, $\Delta = a_{11}a_{22} - a_{12}a_{21} = k_1 k_{-3} + (k_3 + k_{-3})(k_{-1} + k_2 z^2) - 2k_2 k_{-3} z(1 - \alpha z)$.

The main bifurcation curves can be written in explicit form for different combinations of parameters. From the stationarity equation (3.34) and the condition $\Delta = 0$ we obtain the expressions for plotting the curves of multiplicity of steady states

$$L_\Delta(k_1, k_2): \quad k_2(z) = \frac{k_{-1}(k_{-3}(1 - \alpha z) + z(k_3 + k_{-3}))}{z^2 [k_{-3}(1 - \alpha z) - z(k_3 + k_{-3})]} ,$$
$$k_1 = k_1(z, k_2(z)) ,$$

$$L_\Delta(k_1, k_{-1}): \quad k_{-1}(z) = \frac{k_2 z^2 (k_{-3}(1 - \alpha z) - z(k_3 + k_{-3}))}{k_{-3}(1 - \alpha z) + z(k_3 + k_{-3})} , \qquad (3.36)$$
$$k_1 = k_1(z, k_{-1}(z)) .$$

The calculated results for the explicit expressions of (3.36) are presented in Figs. 3.4 and 3.5.

We have the expressions for plotting the curves of neutrality of steady states from the stationarity equation (3.34) and the condition $\sigma = 0$:

$$L_\sigma(k_1, k_2): \quad k_2(z) = \frac{k_{-1}(1 - \alpha z + z) + z(k_3 + k_{-3})}{z^2(1 - \alpha z - z)} ,$$
$$k_1 = k_1(z, k_2(z)) ,$$

$$L_\sigma(k_1, k_{-1}): \quad k_{-1}(z) = \frac{k_2 z^2 (1 - \alpha z - z) - z(k_3 + k_{-3})}{(1 - \alpha z + z)} ,$$
$$k_1 = k_1(z, k_{-1}(z)) ,$$

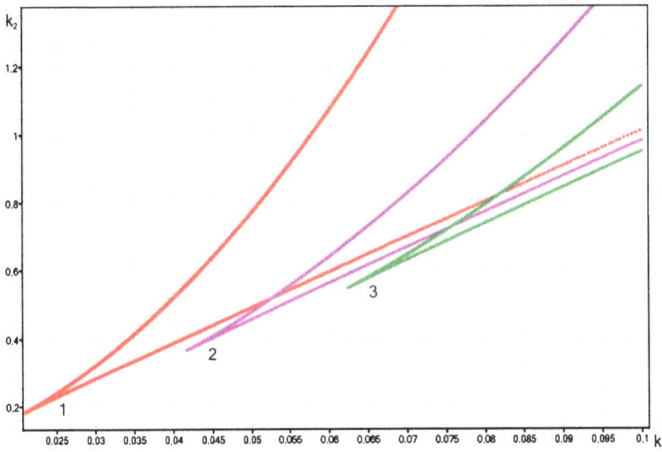

Fig. 3.4: Curves of multiplicity $L_\Delta(k_1, k_2)$ at 1) $k_{-1} = 0.001$; 2) 0.002; 3) 0.003

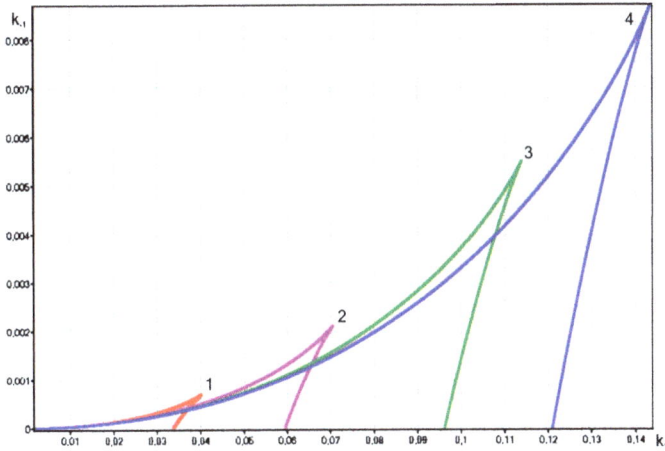

Fig. 3.5: Curves of multiplicity $L_\Delta(k_1, k_{-1})$ at 1) $k_{-3} = 0.0005$; 2) 0.001; 3) 0.002; 4) 0.003

where the functions $k_1(z)$ are taken according to (3.35). As an example the curves of neutrality L_σ are plotted in Fig. 3.6.

One of the possible parametric portraits, which are plotted based on the curves of multiplicity and neutrality, is shown in Fig. 3.7.

The phase trajectories of the system for the case of a single unstable steady state (limit cycle) are shown in Fig. 3.8. The phase trajectories tend to the stable limit cycle from any initial data. The corresponding auto-oscillations of the system in time are given in Fig. 3.9.

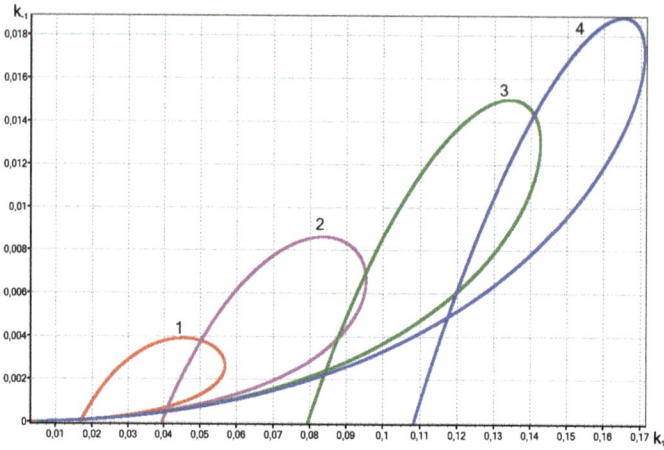

Fig. 3.6: Curves of neutrality $L_\sigma(k_1, k_{-1})$ at 1) $k_{-3} = 0.0005$; 2) 0.001; 3) 0.002; 4) 0.003

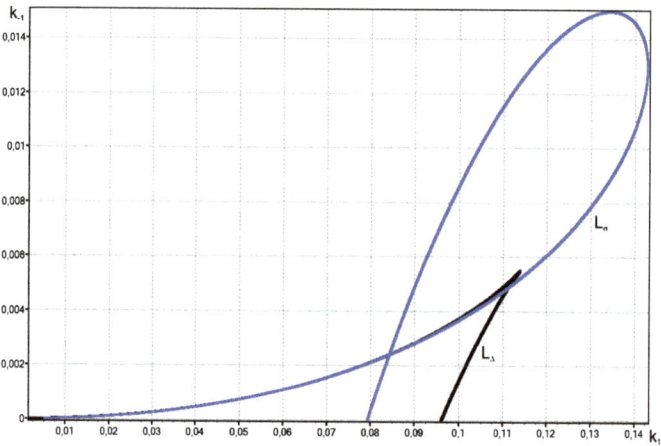

Fig. 3.7: Parametric portrait on the plane (k_1, k_{-1})

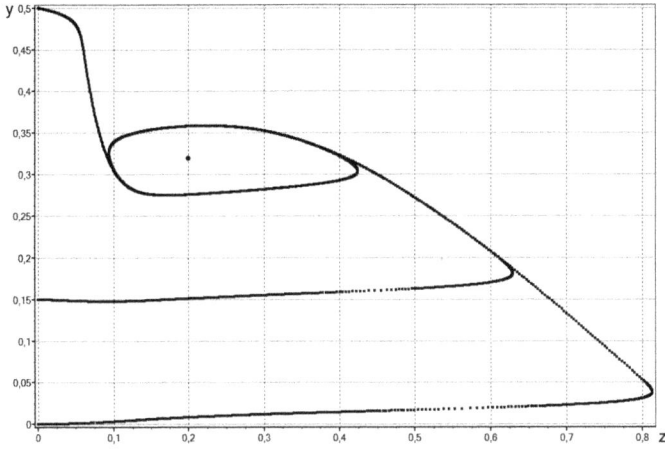

Fig. 3.8: Phase portrait of system (3.33) at $k_1 = 0.12$, $k_{-1} = 0.01$, $k_2 = 1$, $k_3 = 0.0032$, $k_{-3} = 0.002$

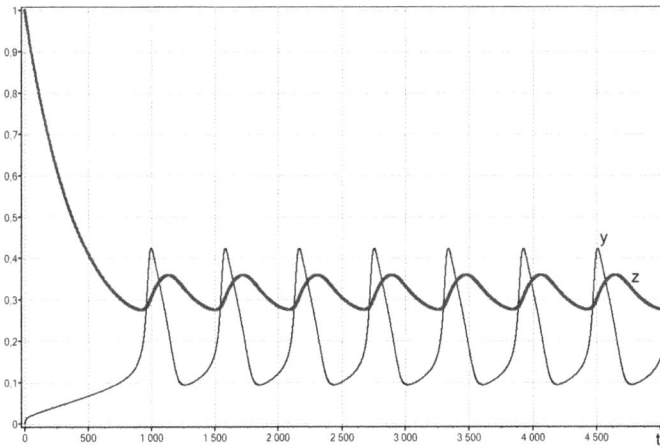

Fig. 3.9: Time dependencies $z(t)$, $y(t)$ at $k_1 = 0.12$, $k_{-1} = 0.01$, $k_2 = 1$, $k_3 = 0.0032$, $k_{-3} = 0.002$

Autocatalytic oscillators with buffer step X ⇌ Y

Replace the third step in scheme of transformations (3.32)

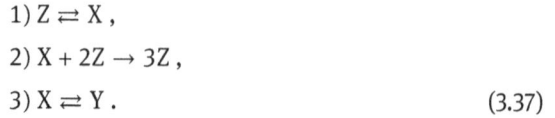

$$1)\, Z \rightleftharpoons X \,,$$
$$2)\, X + 2Z \rightarrow 3Z \,,$$
$$3)\, X \rightleftharpoons Y \,. \tag{3.37}$$

The kinetic model for (3.37) can be written as:

$$\dot{z} = -k_1 z + k_{-1} x + k_2 x z^2 \,,$$
$$\dot{y} = k_3 x - k_{-3} y \,, \tag{3.38}$$

where $x = 1-y-z$. The right sides of system (3.38) are equated to zero and transformed. So we get the equation of steady states depending on one variable z:

$$- k_1 z (k_3 + k_{-3}) + k_{-1} k_{-3} (1 - z) + k_2 k_{-3} z^2 (1 - z) = 0 \,. \tag{3.39}$$

We find the parametric dependencies from (3.39):

$$k_1(z) = \frac{k_{-3}(1 - z)(k_{-1} + k_2 z^2)}{z(k_3 + k_{-3})} \,,$$

$$k_{-1}(z) = \frac{k_1 z (k_3 + k_{-3}) - k_2 k_{-3} z^2 (1 - z)}{k_{-3}(1 - z)} \,,$$

$$k_2(z) = \frac{k_1 z (k_3 + k_{-3}) - k_{-1} k_{-3}(1 - z)}{k_{-3} z^2 (1 - z)} \,,$$

$$k_3(z) = \frac{k_{-3}(1 - z)(k_{-1} + k_2 z^2) - k_1 k_{-3} z}{k_1 z} \,,$$

$$k_{-3}(z) = \frac{k_1 k_3 z}{(1 - z)(k_{-1} + k_2 z^2) - k_1 z} \,. \tag{3.40}$$

As an example one of the parametric dependencies (3.40) is given in Fig. 3.10, following the set of parameters: $k_1 = 0.08$, $k_{-1} = 0.001$, $k_2 = 1$, $k_3 = 0.0032$, $k_{-3} = 0.002$.

The elements of the Jacobian matrix for the right sides of (3.38) have the form:

$$a_{11} = -k_1 - k_{-1} - k_2 z^2 + \frac{2 k_2 k_{-3} z (1 - z)}{k_3 + k_{-3}} \,,$$

$$a_{12} = -k_{-1} - k_2 z^2 \,,$$

$$a_{21} = -k_3 \,,$$

$$a_{22} = -k_3 - k_{-3} \,.$$

The main bifurcation curves can be written explicitly, and in this case for different combinations of parameters. The curves of multiplicity of steady states are obtained

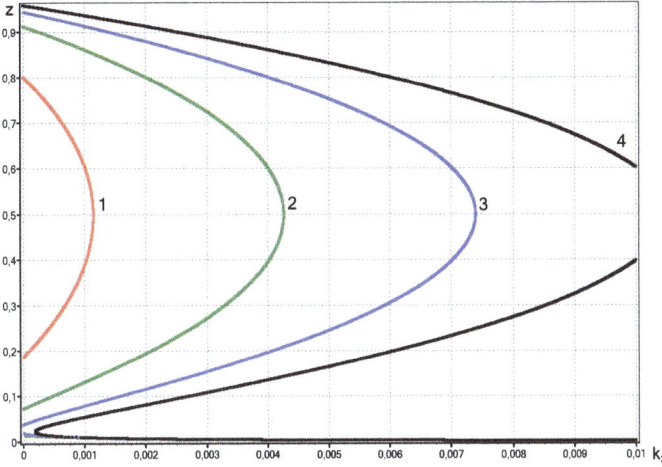

Fig. 3.10: Parametric dependencies $z(k_3)$ at 1) $k_2 = 0.5$; 2) 1; 3) 1.5; 4) 2

from the stationarity equation (3.39) and the condition $\Delta = 0$

$$L_\Delta(k_1, k_2): \quad \begin{aligned} k_2(z) &= \frac{k_{-1}}{z^2(1 - 2z)} \,, \\ k_1 &= k_1(z, k_2(z)) \,, \end{aligned}$$

$$L_\Delta(k_1, k_{-1}): \quad \begin{aligned} k_{-1}(z) &= k_2 z^2(1 - 2z) \,, \\ k_1 &= k_1(z, k_{-1}(z)) \,. \end{aligned} \tag{3.41}$$

From (3.39) and the condition $\sigma = 0$ we have the curves of neutrality

$$L_\sigma(k_1, k_2): \quad \begin{aligned} k_2(z) &= \frac{k_{-1}k_{-3}(1 - z) + z(k_3 + k_{-3})(k_{-1} + k_3 + k_{-3})}{k_{-3}z^2(1 - z) - z^3(k_3 + k_{-3})} \,, \\ k_1 &= k_1(z, k_2(z)) \,, \end{aligned}$$

$$L_\sigma(k_1, k_{-1}): \quad \begin{aligned} k_{-1}(z) &= \frac{k_2 k_{-3} z^2(1 - z) - k_2 z^3(k_3 + k_{-3}) - z(k_3 + k_{-3})^2}{k_3 z + k_{-3}} \,, \\ k_1 &= k_1(z, k_{-1}(z)) \,, \end{aligned} \tag{3.42}$$

where the functions $k_1(z)$ are taken from (3.40). The curves of multiplicity and neutrality according to the formulas (3.41) and (3.42) are plotted in Figs. 3.11 and 3.12.

The parametric portrait of the system is presented in Fig. 3.13. In this case, we have failed to get a single unstable steady state. Numerous calculations have shown that perhaps in this system there is no auto-oscillation.

The phase portrait of system (3.38) for the case of three steady states is presented in Fig. 3.14. Depending on the initial data, the system is stabilized to one of stable steady states (extreme points on the phase portrait). An unstable steady state, lying between them, determines the attraction domain of the stable steady states.

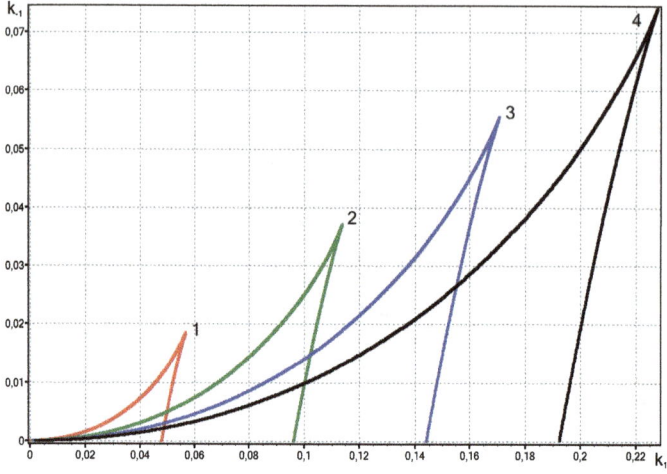

Fig. 3.11: Curves of multiplicity $L_\Delta(k_1, k_{-1})$ at 1) $k_2 = 0.5$; 2) 1; 3) 1.5; 4) 2

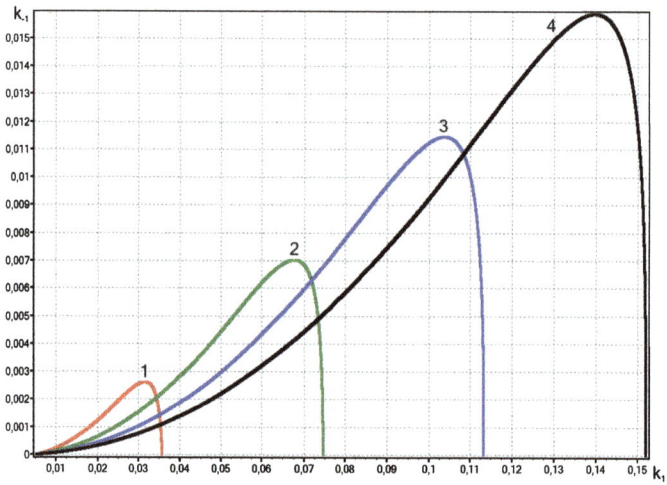

Fig. 3.12: Curves of neutrality $L_\sigma(k_1, k_{-1})$ at 1) $k_2 = 0.5$; 2) 1; 3) 1.5; 4) 2

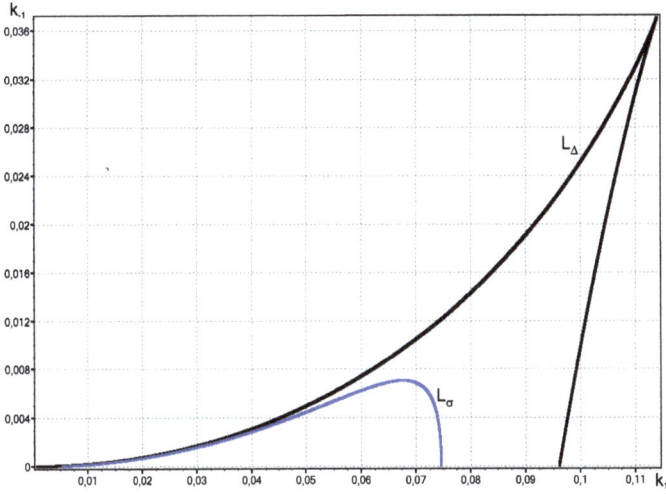

Fig. 3.13: Parametric portrait in (k_1, k_{-1}) at $k_2 = 1$, $k_3 = 0.0032$ and $k_{-3} = 0.002$

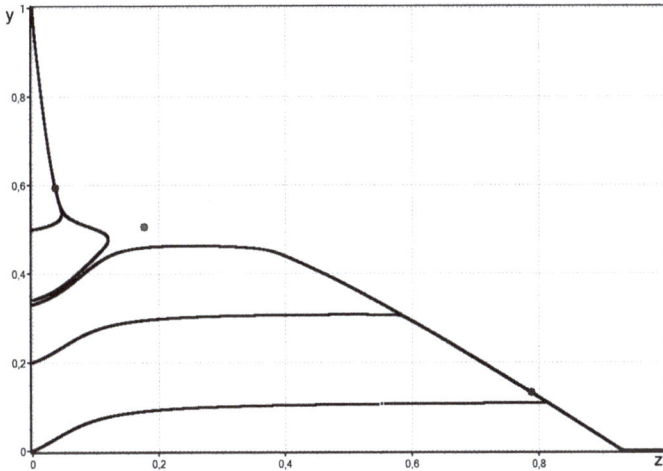

Fig. 3.14: Phase portrait (z, y) for $k_1 = 0.065$, $k_{-1} = 0.005$, $k_2 = 1$, $k_3 = 0.0032$, $k_{-3} = 0.002$

Autocatalytic oscillators with buffer step $2Z \rightleftharpoons Y$

We replace the third step in the scheme of transformations (3.32) and consider the following scheme:

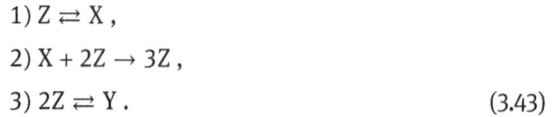

$$1) Z \rightleftharpoons X ,$$

$$2) X + 2Z \rightarrow 3Z ,$$

$$3) 2Z \rightleftharpoons Y . \tag{3.43}$$

The following kinetic model satisfies the scheme (3.43).

$$\dot{z} = -k_1 z + k_{-1} x + k_2 xz^2 - 2k_3 z^2 + 2k_{-3} y ,$$

$$\dot{x} = k_1 z - k_{-1} x - k_2 xz^2 , \tag{3.44}$$

where $y = (1 - x - z)/2$. The right sides of system (3.44) are equated to zero and transformed. We obtain the equation of steady states depending on one variable z:

$$\frac{k_1 z (k_2 z^2 + k_{-1} - k_{-3})}{k_{-1} + k_2 z^2} - k_1 z - 2k_3 z^2 + k_{-3}(1 - z) = 0 . \tag{3.45}$$

If we perform elementary transformations with (3.45) we obtain the following parametric dependencies:

$$k_1(z) = \frac{(k_{-1} + k_2 z^2)(k_{-3}(1 - z) - 2k_3 z^2)}{k_{-3} z} ,$$

$$k_{-1}(z) = \frac{k_1 k_{-3} z}{k_{-3}(1 - z) - 2k_3 z^2} - k_2 z^2 ,$$

$$k_2(z) = \frac{k_{-3}(k_1 z - k_{-1}(1 - z)) + 2k_{-1} k_3 z^2}{z^2 (k_{-3}(1 - z) - 2k_3 z^2)} ,$$

$$k_3(z) = \frac{k_{-3}(1 - z)(k_{-1} + k_2 z^2) - k_1 k_{-3} z}{2z^2 (k_{-1} + k_2 z^2)} ,$$

$$k_{-3}(z) = \frac{2k_3 z^2 (k_{-1} + k_2 z^2)}{(1 - z)(k_{-1} + k_2 z^2) - k_1 z} . \tag{3.46}$$

The parametric dependencies $z(k_2)$ for changing the second parameter k_{-3} are shown in Fig. 3.15. The set of parameter values is $k_1 = 0.12$, $k_{-1} = 0.001$, $k_2 = 1$, $k_3 = 0.0032$, and $k_{-3} = 0.002$.

The elements of the Jacobian matrix for the right sides of (3.44) are represented by:

$$a_{11} = -k_1 + 2k_2 xz - 4k_3 z - k_{-3} ,$$

$$a_{12} = k_{-1} + k_2 z^2 - k_{-3} ,$$

$$a_{21} = -k_1 - 2k_2 xz ,$$

$$a_{22} = -k_{-1} - k_2 z^2 ,$$

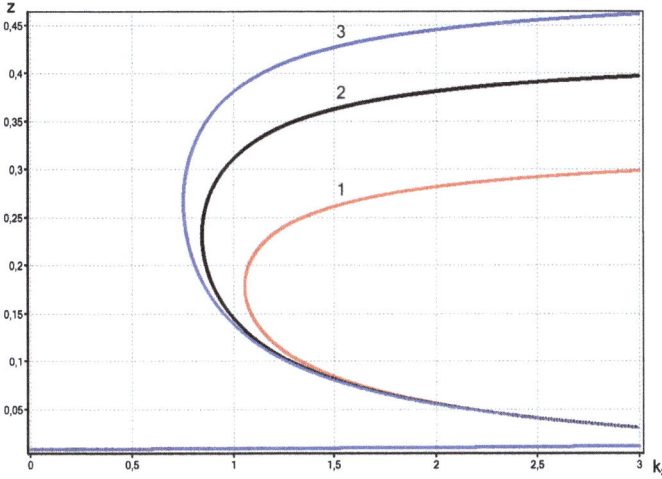

Fig. 3.15: Parametric dependencies $z(k_2)$ at 1) $k_{-3} = 0.001$; 2) 0.002; 3) 0.003

where $x = k_1 z/(k_{-1} + k_2 z^2)$. The main bifurcation curves can be written explicitly for different combinations of parameters. From the stationarity equation (3.45) and the condition $\Delta = 0$, the curves of multiplicity of steady states can be obtained:

$$L_\Delta(k_1, k_2): \qquad k_2(z) = \frac{k_{-1}(k_{-3} + 2k_3 z^2)}{k_{-3}z^2(1 - 2z) - 6k_3 z^4} \, ,$$

$$k_1 = k_1(z, k_2(z)) \, ,$$

$$L_\Delta(k_1, k_{-1}): \qquad k_{-1}(z) = \frac{2k_2 z^2 (k_{-3}(1 - z) - 2k_3 z^2)}{2k_3 z^2 + k_{-3}} - k_2 z^2 \, ,$$

$$k_1 = k_1(z, k_{-1}(z)) \, .$$

We can obtain the curves of neutrality from (3.45) and the condition $\sigma = 0$:

$$L_\sigma(k_1, k_2): \qquad k_2(z) = \frac{k_1(2k_3 z^2 + k_{-3}) + k_{-3}^2 z(1 + 4z)}{k_{-3}z^2 - 2z^3(k_3 z + k_{-3})} \, ,$$

$$k_1 = k_1(z, k_2(z)) \, ,$$

$$L_\sigma(k_1, k_{-1}): \qquad k_{-1}(z) = \frac{k_{-3}z^2(k_2 - 4k_{-3}) - 2k_2 z^3(k_3 z + k_{-3}) - k_{-3}^2 z}{2k_3 z^2 + k_{-3}} \, ,$$

$$k_1 = k_1(z, k_{-1}(z)) \, ,$$

where the functions $k_1(z)$ are determined in (3.46).

Examples of the curves of multiplicity and neutrality for changing the third parameter are shown in Figs. 3.16 and 3.17. It can be seen from the figures that, for example, with an increase of the parameter value k_{-3} the region of multiplicity of steady states and the region of neutrality of steady states increase. Similarly the character-

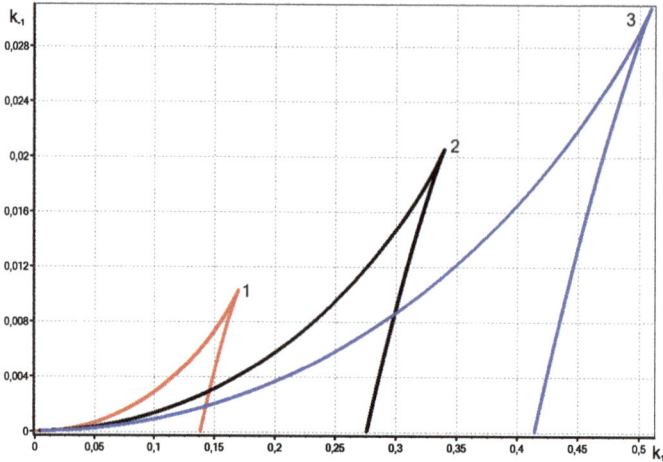

Fig. 3.16: Curves of multiplicity $L_\Delta(k_1, k_{-1})$ at 1) $k_2 = 1$; 2) $k_2 = 2$; 3) $k_2 = 3$

istics of regions of multiplicity of steady states and their stability can be obtained by varying other parameters.

The mutual arrangement of curves L_Δ: $\Delta = 0$ and L_σ: $\sigma = 0$, divides the plane of parameters into a few regions, differing in the number and type of stability of steady states (see Fig. 3.18). The parametric portrait of system (3.44) in the $(k_1, k_a - 1)$ plane is plotted for the following values of the parameters: $k_1 = 0.12$, $k_{-1} = 0.001$, $k_2 = 1$, $k_3 = 0.0032$, $k_{-1} = 0.002$.

Direct calculation of model (3.44) allows you to get the solutions $z(t)$, $x(t)$ and investigate the dependence of the oscillation period on the parameters. For example, with an increase in the parameter k_{-3}, the period of oscillation decreases (Fig. 3.19 and 3.20). A distinctive feature of time dependence corresponding to the auto-oscillation modes is a relaxing character. Sections of slow change of concentrations are replaced by sections with sharp jumps in time. The system is rebuilt from one mode to another.

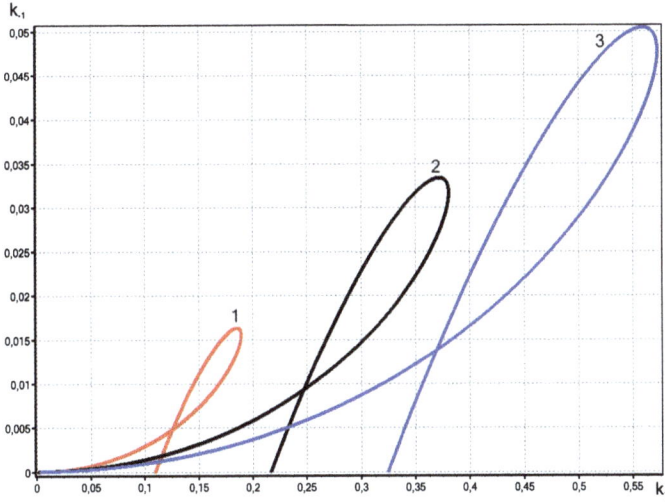

Fig. 3.17: Curves of neutrality $L_\sigma(k_1, k_{-1})$ at 1) $k_2 = 1$; 2) $k_2 = 2$; 3) $k_2 = 3$

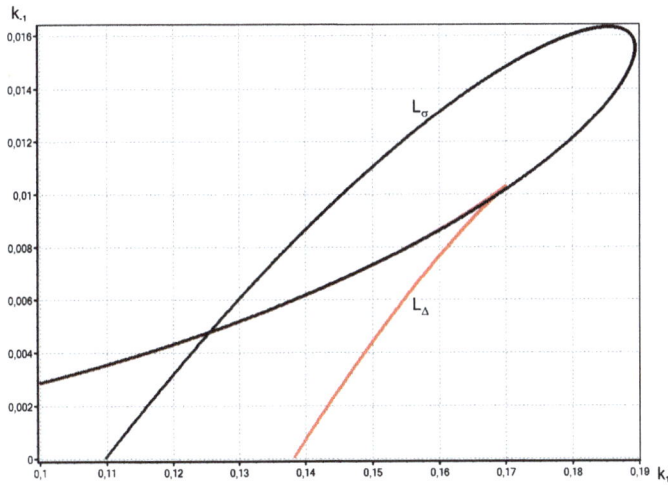

Fig. 3.18: Parametric portrait in (k_1, k_{-1})

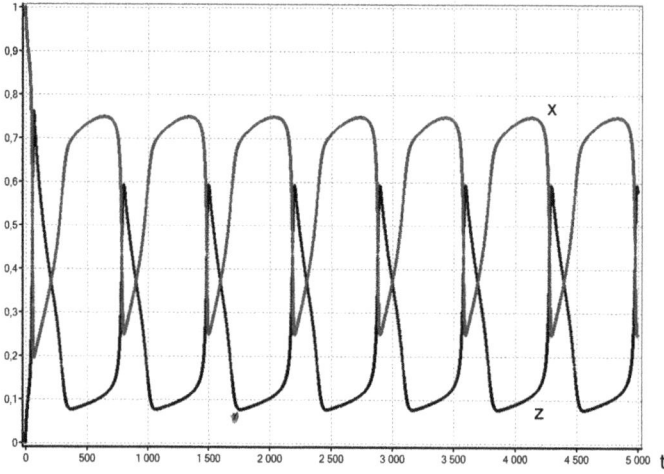

Fig. 3.19: Time dependencies $z(t)$, $x(t)$ at $k_1 = 0.15$, $k_{-1} = 0.01$, $k_2 = 1$, $k_3 = 0.0032$, $k_{-3} = 0.002$

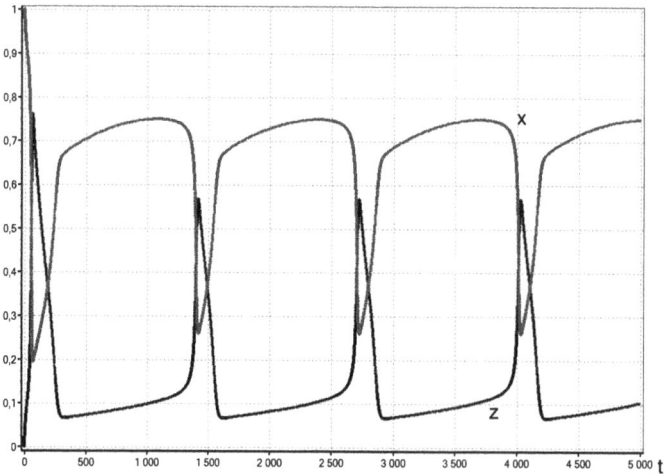

Fig. 3.20: Time dependencies $z(t)$, $x(t)$ at $k_1 = 0.15$, $k_{-1} = 0.01$, $k_2 = 1$, $k_3 = 0.0032$, $k_{-3} = 0.001$

Autocatalytic oscillators with buffer step $Z + X \rightleftharpoons Y$

Similarly to (3.32) we consider the scheme of transformations:

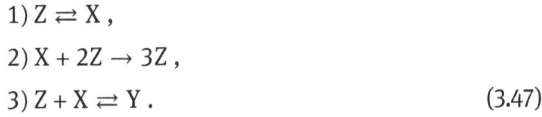

$$1)\ Z \rightleftharpoons X\,,$$
$$2)\ X + 2Z \rightarrow 3Z\,,$$
$$3)\ Z + X \rightleftharpoons Y\,. \tag{3.47}$$

For scheme (3.47) we can write the following kinetic model:

$$\dot{z} = -k_1 z + k_{-1} x + k_2 x z^2 - k_3 x z + k_{-3} y\,,$$
$$\dot{y} = k_3 x z - k_{-3} y\,,$$

where $x = 1 - 2y - z$. Similarly to the models above we obtain the equation of steady states depending on one variable z:

$$-k_1 z + \frac{k_3(1 - z)(k_{-1} + k_2 z^2)}{2k_3 z + k_{-3}} = 0\,. \tag{3.48}$$

The parametric dependencies can be obtained from equation (3.48):

$$k_1(z) = \frac{k_{-3}(1 - z)(k_{-1} + k_2 z^2)}{z(2k_3 z + k_{-3})}\,,$$

$$k_{-1}(z) = \frac{k_1 z(2k_3 z + k_{-3}) - k_2 k_{-3} z^2(1 - z)}{k_{-3}(1 - z)}\,,$$

$$k_2(z) = \frac{k_1 z(2k_3 z + k_{-3}) - k_{-1} k_{-3}(1 - z)}{k_{-3} z^2(1 - z)}\,,$$

$$k_3(z) = \frac{k_{-3}(1 - z)(k_{-1} + k_2 z^2) - k_1 k_{-3} z}{2k_1 z^2}\,,$$

$$k_{-3}(z) = \frac{2k_1 k_3 z^2}{(1 - z)(k_{-1} + k_2 z^2) - k_1 z}\,. \tag{3.49}$$

As an example one of the parametric dependencies (3.48) has been plotted in Fig. 3.21 for the following values of the parameters: $k_1 = 0.12$, $k_{-1} = 0.001$, $k_2 = 1$, $k_3 = 0.0032$, $k_{-3} = 0.002$.

The elements of the Jacobian matrix for right sides of the system have the form:

$$a_{11} = -k_1 - k_{-1} - k_2 z^2 + 2k_2 x z + k_3 z - k_3 x\,,$$
$$a_{12} = -2k_{-1} - 2k_2 z^2 - 2k_3 z + k_{-3}\,,$$
$$a_{21} = k_3 x - k_3 z\,,$$
$$a_{22} = -2k_3 z - k_{-3}\,.$$

Further, we get the equations for the curves of multiplicity from the stationarity equation (3.48) and the condition $\Delta = 0$. Such equations of the bifurcation curves can be

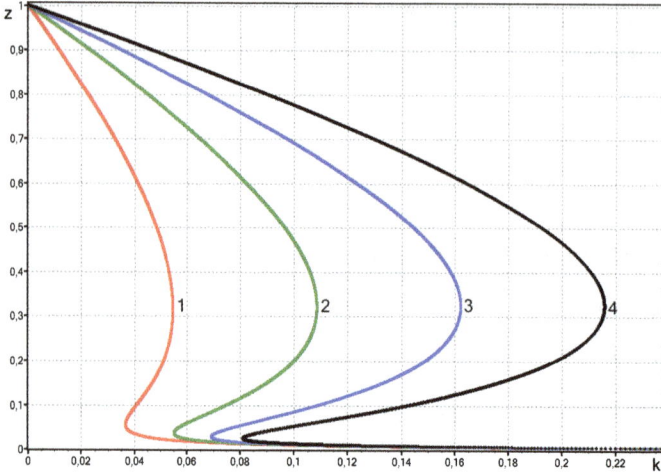

Fig. 3.21: Parametric dependencies $z(k_1)$ at 1) $k_2 = 0.5$; 2) 1; 3) 1.5; 4) 2

written for different combinations of parameters.

$$L_\Delta(k_1, k_2): \quad k_2(z) = \frac{4k_3^2 z^2 (z - x) - k_{-1}(2k_3 xz + k_{-3})}{z^2(k_{-3}(1 - 2x) - 2k_3 xz)},$$

$$k_1 = k_1(z, k_2(z)),$$

$$L_\Delta(k_1, k_{-1}): \quad k_{-1}(z) = \frac{4k_3^2 z^2 (z - x) + 2k_2 xz^2 (k_3 z + k_{-3}) - k_2 k_{-3} z^2}{2k_3 xz + k_{-3}},$$

$$k_1 = k_1(z, k_{-1}(z)).$$

Similarly the equations for the curves of neutrality can be obtained from the stationarity equation (3.48) and the condition $\sigma = 0$:

$$L_\sigma(k_1, k_2): \quad k_2(z) = \frac{z(2k_3 z^2 + k_{-3})[k_3(x + z) + k_{-1} + k_{-3}] + k_{-1}k_{-3}(1 - z)}{z^2(2k_3 z + k_{-3})(2x - z) + k_{-3}z^2(1 - z)},$$

$$k_1 = k_1(z, k_2(z)),$$

$$L_\sigma(k_1, k_{-1}): \quad k_{-1}(z) = \frac{z(2k_3 z + k_{-3})[2k_3 xz - k_3(x + z) - k_{-3}]}{k_{-3}(1 - z) + z(2k_3 z + k_{-3})},$$

$$k_1 = k_1(z, k_{-1}(z)),$$

where $x = 1 - z - 2k_3 z(1 - z)/(2k_3 z + k_{-3})$. The functions $k_1(z)$ were determined in (3.49). Some of the curves of multiplicity and neutrality are plotted in Figs. 3.22 and 3.23.

The variation of the parameter k_{-3} has shown that the region of multiplicity of steady states is increased with an increase in k_{-3}. Similarly a study of the dependence of the region of multiplicity of steady states on the other parameters can be performed.

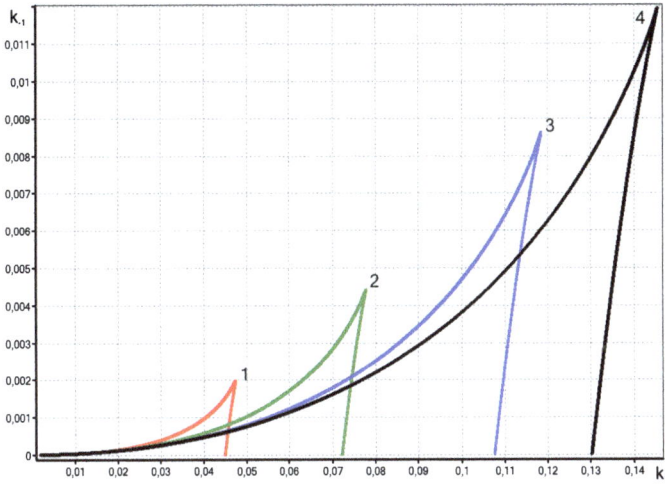

Fig. 3.22: Curves of multiplicity $L_\Delta(k_1, k_{-1})$ at 1) $k_{-3} = 0.001$; 2) 0.002; 3) 0.003

It is easy to do based on explicit expressions of the parametric dependencies for steady states and bifurcation curves. The curves of neutrality are plotted in Fig. 3.23. Here it is shown that the characteristic loop of the bifurcation curve significantly changes its position in the plane of the selected parameters at increasing values of k_{-3}.

The parametric portrait of the system, which is based on curves of multiplicity and neutrality of steady states, is presented in Fig. 3.24. The results of the initial step of the parametric analysis of a base model are accumulated here. It is a kind of skeleton which can be expanded by other details in a deeper parametric analysis of the appropriate dynamic models.

The phase portrait of the dynamical system is presented in Figure 3.25. The set of parameters was taken from the region of three unstable steady states in Fig. 3.24. All three steady states are inside a stable limit cycle.

Phase trajectories from any initial data tend to the limit cycle, which is characterized by undamped auto-oscillations of the concentrations of x, y, z.

Time dependencies of the corresponding phase portrait in Fig. 3.25 are shown in Fig. 3.26 for the same values of the parameters.

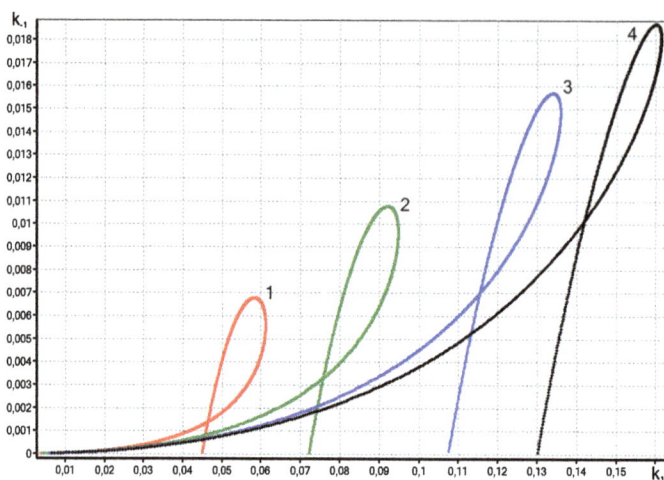

Fig. 3.23: Curves of neutrality $L_\sigma(k_1, k_{-1})$ at 1) $k_{-3} = 0.0005$; 2) 0.001; 3) 0.002; 4) 0.003

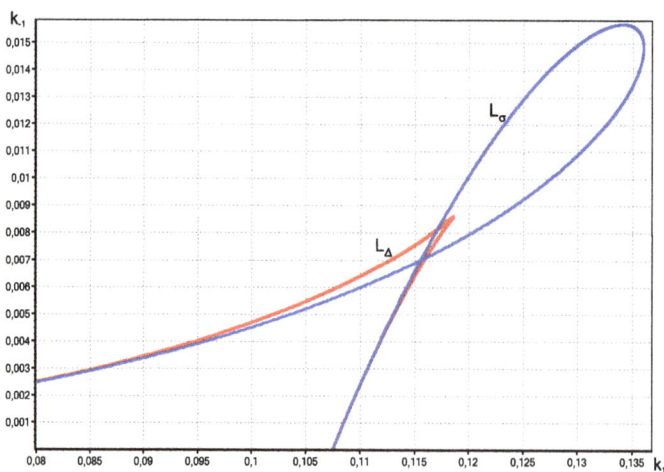

Fig. 3.24: Parametric portrait in (k_1, k_{-1})

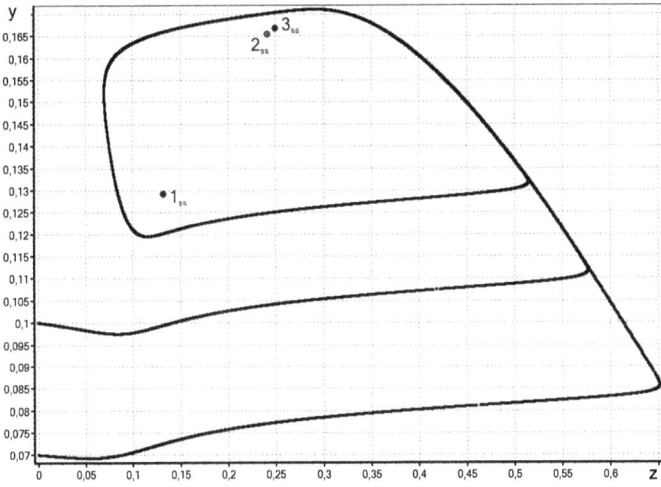

Fig. 3.25: Phase portrait (z, y) at $k_1 = 0.1175$, $k_{-1} = 0.008$, $k_2 = 1$, $k_3 = 0.0032$, $k_{-3} = 0.002$

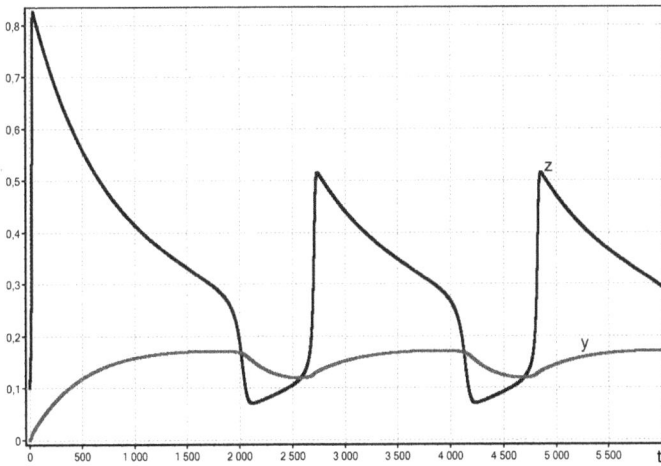

Fig. 3.26: Time dependencies $z(t)$ and $y(t)$

Autocatalytic oscillators with buffer step $2Z \rightleftharpoons 2Y$

Finally we consider one more scheme of transformations:

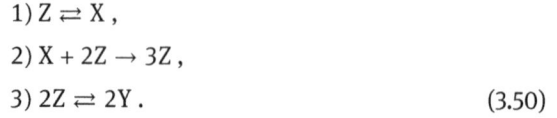

$$1)\, Z \rightleftharpoons X,$$
$$2)\, X + 2Z \rightarrow 3Z,$$
$$3)\, 2Z \rightleftharpoons 2Y. \tag{3.50}$$

A kinetic model corresponding to (3.50) is:

$$\dot{z} = -k_1 z + k_{-1} x + k_2 x z^2 - 2k_3 z^2 + 2k_{-3} y^2,$$
$$\dot{y} = 2k_3 z^2 - 2k_{-3} y^2, \tag{3.51}$$

where $x = 1 - y - z$. Similarly to the schemes we considered above we obtain the stationarity equation depending on variable z:

$$- k_1 z + (1 - z - \alpha z)(k_{-1} + k_2 z^2) = 0, \tag{3.52}$$

where $\alpha = \sqrt{k_3 / k_{-3}}$. From the stationarity equation (3.52) the parametric dependencies can be written:

$$k_1(z) = \frac{(1 - z - \alpha z)(k_{-1} + k_2 z^2)}{z},$$
$$k_{-1}(z) = \frac{k_1 z - k_2 z^2 (1 - z - \alpha z)}{1 - z - \alpha z},$$
$$k_2(z) = \frac{k_1 z - k_{-1}(1 - z - \alpha z)}{z^2 (1 - z - \alpha z)},$$
$$\alpha(z) = \frac{(1 - z)(k_{-1} + k_2 z^2) - k_1 z}{z(k_{-1} + k_2 z^2)}. \tag{3.53}$$

Examples of the calculation of some parametric dependencies (3.53) are given in Fig. 3.27 and Fig. 3.28 for the following values of the parameters: $k_1 = 0.12$, $k_{-1} = 0.001$, $k_2 = 1$, $k_3 = 0.0032$, $k_{-3} = 0.002$.

The elements of the Jacobian matrix for the right sides of system (3.51) are

$$a_{11} = -k_1 - k_{-1} - k_2 z^2 + 2k_2 xz - 4k_3 z, \quad a_{21} = 4k_3 z,$$
$$a_{12} = -k_{-1} - k_2 z^2 + 4k_{-3} y, \qquad\qquad a_{22} = -4k_{-3} y.$$

The equations for the curves of multiplicity are obtained according to the condition $\Delta = 0$ and the stationarity equation (3.52). Some of them are

$$L_\Delta(k_1, k_2): \quad k_2(z) = \frac{k_{-1}[\alpha k_{-3}(1 - z - \alpha z) + z(\alpha k_{-3} + k_3)]}{z^2 (\alpha k_{-3}(1 - z - \alpha z) - z(\alpha k_{-3} + k_3))},$$
$$k_1 = k_1(z, k_2(z)),$$

$$L_\Delta(k_1, k_{-1}): \quad k_{-1}(z) = \frac{k_2 z^2 (\alpha k_{-3}(1 - z - \alpha z) - z(\alpha k_{-3} + k_3))}{\alpha k_{-3}(1 - z - \alpha z) + z(\alpha k_{-3} + k_3)},$$
$$k_1 = k_1(z, k_{-1}(z)).$$

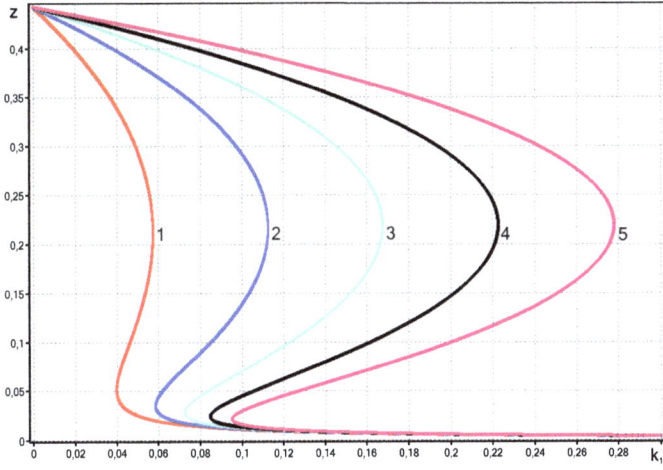

Fig. 3.27: Parametric dependencies $z(k_1)$ at 1) $k_2 = 0.5$; 2) 1; 3) 1.5; 4) 2; 5) 2.5

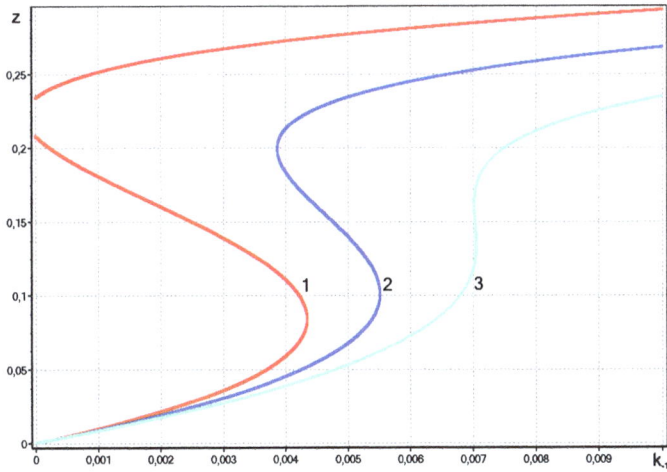

Fig. 3.28: Parametric dependencies $z(k_{-1})$ at 1) $k_1 = 0.11$; 2) 0.12; 3) 0.13

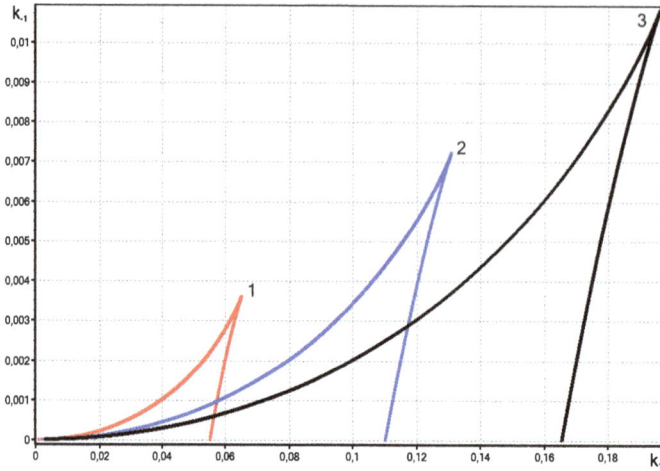

Fig. 3.29: Curves of multiplicity $L_\Delta(k_1, k_{-1})$ at 1) $k_2 = 0.5$; 2) 1; 3) 1.5

An example of the curves of multiplicity is given in Fig. 3.29 where we vary the third parameter k_2. Varying the values of the parameter k_2 significantly changes the sizes of the regions of the multiplicity of steady states.

Similarly, from (3.52) and the condition $\sigma = 0$ we have the equations for the curves of neutrality:

$$L_\sigma(k_1, k_2): \quad k_2(z) = \frac{k_{-1}(1 - \alpha z) + 4z^2(\alpha k_{-3} + k_3)}{z^2(1 - z - \alpha z) - z^3},$$

$$k_1 = k_1(z, k_2(z)),$$

$$L_\sigma(k_1, k_{-1}): \quad k_{-1}(z) = \frac{z^2(k_2(1 - 2z - \alpha z) - 4(\alpha k_{-3} + k_3))}{1 - \alpha z},$$

$$k_1 = k_1(z, k_{-1}(z)),$$

where $k_1(z)$ was determined in (3.53). More equations of the bifurcation curves can be written for different combinations of model parameters.

The curves of neutrality are shown in Fig. 3.30. The sizes of the regions of instability of steady states increase at growing values of k_2.

Using the specifics of the mathematical model (3.51), the equations of the bifurcation curves were obtained explicitly in the plane of parameters (k_1, k_{-1}). The analysis of their mutual arrangement allows us to classify the number and stability of the steady states (see Fig. 3.31).

The behavior of system (3.51) in time has the character of relaxation oscillations for sufficiently small values k_3, k_{-3}. During the period, there are two regions of motion: fast (motion along the initial trajectory) and slow (motion to a steady state). The region of fast motion corresponds to a transition from one branch of monotonicity of the function $z(t)$ to another. The dependence of period and amplitude of the oscilla-

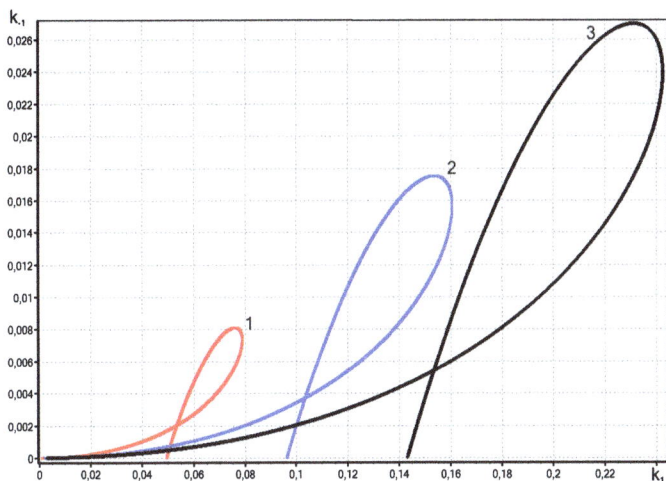

Fig. 3.30: Curves of neutrality $L_\sigma(k_1, k_{-1})$ at 1) $k_2 = 0.5$; 2) 1; 3) 1.5

tions on k_3, for example, can be numerically studied. An increase of k_3 reduces both the amplitude and the period of oscillations.

Graphs of time dependencies with a single unstable steady state and the corresponding phase portrait with a limit cycle are shown in Figs. 3.32 and 3.33.

Comparative analysis of the regions of the parametric portrait for the mechanisms of type (3.32) with different buffer stages shows that in the plane (k_1, k_{-1}), the mechanisms with buffer stages $2Z \leftrightarrow Y$ and $2Z \leftrightarrow 2Y$ have the largest regions where there is a single and unstable steady state (and hence oscillations). In this sense, the least probability of occurrence of undamped oscillations is characterized by the reaction scheme with a buffer step $XZ \leftrightarrow Y$.

This conclusion is hardly a common result, because in other planes of parameters the ratio of the size of these regions may be different.

A general property of plotted curves is that by varying the parameters k_3, k_{-3} $(k_3/k_{-3} = \mathrm{const})$ the multiplicity region of steady states is not changed. The region bounded by the curve L_σ decreases with an increase in k_3, k_{-3}. Further this region disappears at definite values of k_3 and k_{-3}, and it tends to maximum size when k_3 and k_{-3} tend to zero. However, note that if $k_3 = k_{-3} = 0$ (step 3) there are no auto-oscillations. At arbitrarily small k_3, k_{-3} the system is characterized by an arbitrarily large period of oscillations. Their frequency is set by the values of k_3, k_{-3}. The lower boundary of the oscillation period is already determined by another set of parameters.

The models considered above can serve as a base for building more difficult realistic models. The results of numerical and qualitative analysis allow us to obtain some a priori estimates of the dynamics of complex catalytic processes [1, 2, 8, 89–96].

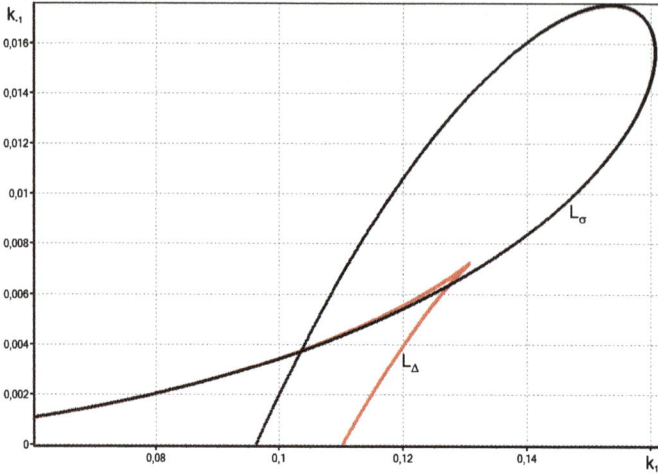

Fig. 3.31: Parametric portrait in (k_1, k_{-1}) at $k_2 = 1$, $k_3 = 0.0032$, $k_{-3} = 0.002$

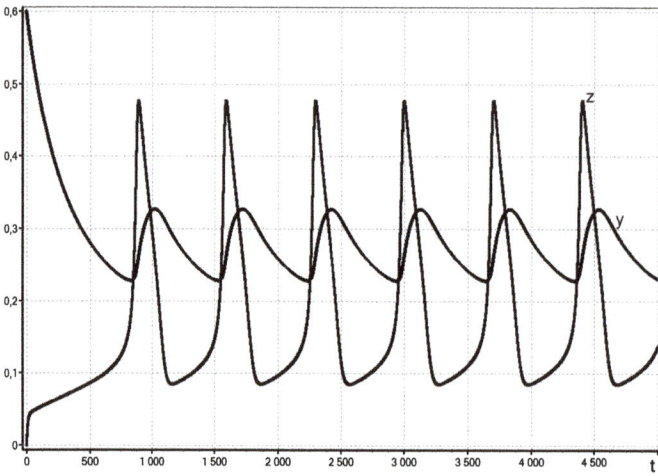

Fig. 3.32: Time dependencies $z(t)$, $y(t)$ at $k_1 = 0.13$, $k_{-1} = 0.01$, $k_2 = 1$, $k_3 = 0.0032$, $k_{-3} = 0.002$

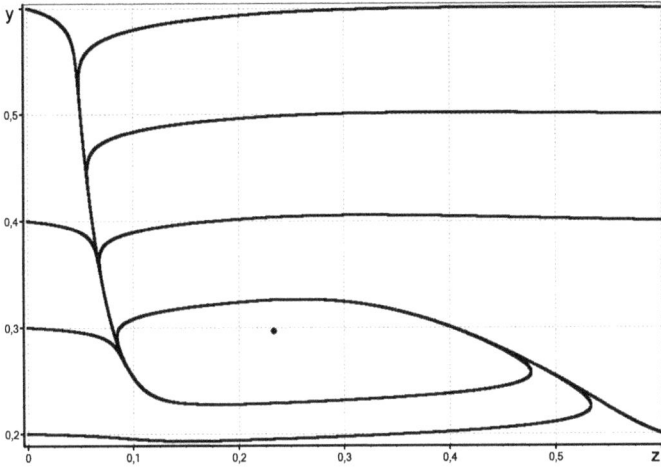

Fig. 3.33: Phase portrait of system (3.51) in the plane (z, y)

3.2.3 Association reaction

In the simplest case, the phenomenon of association of some substances X (monomer) is the formation of complexes: X_2 (dimer), X_3 (trimer), etc. The formation of such complexes is characteristic of liquids and adsorbed substances on a solid surface. As shown in [91], kinetic models of the association processes can have auto-oscillation regimes

In this section, we will do a parametric analysis of two basic models of an association reaction. As above, a set of kinetic parameters will be found in which exist a single unstable steady state. It guarantees the existence of a limit cycle in the corresponding mathematical model.

Three-step mechanism

Consider a scheme of transformations:

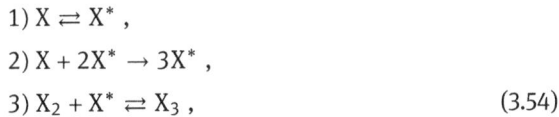

$$1)\ X \rightleftarrows X^* ,$$
$$2)\ X + 2X^* \rightarrow 3X^* ,$$
$$3)\ X_2 + X^* \rightleftarrows X_3 , \tag{3.54}$$

where X, X_2, X_3 are a monomer, a dimer, and a trimer, respectively. X^* is an excited (active) monomer form. The mechanism (3.54) is similar to (3.37), but the buffer step 3) is somewhat different.

Two linear conservation laws corresponds to scheme (3.54):

$$[X_2] + [X_3] = \mathrm{const}_1 ,$$
$$[X^*] + [X] + 2[X_2] + 3[X_3] = \mathrm{const}_2 ,$$

where the square brackets indicate the concentration of the relevant substance. Without loss of generality we accept that $const_1 = \alpha < 1$, $const_2 = 1$. A nonstationary kinetic model for scheme (3.54) has the form

$$\dot{x} = -k_1 x + k_{-1} x^* - k_2 x x^{*2} ,$$

$$\dot{x}_3 = k_3 x_2 x^* - k_{-3} x_3 , \qquad (3.55)$$

where x, x_3 are the dimensionless concentrations of X, X_3 respectively, and k_i are the rate constants of the stages. The values of x^*, x_2 are determined from conservation laws

$$x^* = 1 - 2\alpha - x - x_3 ,$$

$$x_2 = \alpha - x_3 . \qquad (3.56)$$

Note that scheme (3.54) can only serve as a fragment of a more detailed scheme of an association reaction. For example, there is no the step of the dimer formation in (3.54). However, this simple scheme allows us to interpret auto-oscillations in a system. Therefore, considering a real scheme of conversion, mechanism (3.54) can be used as a base scheme. Starting from the simplest model studied, it can be built into a more detailed model with minimum cost.

The solutions (3.55) are defined in the polyhedron of reaction

$$S_r = \{(x, x_3) : x \geq 0, \ 0 \leq x_3 \leq \alpha, \ x + x_3 \leq 1 - 2\alpha\} .$$

Stationary and nonstationary solutions of system (3.55) are searched in the ω-invariant set S_r.

Steady states are determined from the stationarity equation. It can be written as an equality of isoclines $f(x^*) = g(x^*)$:

$$x_3 = 1 - 2\alpha - x^* - \frac{k_{-1} x^*}{k_1 + k_2 x^{*2}} ,$$

$$x_3 = \frac{\alpha x^*}{k_{-3}/k_3 + x^*} ,$$

where we can accept $k_2 = 1$. The inequalities for the parameters (at $k_{-3} = \sqrt{3 k_1}$) can be obtained from the requirement of uniqueness of the intersection point of isoclines

$$8 < k_{-1}/k_1 < \frac{8 - 18\alpha}{1 - 3\alpha} .$$

The steady state will be single and unique in this case.

For (3.55), (3.56) the elements of the Jacobian matrix of the right sides of the equations are calculated in the form

$$a_{11} = -k_1 - k_{-1} - 3x^{*2} + 2x^*(1 - 2\alpha - x_3) - k_3(\alpha - x_3) ,$$

$$a_{12} = -k_1 - x^{*2} + k_3 x^* + k_{-3} , \quad a_{21} = k_3(\alpha - x_3) , \quad a_{22} = -k_3 x^* - k_{-3} .$$

It can be shown that the sign of the free member of the characteristic equation Δ matches with the sign of the difference of the derivatives f' and g'. For sufficiently small k_3, k_{-3} the value $\Delta < 0$, which means instability of the steady states.

From the stationarity equation $f(x^*) = g(x^*)$ the parametric dependencies can be written

$$k_{-1} = \frac{((k_{-3} + k_3 x^*)(1 - 2\alpha - x^*) - \alpha x^*)(k_1 + x^{*2})}{(k_{-3} + x^*)x^*},$$

$$k_1 = \frac{\alpha x^{*3} - ((1 - 2\alpha - x^*)x^{*2} - k_{-1}x^*)(k_{-3} + x^{*2})}{(1 - 2\alpha - x^*)(k_{-3} + k_3 x^*) - \alpha x^*}. \tag{3.57}$$

As above, the region of multiplicity of steady states can be plotted on the basis of explicit expressions (3.57). Studying the influence of different parameters on the sizes of the region of multiplicity of steady states (its boundaries are defined by the curves L_Δ) showed this region is reduced in size with growth in α and k_{-1}/k_1. The neutrality curve L_σ can be plotted on the basis of (3.57) and the condition $\sigma = a_{11}a_{22} = 0$. Variation of the parameters α, k_1, k_{-1}, k_3, k_{-3} has shown that as a rule as they grow the region of single and unstable steady states decreases. In particular, at $\alpha = 0.1$, $k_3 = 10^{-4}$, $k_{-3} = 0.24 \cdot 10^{-4}$, $k_{-1}/k_1 = 8.5$ a single steady state can be unstable.

Four-step mechanism

Let us add to scheme (3.54) one more step 4):

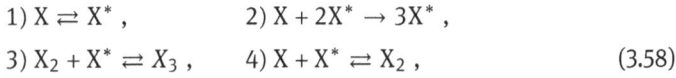

$$\begin{array}{ll} 1)\ X \rightleftarrows X^*, & 2)\ X + 2X^* \rightarrow 3X^*, \\ 3)\ X_2 + X^* \rightleftarrows X_3, & 4)\ X + X^* \rightleftarrows X_2, \end{array} \tag{3.58}$$

where the notation is similar to (3.54). The following nonstationary kinetic model matches scheme (3.58):

$$\dot{x} = -k_1 x + k_{-1} x^* - k_2 x x^{*2},$$

$$\dot{x}_2 = -k_3 x_2 x^* + k_{-3} x_3 + k_4 x x^* - k_{-4} x_2,$$

$$\dot{x}_3 = k_3 x_2 x^* - k_{-3} x_3, \tag{3.59}$$

where $x^* = 1 - x - 2x_2 - 3x_3$. The solutions $x(t)$, $x_2(t)$, $x_3(t)$ of system (3.59) are considered in the field

$$S_r = \{(x, x_2, x_3):\ x, x_2, x_3 \geq 0,\ x + 2x_2 + 3x_3 \leq 1\}.$$

The simplex of the reaction S_r is the ω-invariant set for (3.59), i.e., for any initial data $x(0)$, $x_2(0)$, $x_3(0) \in S_r$ the solutions $x(t)$, $x_2(t)$, $x_3(t) \in S_r$ for any $t > 0$. This guarantees the existence of S_r and at least one steady state for (3.59).

Steady states for (3.59) are solutions of a system of algebraic equations

$$k_1 x - k_{-1} x^* + k_2 x x^{*2} = 0,$$

$$k_3 x_2 x^* - k_{-3} x_3 - k_4 x x^* + k_{-4} x_2 = 0,$$

$$k_3 x_2 x^* - k_{-3} x_3 = 0,$$

$$x + x^* + 2x_2 + 3x_3 = 1. \tag{3.60}$$

After elementary transformations, system (3.60) can be rewritten in the form

$$x_2 = \frac{k_{-3}x_3}{k_2x^*} \, ,$$

$$x_3 = \frac{k_3k_4x^{*2}(1-x^*)}{3k_3k_4x^{*2} + k_{-3}(k_{-4} + 2k_4x^*)} = f(x^*) \, ,$$

$$x_3 = \frac{k_{-4}((1-x^*)(k_1 + k_2x^{*2}) - k_{-1}x^*) - 2k_4k_{-1}x^{*2}}{k_{-4}(k_1 + k_2x^{*2})} = g(x^*) \, . \tag{3.61}$$

Steady states are solutions of the equality $f(x^*) = g(x^*)$, i.e., the equation in one variable x^*. Knowing the stationary value of x^*, it is easy to find appropriate values for x_2, x_3, and x.

We select the two parameters k_1, k_{-1} in (3.61). They are linear in the stationarity equation

$$F(x^*, k_1, k_{-1}) = f(x^*) - g(x^*) = 0 \, . \tag{3.62}$$

Therefore, we can obtain the parametric dependencies from (3.62):

$$k_{-1} = \frac{k_{-3}k_4(1-x^*)(k_1 + k_2x^{*2})}{x^*(3k_3k_4x^{*2} + k_{-3}(k_{-4} + 2k_4x^*))} \, ,$$

$$k_1 = \frac{k_{-1}x^*((k_{-4} + 2k_4x^*)k_{-3} + 3k_3k_4x^{*2}) - k_{-3}k_{-4}k_2(1-x^*)x^{*2}}{k_{-3}k_{-4}(1-x^*)} \, , \tag{3.63}$$

where x^* is changed as a parameter in the range $(0, 1)$.

The explicit expressions of the parametric dependencies (3.63) allow us to relatively easily plot the bifurcation curves L_Δ and L_σ (the curves of multiplicity and neutrality of steady states) by using, for example, the procedure of direct calculation of the determinant and trace of the Jacobian matrix for the system of stationarity equations. One of the possible parametric portraits of system (3.60) formed by the curves L_Δ and L_σ can be found in [91].

Numerical study of the mutual arrangement of curves of multiplicity L_Δ and neutrality L_σ of steady states showed at $k_2 = 1$, $k_3 = 10^{-4}$, $k_{-3} = 0.24 \cdot 10^{-4}$, $k_4 = 10^{-5}$, $k_{-4} = 2.45 \cdot 10^{-5}$, the plane of parameters (k_{-1}, k_1) is divided into six subareas, distinguished by the number and stability of steady states.

The oscillation amplitude can be evaluated a priori from the analysis of diagrams of isoclines $f(x^*)$, $g(x^*)$. An a priori estimate can be obtained for the oscillation period. Knowing the value of the eigenvalues in a steady state, it is located on the border of the stability region (in a neighborhood of a point of Andronov–Hopf bifurcation), and the oscillation period is calculated by the formula $\tau = 2\pi/|\lambda|$ [42, 43, 97].

Comparative analysis of the two models (3.55) and (3.59) shows that the resulting parametric dependencies and the conditions for the occurrence of undamped oscillations are close enough. It allows us to use a simplified model (3.55) for a preliminary assessment of critical phenomena in real systems. Knowing the values of the parameters of the basic model, it is much easier to carry out the parametric analysis of more realistic models of specific processes of association.

3.3 Catalytic schemes of transformations

The traditional objects of study in heterogeneous catalysis are oxidation reactions by Pt and especially CO [1, 2, 43, 45–68, 98–151]. Experimental data confirm the possibility of complicated dynamic behaviour in these reactions. In this section the general procedure of parametric analysis is demonstrated with the example of two model mechanisms of catalytic oxidation reactions (without autocatalytic stages) with various "buffer" stages. The use of "buffer" steps allows to describe critical phenomena such as a multiplicity of steady states and auto-oscillations.

3.3.1 Catalytic triggers

Parallel catalytic trigger 1
The simplest formal scheme of the catalytic reaction, allowing three steady states, is as follows:

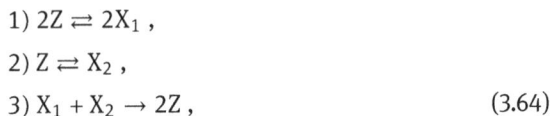

$$1)\ 2Z \rightleftarrows 2X_1 \,,$$
$$2)\ Z \rightleftarrows X_2 \,,$$
$$3)\ X_1 + X_2 \rightarrow 2Z \,, \tag{3.64}$$

where Z is a catalyst and X_1, X_2 are intermediate substances. A three-step mechanism (3.64) may be interpreted as known in the catalysis mechanism of Langmuir–Hinshelwood. For example, CO oxidation on Pt:

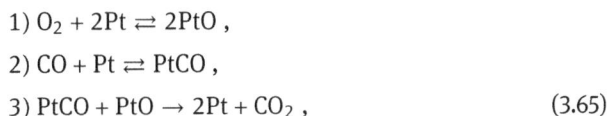

$$1)\ O_2 + 2Pt \rightleftarrows 2PtO \,,$$
$$2)\ CO + Pt \rightleftarrows PtCO \,,$$
$$3)\ PtCO + PtO \rightarrow 2Pt + CO_2 \,, \tag{3.65}$$

or as the mechanism of oxidation of some substance A on metal:

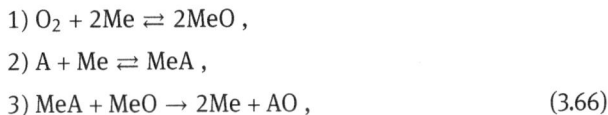

$$1)\ O_2 + 2Me \rightleftarrows 2MeO \,,$$
$$2)\ A + Me \rightleftarrows MeA \,,$$
$$3)\ MeA + MeO \rightarrow 2Me + AO \,, \tag{3.66}$$

In the reactions (3.65) and (3.66) the substances in the gas phase are often accepted as constant. Their partial pressures are included as cofactors in the expression for the rates of the corresponding stages. Therefore, the variation of the rate constants for the formal scheme (3.64) can be understood as the modification of these partial pressures. Other rate constants can be changed with temperature variation.

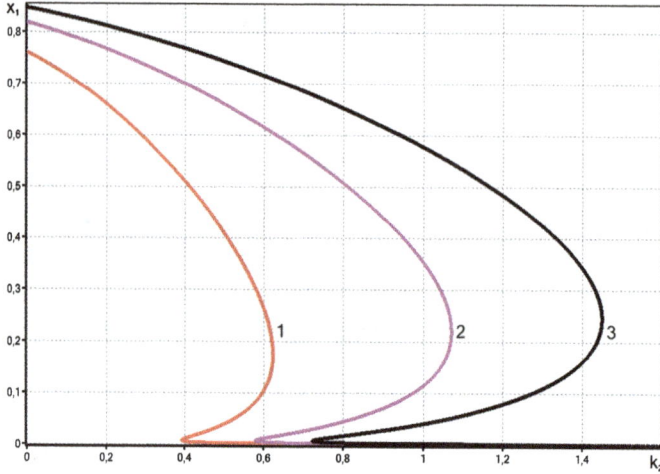

Fig. 3.34: Parametric dependencies $x_1(k_2)$ at 1) $k_1 = 0.5$; 2) 1; 3) 1.5

The kinetic model corresponding to scheme (3.64) is

$$\dot{x}_1 = 2k_1 z^2 - 2k_{-1} x_1^2 - k_3 x_1 x_2 ,$$
$$\dot{x}_2 = k_2 z - k_{-2} x_2 - k_3 x_1 x_2 , \qquad (3.67)$$

where $z = 1 - x_1 - x_2$. From the stationarity conditions for (3.67) we get:

$$x_2 = \frac{k_2(1 - x_1)}{k_2 + k_{-2} + k_3 x_1} ,$$

$$2k_1 \left(1 - x_1 - \frac{k_2(1 - x_1)}{k_2 + k_{-2} + k_3 x_1}\right) - 2k_{-1} x_1^2 - \frac{k_2 k_3 x_1(1 - x_1)}{k_2 + k_{-2} + k_3 x_1}) = 0 , \qquad (3.68)$$

From the stationarity equation (3.68) the parametric dependencies can be expressed

$$k_1(z) = \frac{(2k_{-1} x_1^2(k_2 + k_{-2} + k_3 x_1) + k_2 k_3 x_1(1 - x_1))(k_2 + k_{-2} + k_3 x_1)}{2(1 - x_1)^2(k_{-2} + k_3 x_1)^2} ,$$

$$k_{-1}(z) = \frac{2k_1(1 - x_1)^2(k_{-2} + k_3 x_1)^2 - k_2 k_3 x_1(1 - x_1)(k_2 + k_{-2} + k_3 x_1)}{2x_1^2(k_2 + k_{-2} + k_3 x_1)^2} ,$$

$$k_2(z) = \frac{-B + \sqrt{B^2 - 4AC}}{2A} , \qquad (3.69)$$

where $A = 2k_{-1} x_1^2 + k_3 x_1(1 - x_1)$, $B = (k_{-2} + k_3 x_1)(4k_{-1} x_1^2 + k_3 x_1(1 - x_1))$, $C = 2(k_{-2} + k_3 x_1)^2(k_{-1} x_1^2 - k_1(1 - x_1)^2)$.

The example of the plotted parametric dependencies of (3.69) are shown in Fig. 3.34.

Here we only note that if k_{-1}, k_{-2} are not equal to 0, there are either one or three steady states. So if $k_1 = k_2 = 1$, $k_3 = 10$, $k_{-1} = k_{-2} = 0.01$, system (3.67) has three

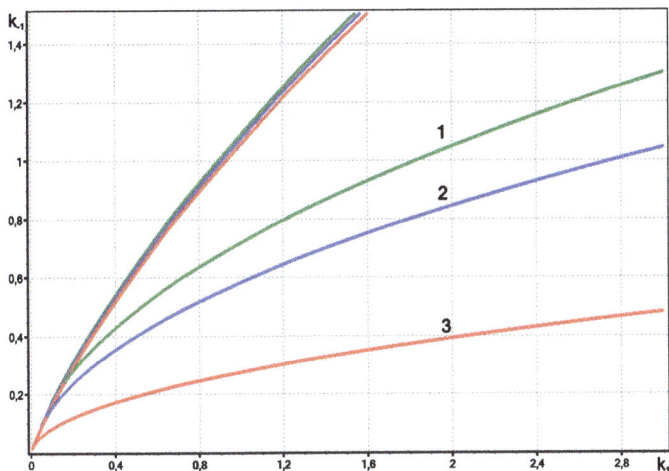

Fig. 3.35: Curves of multiplicity $L_\Delta(k_1, k_2)$ at 1) $k_{-2} = 0.01$; 2) 0.05; 3) 0.08

steady states. Two of them are stable and one is unstable. The system cannot make auto-oscillations, because the criterion of Bendixson (the criterion of absence of limit cycles) is met. The function is $P'_x + Q'_y < 0$, where P'_x, Q'_y are the derivatives of the corresponding right sides of (3.67).

The elements of the Jacobian matrix for (3.67) are of the form:

$$a_{11} = -4k_1(1 - x_1 - x_2) - 4k_{-1}x_1 - k_3x_2 \ ,$$
$$a_{12} = -4k_1(1 - x_1 - x_2) - k_3x_1 \ ,$$
$$a_{21} = -k_2 - k_3x_2 \ ,$$
$$a_{22} = -k_2 - k_{-2} - k_3x_1 \ .$$

As before, the stability of the steady state is determined by the roots of the character-istic equation where $\sigma < 0$. In this case, the curve of the multiplicity ($\Delta = 0$) is a main bifurcation curve. An example of plotted curves of multiplicity is shown in Fig. 3.35 at various values of the parameter k_{-2}.

Parallel catalytic trigger 2

Consider the following scheme of reactions which consists of three stages:

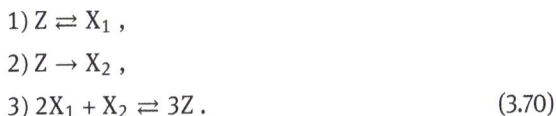

$$1) \ Z \rightleftarrows X_1 \ ,$$
$$2) \ Z \rightarrow X_2 \ ,$$
$$3) \ 2X_1 + X_2 \rightleftarrows 3Z \ . \tag{3.70}$$

A kinetic model for scheme (3.70) can be written in the form

$$\dot{x}_1 = k_1z - k_{-1}x_1 - 2k_3x_1^2x_2 \ ,$$
$$\dot{x}_2 = k_2z - k_{-2}x_2 - k_3x_1^2x_2 \ , \tag{3.71}$$

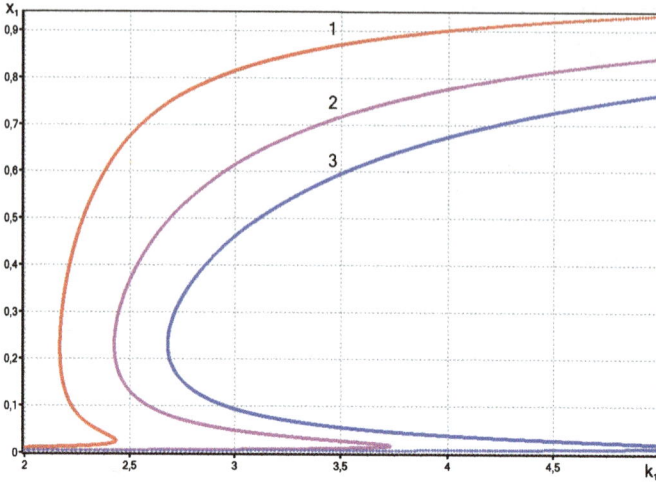

Fig. 3.36: Parametric dependencies $x(k_1)$ at 1) $k_{-1} = 0.2$; 2) 0.5; 3) 0.8

where $z = 1 - x_1 - x_2$. From the stationarity conditions for (3.71) we get:

$$x_2 = \frac{k_1(1 - x_1) - k_{-1}x_1}{k_1 + 2k_3x_1^2},$$

$$2k_2k_3x_1^2(1-x_1)+k_{-1}x_1(k_2+k_{-2})-k_1k_{-2}(1-x_1)-k_1k_3x_1^2(1-x_1)+k_{-1}k_3x_1^3 = 0. \quad (3.72)$$

Parametric dependencies are expressed from (3.72) as

$$k_1(z) = \frac{k_3x_1^2(2k_2(1 - x_1) + k_{-1}x_1) + k_{-1}x_1(k_2 + k_{-2})}{(1 - x_1)(k_{-2} + k_3x_1^2)},$$

$$k_2(z) = \frac{(k_{-2} + k_3x_1^2)(k_1(1 - x_1) - k_{-1}x_1)}{x_1(2k_3x_1(1 - x_1) + k_{-1})}. \quad (3.73)$$

Examples of some parametric dependencies (3.73) are shown in Fig. 3.36 with changes in the second parameter k_{-1}.

The elements of the Jacobian matrix for (3.71) take the form:

$$a_{11} = -k_1 - k_{-1} - 4k_3x_1x_2, \quad a_{12} = -k_1 - 2k_3x_1^2,$$
$$a_{21} = -k_2 - 2k_3x_1x_2, \quad a_{22} = -k_2 - k_{-2} - k_3x_1^2.$$

An example of plotting the curves of multiplicity L_Δ is shown in Fig. 3.37, varying the value of the parameter k_2.

One of the possible phase portraits of the dynamic system (3.71) for the case of three steady states is given in Fig. 3.38. In this special case the region of attraction of the stable steady state covers almost the whole reaction simplex of S. The system is stabilized to a stable steady state from almost all initial data, excepting a small neighbourhood of the point $(0, 1)$.

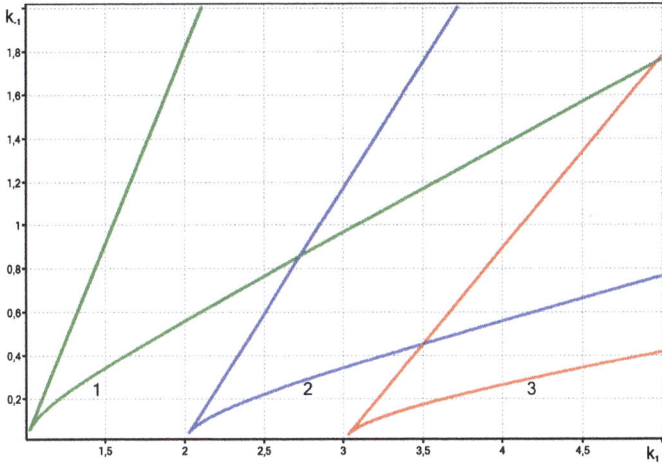

Fig. 3.37: Curves of multiplicity $L_\Delta(k_1, k_{-1})$ at 1) $k_2 = 0.5$; 2) 1; 3) 1.5

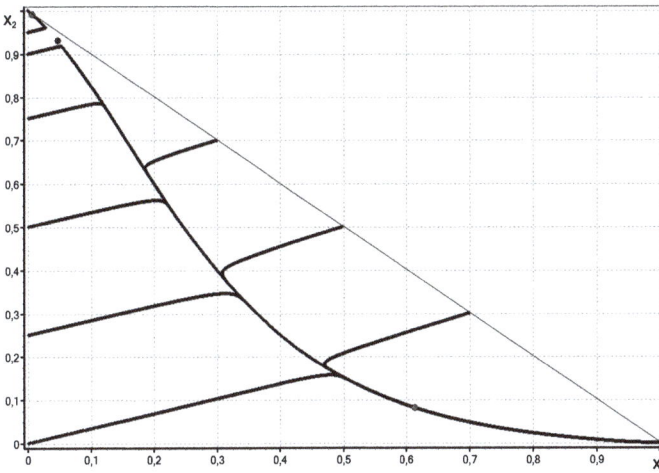

Fig. 3.38: Phase portrait in (x_1, x_2) at $k_1 = 3$, $k_{-1} = 0.5$, $k_2 = 1$, $k_{-2} = 0.001$, $k_3 = 10$

Sequential catalytic trigger

Consider the following scheme of transformations:

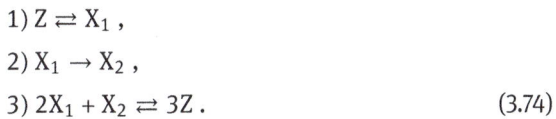

$$1)\ Z \rightleftarrows X_1 ,$$
$$2)\ X_1 \rightarrow X_2 ,$$
$$3)\ 2X_1 + X_2 \rightleftarrows 3Z . \tag{3.74}$$

The kinetic model corresponding to (3.74) is

$$\dot{x}_1 = k_1 z - k_{-1}x_1 + k_{-2}x_2 - k_2 x_1 - 2k_3 x_1^2 x_2 ,$$
$$\dot{x}_2 = k_2 x_1 - k_{-2}x_2 - k_3 x_1^2 x_2 , \tag{3.75}$$

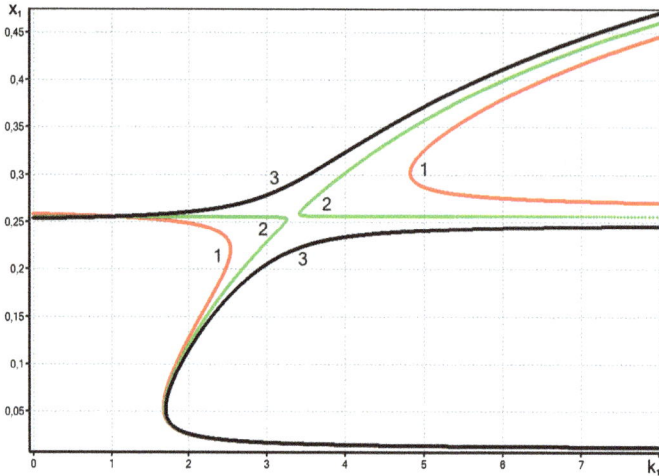

Fig. 3.39: Parametric dependencies $x_1(k_1)$ at $k_{-1} = k_{-2} = 0.01$, $k_2 = 1$, 1) $k_3 = 5$; 2) 5.11; 3) 5.2

where $z = 1 - x_1 - x_2$. In contrast to (3.64) the scheme (3.74) can be called sequential, because the first two stages are sequential. From the stationarity conditions for (3.75) we get:

$$x_2 = \frac{k_2 x_1}{k_{-2} + k_3 x_1^2},$$

$$k_1(1 - x_1)(k_{-2} + k_3 x_1^2) - k_1 k_2 x_1 - x_1(k_2 + k_{-1})(k_{-2} + k_3 x_1^2) + k_2 x_1(k_2 + k_{-2}) - 2k_2 k_3 x_1^3 = 0.$$
$$(3.76)$$

Parametric dependencies are found from the stationarity of equation (3.76):

$$k_1(z) = \frac{x_1\left((k_{-2} + k_3 x_1^2)(k_2 + k_{-1}) - k_2(k_2 + k_{-2}) + 2k_2 k_3 x_1^2\right)}{(1 - x_1)(k_{-2} + k_3 x_1^2) - k_2 x_1},$$

$$k_2(z) = \frac{-B + \sqrt{B^2 - 4AC}}{2A},$$

where $A = x_1$, $B = -x_1(k_1 + 3k_3 x_1^2)$, $C = (k_{-2} + k_3 x_1^2)(k_1(1 - x_1) - k_{-1} x_1)$.

Examples of typical parametric dependencies are shown in Figs. 3.39 and 3.40. The elements of the Jacobian matrix for (3.75) are written:

$$a_{11} = -k_1 - k_{-1} - k_2 - 4k_3 x_1 x_2,$$
$$a_{12} = -k_1 + k_{-2} - 2k_3 x_1^2,$$
$$a_{21} = k_2 - 2k_3 x_1 x_2,$$
$$a_{22} = -k_{-2} - k_3 x_1^2.$$

As above, we compute a determinant and a trace of the Jacobian matrix. Next, we plot the basic bifurcation curves of L_Δ and L_σ, which allow us to plot the corresponding parametric portraits (see Fig. 3.41).

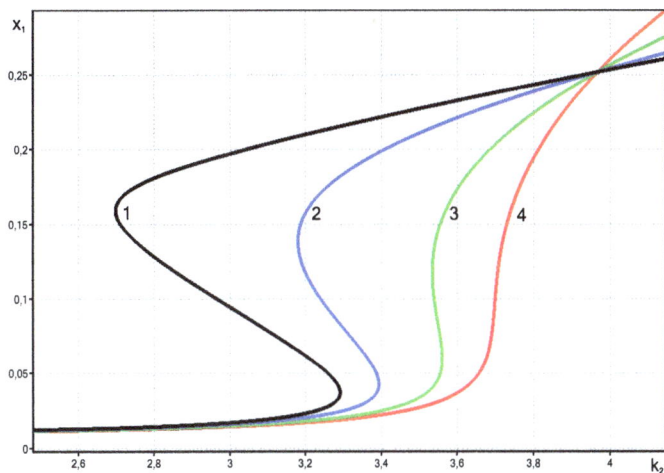

Fig. 3.40: Parametric dependencies $x_1(k_2)$ at $k_1 = 4$, $k_{-1} = k_{-2} = 0.01$, 1) $k_3 = 10$; 2) 8; 3) 5; 4) 3

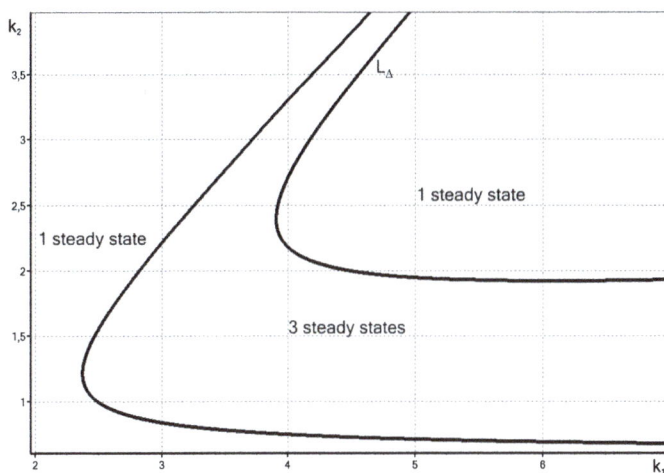

Fig. 3.41: Curves of multiplicity $L_\Delta(k_1, k_{-1})$ at $k_{-2} = 0.01$, $k_2 = 1$, 1) $k_3 = 5.12$; 2) 5.5

Three-step catalytic scheme in general form

Here we present in a compact form the results of the analysis of a series of mechanisms, which are given in detail in [85, 86]. Consider a three-step catalytic scheme in general form:

$$1)\ mZ \rightleftarrows mX\ ,$$

$$2)\ nZ \rightarrow nY\ ,$$

$$3)\ pX + qY \rightarrow (p + q)Z\ , \tag{3.77}$$

where m, n, p, q are the stoichiometric coefficients (positive integers) and X, Y, Z are intermediate substances. Special cases (3.77) were discussed above (for example, in (3.64) $m = 2$, $n = 1$, $p = 1$, $q = 1$).

Let us write a kinetic model for scheme (3.77)

$$\dot{x} = mk_1 z^m - mk_{-1}x^m - pk_3 x^p y^q ,$$
$$\dot{y} = nk_2 z^n - nk_{-2}y^n - qk_3 x^p y^q , \tag{3.78}$$

where $z = 1 - x - y$, x, y are concentrations of Z, X, Y respectively.

As usual, the solution of system (3.78) is defined in the reaction simplex

$$S = \{(x, y): \ x \geq 0, \ y \geq 0, \ x + y \leq 1\} .$$

The reaction simplex is an invariant set for a dynamical system (3.78). This guarantees the existence in S of at least one steady state. Steady states for (3.78) will be divided into internal and boundary. The internal steady states lie strictly inside of a region S, i.e., $x, y > 0$, $x + y < 1$. The extreme steady states lie on the boundary S at $x = 0$ or $y = 0$ or $x + y = 1$.

Steady states are determined from the system of algebraic equations

$$mk_1 z^m - mk_{-1}x^m - pk_3 x^p y^q = 0 ,$$
$$nk_2 z^n - nk_{-2}y^n - qk_3 x^p y^q = 0 ,$$
$$x + y + z = 1 . \tag{3.79}$$

The solutions of the nonlinear system (3.79) are located in S. We review and consider such solutions here. At different values of parameters the system can have one, three, or five steady states in S. Briefly we describe the most important variants of system (3.79). For example, at $m = n$ and $m \geq p$, q there is only a single steady state, i.e., the inequality of stoichiometric coefficients m and n is a necessary condition of a multiplicity of steady states. In particular, only internal steady states exist for the first two stages ($k_{-1}, k_{-2} > 0$), which are reversible.

1) $m = 1$, $n = 2$, $p = q = 1$: one or three steady states exist
2) $m = n = 1$, $p = 2$, $q = 1$: one or three steady states exist
3) $m = 1$, $n = 2$, $p = 2$, $q = 1$: one, three or five steady states exist (but no more)
4) $m = n = p = q$: only one steady state.

Note that a case of five steady states seems to be rather exotic for model (3.79). Moreover, it seems plausible that the number of steady states for (3.79) does not exceed five for any values of m, n, p, q. So, the most typical nonlinear situation for a system (3.78) is the presence of three steady states.

Together with (3.77) we consider another three-step catalytic mechanism

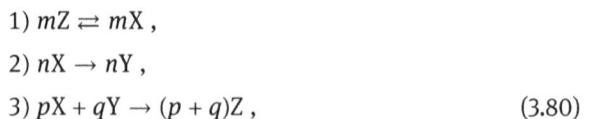

$$1) \ mZ \rightleftarrows mX ,$$
$$2) \ nX \rightarrow nY ,$$
$$3) \ pX + qY \rightarrow (p + q)Z , \tag{3.80}$$

where Z is a catalyst and X, Y are intermediate substances on the catalyst surface. As above, the observed substances (e.g., substances in the gas phase) are assumed to be constant (their partial pressures are included as cofactors in the rate constants of the corresponding reactions of scheme (3.80)).

The scheme (3.80), like (3.77), contains the step of interaction between various intermediate substances X and Y (step 3). Its presence in the reaction mechanism is a necessary condition of a multiplicity of steady states, but its still not enough for the emergence of a multiplicity of steady states. Study of mechanism (3.77) shows that a certain ratio between kinetic parameters of its stages is required. Therefore, a multiplicity of steady states is the result of competition between the first two stages and the interaction of two different substances in the third step.

A nonstationary kinetic model follows from catalytic mechanism (3.80):

$$\dot{y} = nk_2 z^n - nk_{-2} y^n - qk_3 x^p y^q \,,$$
$$\dot{z} = -mk_1 z^m + mk_{-1} x^m + (p+q)k_3 x^p y^q \,, \tag{3.81}$$

where $x = 1 - y - z$, y, z are the concentrations of the corresponding substances. The steady states of system (3.81) are defined in the reaction simplex S as the solutions of the nonlinear system of algebraic equations

$$nk_2 z^n - nk_{-2} y^n - qk_3 x^p y^q = 0 \,,$$
$$mk_1 z^m - mk_{-1} x^m - (p+q)k_3 x^p y^q = 0 \,,$$
$$x + y + z = 1 \,. \tag{3.82}$$

A detailed analysis of the number of solutions of system (3.81) in S is performed in [129]. Here we only note the following major conclusions:

1. at $p \le n$ a single internal steady state exists;
2. for reversible stages ($k_{-1}, k_{-2} > 0$) at $m = n = q = 1$, $p = 2$ one or three steady states exist;
3. at $m = p = 2$, $n = q = 1$, one, three or five steady states exist, but no more than five steady states (the question of five steady states in the general case is still open).

Thus, for a multiplicity of steady states for (3.81), we need to search among mechanisms for which $p > n$. Comparing schemes (3.80) and (3.77) showed that the main necessary condition of multiplicity of steady states for (3.77) was the inequality $m \ne n$. Here we have $p > n$.

As above, it is possible to suggest that scheme (3.80) for arbitrary stoichiometric coefficients m, n, p, q can be characterized by not more than five steady states. In this case a situation of five steady states seems unlikely. At least, it can be realized in a very narrow region of parameters k_i. So, if more than three steady states were detected in the experiment, for their description it is necessary to involve schemes more complex than (3.77) or (3.80). For example, mechanisms with a large number of stages and reagents can be considered. For a qualitative interpretation of three steady states, it may be sufficient to use a simple three-step scheme as considered above.

3.3.2 Catalytic oscillators

According to the scheme which was shown for autocatalytic mechanisms we will conduct a parametric analysis of kinetic models in this section. Catalytic triggers are added by one or another buffer step. We will call such mechanisms catalytic oscillators.

Mechanism 1

Consider a catalytic scheme of transformations

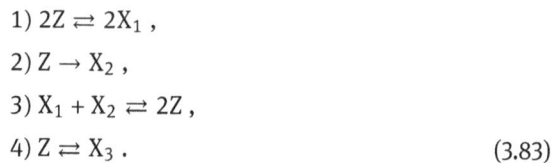

$$1)\ 2Z \rightleftharpoons 2X_1 \,,$$
$$2)\ Z \rightarrow X_2 \,,$$
$$3)\ X_1 + X_2 \rightleftharpoons 2Z \,,$$
$$4)\ Z \rightleftharpoons X_3 \,. \tag{3.83}$$

Stages 1)–3) correspond to the mechanism of Langmuir–Hinshelwood, and step 4) is a buffer. The formal scheme (3.83) may be interpreted, for example, as one of the possible schemes of hydrogen oxidation on metal:

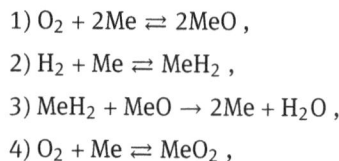

$$1)\ O_2 + 2Me \rightleftharpoons 2MeO \,,$$
$$2)\ H_2 + Me \rightleftharpoons MeH_2 \,,$$
$$3)\ MeH_2 + MeO \rightarrow 2Me + H_2O \,,$$
$$4)\ O_2 + Me \rightleftharpoons MeO_2 \,,$$

where MeO_2 is an inactive form of oxygen on the metal surface. A scheme similar to (3.83) can be written, for example, for the simplest mechanism of CO oxidation on Pt:

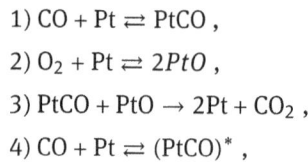

$$1)\ CO + Pt \rightleftharpoons PtCO \,,$$
$$2)\ O_2 + Pt \rightleftharpoons 2PtO \,,$$
$$3)\ PtCO + PtO \rightarrow 2Pt + CO_2 \,,$$
$$4)\ CO + Pt \rightleftharpoons (PtCO)^* \,,$$

where $(PtCO)^*$ is a second (more tightly bound) form of adsorption of CO on the surface Pt.

The kinetic model corresponding to scheme (3.83) is

$$\dot{x}_1 = 2k_1 z^2 - 2k_{-1} x_1^2 - k_3 x_1 x_2 \,,$$
$$\dot{x}_2 = k_2 z - k_{-2} x_2 - k_3 x_1 x_2 \,,$$
$$\dot{x}_3 = k_4 z - k_{-4} x_3 \,, \tag{3.84}$$

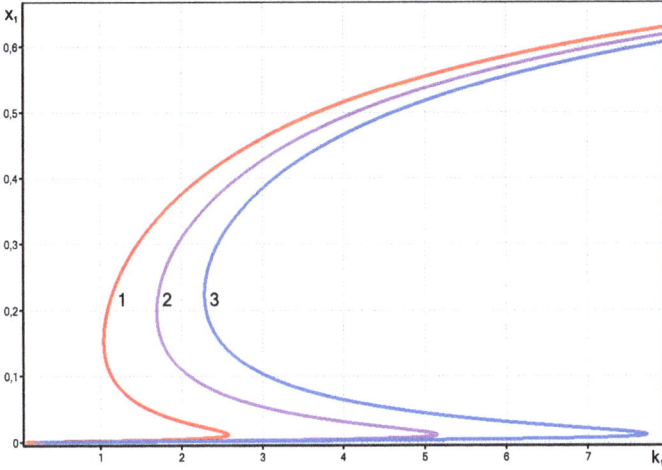

Fig. 3.42: Parametric dependencies $x_1(k_1)$ at $k_{-1} = 1$, $k_{-2} = 0.1$, $k_3 = 10$, $k_4 = 0.0675$, $k_{-4} = 0.022$, 1) $k_2 = 0.5$; 2) 1; 3) 1.5

where $z = 1 - x_1 - x_2 - x_3$. The variables x_2, x_3 can be expressed from the stationarity conditions for (3.84)

$$x_2 = \frac{k_2(1 - x_1)(1 - \alpha)}{k_{-2} + k_3 x_1 + k_2(1 - \alpha)},$$

$$x_3 = \alpha(1 - x_1 - x_2), \tag{3.85}$$

where $\alpha = k_4/(k_4 + k_{-4})$. Using the expression (3.85), we write the stationarity equation in the form:

$$2k_1\left(\frac{(1 - x_1)(1 - \alpha)(k_{-2} + k_3 x_1)}{k_{-2} + k_3 x_1 + k_2(1 - \alpha)}\right)^2 - 2k_{-1}x_1^2 - \frac{k_2 k_3 x_1(1 - x_1)(1 - \alpha)}{k_{-2} + k_3 x_1 + k_2(1 - \alpha)} = 0. \tag{3.86}$$

From equation (3.86) we get the desired parametric dependencies. For example,

$$k_1(x_1) = \frac{A\left(2Ak_{-1}x_1^2 + k_2 k_3 x_1(1 - x_1)(1 - \alpha)\right)}{2\left((1 - x_1)(1 - \alpha)(k_{-2} + k_3 x_1)\right)^2},$$

$$k_{-1}(x_1) = \frac{2k_1\left((1 - x_1)(1 - \alpha)(k_{-2} + k_3 x_1)\right)^2 - Ak_2 k_3 x_1(1 - x_1)(1 - \alpha)}{2A^2 x_1^2}, \tag{3.87}$$

where $A = k_{-2} + k_3 x_1 + k_2(1 - \alpha)$. Examples of plots of the parametric dependencies (3.87) are shown in Fig. 3.42.

The phase space of a system has a dimension of three. Therefore, the procedure of stability analysis of steady states and their bifurcations becomes more time-consuming compared to previous models. The elements of the Jacobian matrix for

(3.84) have the form:

$$a_{11} = -4k_1z - 4k_{-1}x_1 - k_3x_2, \qquad a_{12} = -4k_1z - k_3x_1, \qquad a_{13} = -4k_1z,$$
$$a_{21} = -k_2 - k_3x_2, \qquad a_{22} = -k_2 - k_{-2} - k_3x_1, \qquad a_{23} = -k_2,$$
$$a_{31} = -k_4, \qquad a_{32} = -k_4, \qquad a_{33} = -k_4 - k_{-4}.$$

The characteristic equation in the analysis of stability of steady states is cubic:

$$\lambda^3 + \sigma\lambda^2 + \theta\lambda + \Delta = 0,$$

where

$$\sigma = -(a_{11} + a_{22} + a_{33}),$$

$$\theta = \begin{vmatrix} a_{11} & a_{12} \\ a_{21} & a_{22} \end{vmatrix} + \begin{vmatrix} a_{11} & a_{13} \\ a_{31} & a_{33} \end{vmatrix} + \begin{vmatrix} a_{22} & a_{23} \\ a_{32} & a_{33} \end{vmatrix},$$

$$\Delta = -\begin{vmatrix} a_{11} & a_{12} & a_{13} \\ a_{21} & a_{22} & a_{23} \\ a_{31} & a_{32} & a_{33} \end{vmatrix}.$$

Stability of steady states is defined by a sequence of Routh–Hurwitz σ, $\xi = \sigma\theta - \Delta$, Δ [171]. Since we always have $\sigma > 0$, the number of bifurcations and the stability of steady states are determined from setting the variables Δ and $\sigma\theta - \Delta$ to zero. Knowing the parametric dependencies, we can numerically plot the corresponding bifurcation curves L_Δ, L_ξ. Examples of computed results are given in Fig. 3.43.

The parametric portrait of system (3.84) in the (k_1, k_2) plane is given in Fig. 3.44. In this particular case, the curve L_ξ lies completely inside the region defined by the curve L_Δ. All three steady states are unstable with such a mutual arrangement of bifurcation curves. This ensures the existence of auto-oscillations.

Time dependencies corresponding to auto-oscillations are given in Fig. 3.45.

Projections of the phase trajectories of system (3.84) in the plane (x_2, x_3) are shown in Fig. 3.46. The system stabilizes to the stable limit cycle from any initial data. All three steady states are unstable.

Note that projections of three-dimensional phase trajectories on any plane can intersect. In addition, here we must not say that unstable steady states lies inside the limit cycle.

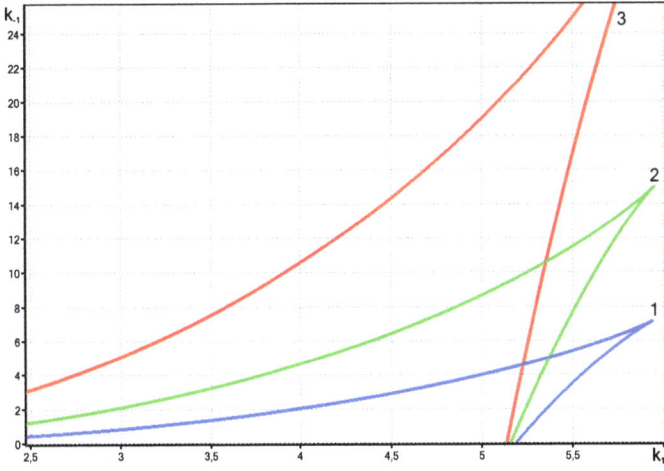

Fig. 3.43: Curves of multiplicity $L_\Delta(k_1, k_{-1})$ at $k_{-2} = 0.1$, $k_3 = 10$, $k_4 = 0.0675$, $k_{-4} = 0.022$, 1) $k_3 = 5$; 2) $k_3 = 7$; 3) $k_3 = 10$

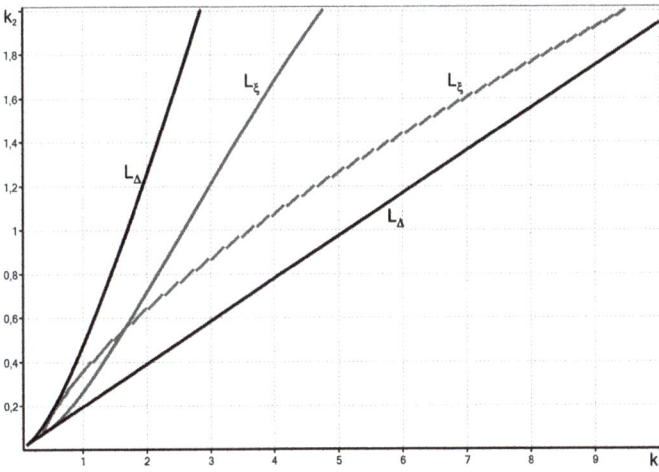

Fig. 3.44: Parametric portrait L_Δ, L_ξ in the plane (k_1, k_2) at $k_{-1} = 1$, $k_{-2} = 0.1$, $k_3 = 10$, $k_4 = 0.0675$, $k_{-4} = 0.022$

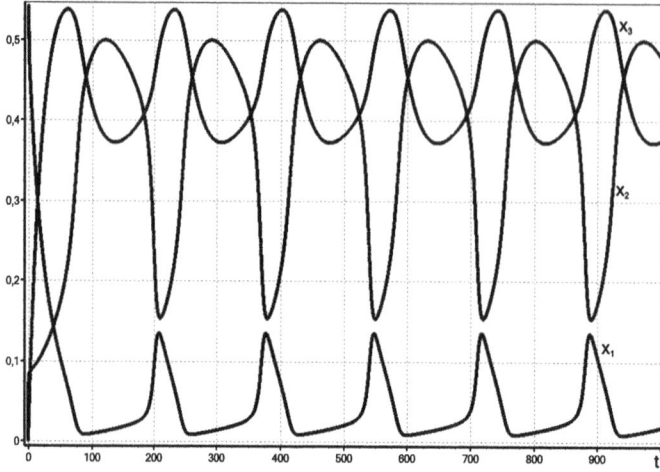

Fig. 3.45: Time dependencies $x_1(t)$, $x_2(t)$, $x_3(t)$ at $k_1 = 2.5$, $k_{-1} = k_2 = 1$, $k_{-2} = 0.1$, $k_3 = 10$, $k_4 = 0.0675$, $k_{-4} = 0.022$

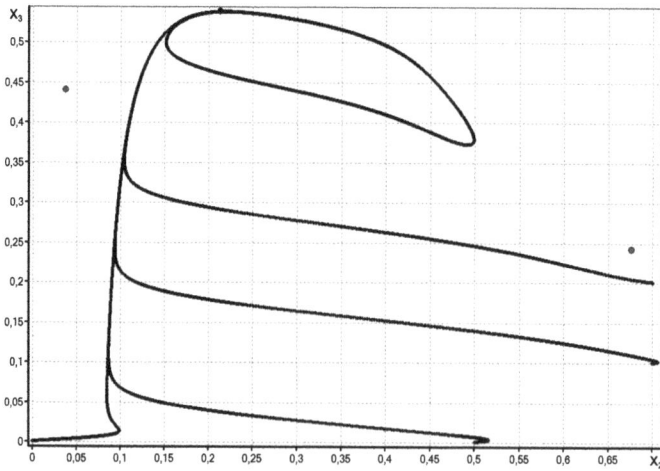

Fig. 3.46: Projections of the phase trajectories to the plane (x_2, x_3) at $k_1 = 2.5$, $k_{-1} = k_2 = 1$, $k_{-2} = 0.1$, $k_3 = 10$, $k_4 = 0.0675$, $k_{-4} = 0.022$

Mechanism 2

Let us consider one more reaction scheme which does not contain autocatalytic stages:

$$1)\ Z \rightleftharpoons X_1\ ,$$
$$2)\ X_1 \rightleftharpoons X_2\ ,$$
$$3)\ 2X_1 + X_2 \rightarrow 3Z\ ,$$
$$4)\ Z \rightleftharpoons X_3\ . \tag{3.88}$$

A nonstationary kinetic model can be written:

$$\dot{x}_1 = k_1 z - k_{-1}x_1 - k_2 x_1 + k_{-2}x_2 - 2k_3 x_1^2 x_2\ ,$$
$$\dot{x}_2 = k_2 x_1 - k_{-2}x_2 - k_3 x_1^2 x_2\ ,$$
$$\dot{x}_3 = k_4 z - k_{-4}x_3\ , \tag{3.89}$$

where $z = 1 - x_1 - x_2 - x_3$. Equating the right sides of system (3.89) to zero, we find x_2 and x_3:

$$x_2 = \frac{k_2 x_1}{k_{-2} + k_3 x_1^2}\ ,$$

$$x_3 = \alpha(1 - x_1 - x_2) = \alpha(1 - x_1) - \frac{\alpha k_2 x_1}{k_{-2} + k_3 x_1^2}\ , \tag{3.90}$$

where $\alpha = k_4/(k_4 + k_{-4})$. Substituting the expressions (3.90) in the first equation of system (3.87), we obtain the following equation for determining steady states:

$$k_1(1 - \alpha)((1 - x_1)(k_{-2} + k_3 x_1^2) - k_2 x_1) - x_1(k_{-2} + k_3 x_1^2)(k_2 +$$
$$+ k_{-1}) + k_2 x_1(k_{-2} - 2k_3 x_1^2) = 0\ . \tag{3.91}$$

From the stationarity equation (3.91) we can express some parametric dependencies:

$$k_1(x_1) = \frac{x_1(k_{-2} + k_3 x_1^2)(k_2 + k_{-1}) - k_2 x_1(k_{-2} - 2k_3 x_1^2)}{(1 - \alpha)((1 - x_1)(k_{-2} + k_3 x_1^2) - k_2 x_1)}\ ,$$

$$k_{-1}(x_1) = \frac{k_1(1 - \alpha)((1 - x_1)(k_{-2} + k_3 x_1^2) - k_2 x_1) - 3k_2 k_3 x_1^3}{x_1(k_{-2} + k_3 x_1^2)}\ ,$$

$$k_2(x_1) = \frac{(k_{-2} + k_3 x_1^2)(k_1(1 - \alpha)(1 - x_1) - k_{-1}x_1)}{x_1(k_1(1 - \alpha) + 3k_3 x_1^2)}\ ,$$

$$k_{-2}(x_1) = \frac{k_1 x_1(1 - \alpha)(k_2 - k_3 x_1(1 - x_1)) + k_3 x_1^3(k_1 + 3k_2)}{k_1(1 - \alpha)(1 - x_1) - k_{-1}x_1}\ ,$$

$$k_3(x_1) = \frac{k_1(1 - \alpha)(k_2 x_1 - k_{-2}(1 - x_1)) + k_{-1}k_{-2}x_1}{x_1^2(k_1(1 - \alpha)(1 - x_1) - x_1(k_{-1} + 3k_2))}\ . \tag{3.92}$$

It is easy to obtain explicit expressions of the type (3.92) and for the other parameters. Examples of plots of some parametric dependencies are shown in Fig. 3.47 at various values of the parameter k_2.

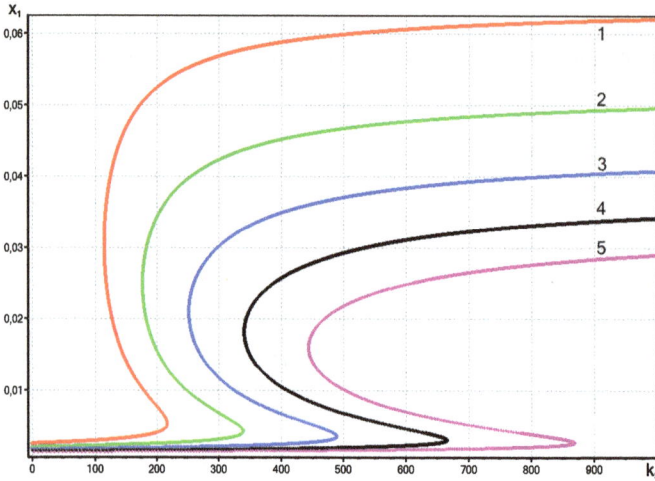

Fig. 3.47: Parametric dependencies $x_1(k_3)$ at $k_1 = 2.5$, $k_{-1} = 0.1$, $k_{-2} = 0.005$, $k_4 = 0.004$, $k_{-4} = 0.0008$, 1) $k_2 = 2$; 2) 2.5; 3) 3; 4) 3.5; 5) 4

Let us write the elements of the Jacobian matrix for (3.89):

$$a_{11} = -k_1 - k_{-1} - k_2 - 4k_3 x_1 x_2 , \qquad a_{12} = -k_1 + k_{-2} - 2k_3 x_1^2 , \qquad a_{13} = -k_1 ,$$
$$a_{21} = -k_2 - 2k_3 x_1 x_2 , \qquad a_{22} = -k_{-2} - k_3 x_1^2 , \qquad a_{23} = 0 ,$$
$$a_{31} = -k_4 , \qquad a_{32} = -k_4 , \qquad a_{33} = -k_4 - k_{-4} .$$

The characteristic equation is written as for mechanism 1. Curves of multiplicity of steady states are given in Fig. 3.48 for various values of k_2.

Bifurcation curves L_ξ are shown in Fig. 4.49. On the border of the selected region the steady states change their type of stability.

If a steady state corresponds to the parameters lying in a characteristic loop L_ξ, then it is unstable. If such a steady state is single, this ensures that the system has oscillating modes.

Time dependencies of the solutions of system (3.89) corresponding to the auto-oscillation mode are presented in Fig. 3.50.

The amplitude and period of oscillations depend strongly on the system parameters.

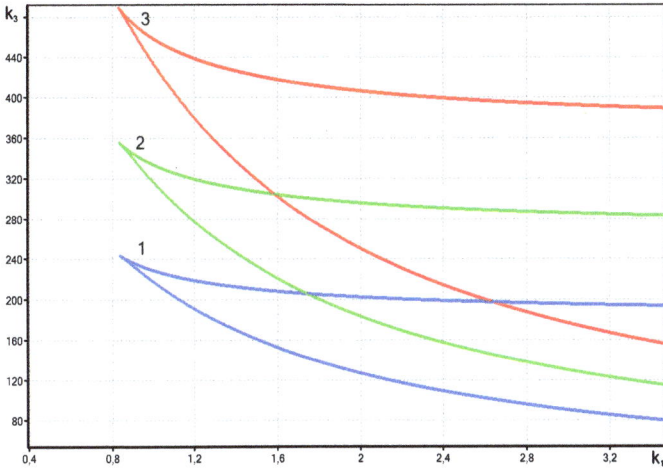

Fig. 3.48: Curves of multiplicity $L_\Delta(k_1, k_3)$ at $k_{-1} = 0.1$, $k_{-2} = 0.005$, $k_4 = 0.004$, $k_{-4} = 0.0008$, 1) $k_2 = 1.9$; 2) 2.3; 3) 2.7

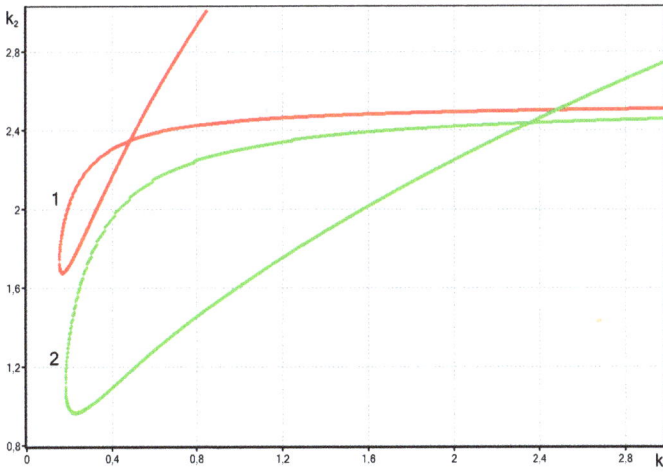

Fig. 3.49: Bifurcation curves $L_\xi(k_1, k_2)$ at $k_{-1} = 0.1$, $k_{-2} = 0.005$, $k_3 = 324$, $k_{-4} = 0.0008$, 1) $k_4 = 0.001$; 2) 0.004

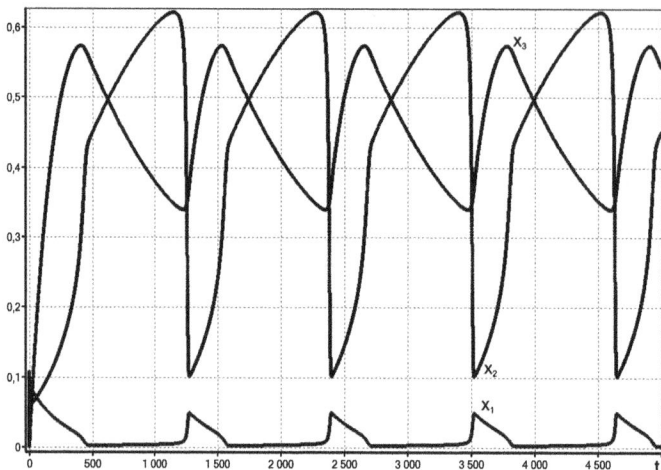

Fig. 3.50: Time dependencies $x_1(t)$, $x_2(t)$ and $x_3(t)$ at $k_1 = 0.5$, $k_{-1} = 0,1$, $k_2 = 1.6$, $k_{-2} = 0.005$, $k_3 = 324$, $k_4 = 0.004$, $k_{-4} = 0.0008$

Mechanism 3
This mechanism is different from the previous mechanism in buffer step 4). The scheme of reactions is

$$1)\ Z \rightleftarrows X_1 \,,$$
$$2)\ X_1 \rightleftarrows X_2 \,,$$
$$3)\ 2X_1 + X_2 \rightarrow 3Z \,,$$
$$4)\ X_1 \rightleftarrows X_3 \,. \tag{3.93}$$

Corresponding to (3.93) a mathematical model looks like this:

$$\dot{x}_1 = k_1 z - k_{-1}x_1 - k_2 x_1 + k_{-2}x_2 - 2k_3 x_1^2 x_2 - k_4 x_1 + k_{-4}x_3 \,,$$
$$\dot{x}_2 = k_2 x_1 - k_{-2}x_2 - k_3 x_1^2 x_2 \,,$$
$$\dot{x}_3 = k_4 x_1 - k_{-4}x_3 \,, \tag{3.94}$$

where $z = 1 - x_1 - x_2 - x_3$. To obtain the steady state equation we equate the right sides of system (3.94) to zero and exclude the variables x_2 and x_3:

$$x_2 = \frac{k_2 x_1}{k_{-2} + k_3 x_1^2} \,, \qquad x_3 = \alpha x_1 \,, \tag{3.95}$$

where $\alpha = k_4/k_{-4}$. Knowing the expression (3.95), let us write the equation of steady state with only the variable x_1:

$$k_1 \left((k_{-2} + k_3 x_1^2)(1 - x_1 - \alpha x_1) - k_2 x_1 \right) - k_{-1}x_1(k_{-2} + k_3 x_1^2) - 3k_2 k_3 x_1^3 = 0 \,. \tag{3.96}$$

For the studied system (3.94) all parametric dependencies have been written from (3.96):

$$k_1(x_1) = \frac{Ax_1(k_2 + k_{-1}) - k_2 x_1(k_{-2} - 2k_3 x_1^2)}{A(1 - x_1 - \alpha x_1) - k_2 x_1} \,,$$

$$k_{-1}(x_1) = \frac{k_1 \left(A(1 - x_1 - \alpha x_1) - k_2 x_1\right) - 3k_2 k_3 x_1^3}{Ax_1} \,,$$

$$k_2(x_1) = \frac{A \left(k_1(1 - x_1 - \alpha x_1) - k_{-1} x_1\right)}{x_1(k_1 + 3k_3 x_1^2)} \,,$$

$$k_{-2}(x_1) = \frac{k_1 x_1 \left(k_2 - k_3 x_1(1 - x_1 - \alpha x_1)\right) + k_3 x_1^3(k_1 + 3k_2)}{k_1(1 - x_1 - \alpha x_1) - k_{-1} x_1} \,,$$

$$k_3(x_1) = \frac{k_1 \left(k_2 x_1 - k_{-2}(1 - x_1 - \alpha x_1)\right) - k_{-1} k_{-2} x_1}{x_1^2[k_1(1 - x_1 - \alpha x_1) - x_1(k_{-1} + 3k_2)]} \,,$$

$$k_4(x_1) = \frac{k_{-4} \left(A(k_1(1 - x_1) - k_{-1} x_1) - k_2 x_1(k_1 + 3k_3 x_1^2)\right)}{Ak_1 x_1} \,,$$

$$k_{-4}(x_1) = \frac{Ak_1 k_4 x_1}{A(k_1(1 - x_1) - k_{-1} x_1) - k_2 x_1(k_1 + 3k_3 x_1^2)} \,.$$

where $A = k_{-2} + k_3 x_1^2$. Examples of plots of some parametric dependencies are given in Fig. 3.51.

Corresponding to system (3.94) the elements of the Jacobian matrix are calculated as follows:

$a_{11} = -k_1 - k_{-1} - k_2 - 4k_3 x_1 x_2 - k_4 \,,$ $\quad a_{12} = -k_1 + k_{-2} - 2k_3 x_1^2 \,,$ $\quad a_{13} = -k_1 + k_{-4} \,,$

$a_{21} = k_2 - 2k_3 x_1 x_2 \,,$ $\qquad\qquad\qquad\qquad a_{22} = -k_{-2} - k_3 x_1^2 \,,$ $\qquad\qquad a_{23} = 0 \,,$

$a_{31} = k_4 \,,$ $\qquad\qquad\qquad\qquad\qquad\qquad a_{32} = 0 \,,$ $\qquad\qquad\qquad\qquad a_{33} = -k_{-4} \,.$

As above, upon calculating the elements of the Jacobian matrix in a steady state, we can relatively easily plot the bifurcation curves L_Δ and L_ξ and the corresponding parametric portraits. Time dependencies corresponding to auto-oscillations are presented in Fig. 3.52.

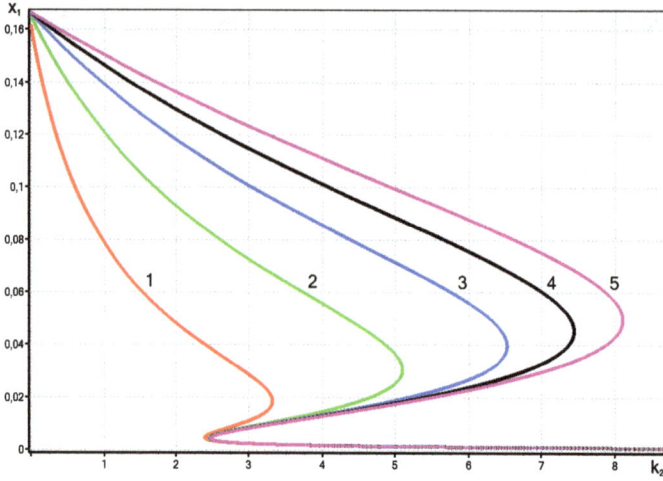

Fig. 3.51: Parametric dependencies $x_1(k_2)$ at $k_{-1} = 0.1$, $k_{-2} = 0.005$, $k_3 = 324$, $k_4 = 0.004$, $k_{-4} = 0.0008$, 1) $k_1 = 0.5$; 2) 1.5; 3) 3; 4) 4.5; 5) 6

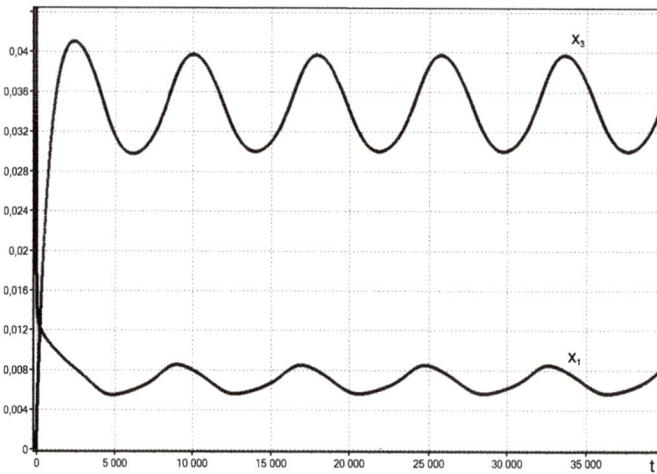

Fig. 3.52: Time dependencies $x_1(t)$ and $x_3(t)$ for $k_1 = 0.15$, $k_{-1} = 0,1$, $k_2 = 2.158$, $k_{-2} = 0.005$, $k_3 = 324$, $k_4 = 0.004$, $k_{-4} = 0.0008$

Mechanism 4

This mechanism is also different in buffer step 4). The scheme of reactions is

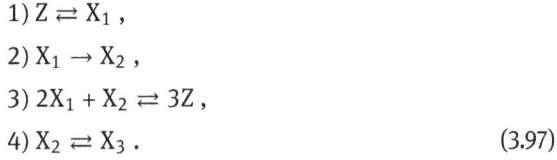

$$1)\ Z \rightleftarrows X_1 \ ,$$
$$2)\ X_1 \rightarrow X_2 \ ,$$
$$3)\ 2X_1 + X_2 \rightleftarrows 3Z \ ,$$
$$4)\ X_2 \rightleftarrows X_3 \ . \tag{3.97}$$

The corresponding kinetic model is written as:

$$\dot{x}_1 = k_1 z - k_{-1}x_1 - k_2 x_1 + k_{-2}x_2 - 2k_3 x_1^2 x_2 \ ,$$
$$\dot{x}_2 = k_2 x_1 - k_{-2}x_2 - k_3 x_1^2 x_2 - k_4 x_2 + k_{-4}x_3 \ ,$$
$$\dot{x}_3 = k_4 x_2 - k_{-4}x_3 \ , \tag{3.98}$$

where $z = 1 - x_1 - x_2 - x_3$. We equate the right sides of system (3.98) to zero and exclude the variables x_2 and x_3:

$$x_2 = \frac{k_2 x_1}{k_{-2} + k_3 x_1^2} \ , \qquad x_3 = \alpha x_2 \ , \tag{3.99}$$

where $\alpha = k_4/k_{-4}$. Using the expression (3.99), let us write the stationarity equation with the variable x_1:

$$k_1 \left((k_{-2} + k_3 x_1^2)(1 - x_1) - k_2 x_1 (1 + \alpha) \right) - k_{-1}x_1(k_{-2} + k_3 x_1^2) - 3k_2 k_3 x_1^3 = 0 \ . \tag{3.100}$$

All parametric dependencies can be written from (3.100) for system (3.98):

$$k_1(x_1) = \frac{Ak_{-1}x_1 + 3k_2 k_3 x_1^3}{A(1 - x_1) - k_2 x_1(1 + \alpha)} \ ,$$

$$k_{-1}(x_1) = \frac{k_1 \left(A(1 - x_1) - k_2 x_1(1 + \alpha) \right) - 3k_2 k_3 x_1^3}{Ax_1} \ ,$$

$$k_2(x_1) = \frac{A \left(k_1(1 - x_1) - k_{-1}x_1 \right)}{k_1 x_1(1 + \alpha) + 3k_3 x_1^3} \ ,$$

$$k_{-2}(x_1) = \frac{k_3 x_1^3(k_{-1} + 3k_2) + k_1 x_1 \left(k_2(1 + \alpha) - k_3 x_1(1 - x_1) \right)}{k_1(1 - x_1) - k_{-1}x_1} \ ,$$

$$k_3(x_1) = \frac{k_1 k_2 x_1(1 + \alpha) + k_{-2}(k_{-1}x_1 - k_1(1 - x_1))}{x_1^2 [k_1(1 - x_1) - x_1(k_{-1} + 3k_2)]} \ ,$$

$$k_4(x_1) = \frac{k_{-4} \left(A(k_1(1 - x_1) - k_{-1}x_1) - k_2 x_1(k_1 + 3k_3 x_1^2) \right)}{k_1 k_2 x_1} \ ,$$

$$k_{-4}(x_1) = \frac{k_1 k_2 k_4 x_1}{A(k_1(1 - x_1) - k_{-1}x_1) - k_2 x_1(k_1 + 3k_3 x_1^2)} \ ,$$

where $A = k_{-2} + k_3 x_1^2$. Examples of plots of parametric dependencies $x_1(k_4)$ are given in Fig. 3.53.

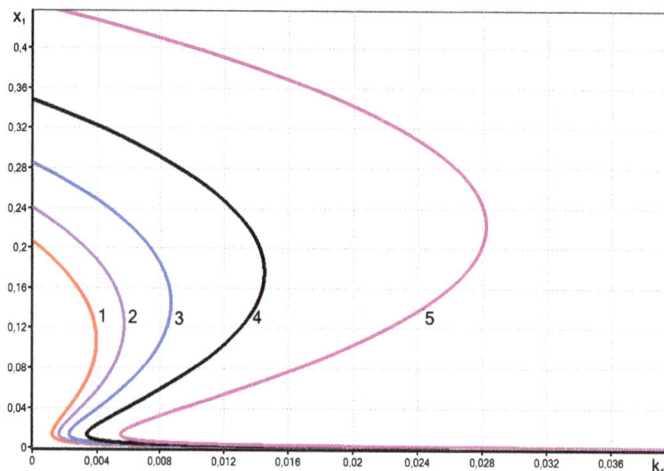

Fig. 3.53: Parametric dependencies $x_1(k_4)$ at $k_1 = 2.5$, $k_{-1} = 0.1$, $k_{-2} = 0.05$, $k_3 = 324$, $k_4 = 0.004$, $k_{-4} = 0.0008$, 1) $k_2 = 3$; 2) 2.5; 3) 2; 4) 1.5; 5) 1

The elements of the Jacobian matrix for system (3.98) are written in the form:

$$a_{11} = -k_1 - k_{-1} - k_2 - 4k_3x_1x_2 , \quad a_{12} = -k_1 + k_{-2} - 2k_3x_1^2 , \quad a_{13} = -k_1 ,$$

$$a_{21} = k_2 - 2k_3x_1x_2 , \qquad\qquad a_{22} = -k_{-2} - k_3x_1^2 - k_4 , \quad a_{23} = k_{-4} ,$$

$$a_{31} = 0 , \qquad\qquad\qquad\qquad a_{32} = k_4 , \qquad\qquad\quad a_{33} = -k_{-4} .$$

First we compute the elements of the Jacobian matrix and the coefficients of the characteristic equation in a steady state. Further, it is possible to plot the bifurcation curves L_Δ, L_ξ, and thus parametric portraits of system (3.98).

Comparative analysis of models has shown that depending on the buffer step, a system may or may not have auto-oscillations.The mechanisms with the greatest region of existence of oscillations in parameter space are found. So, of the two oscillatory systems corresponding to the mechanisms 2 and 3, the highest probability of occurrence of oscillations in the plane (k_1, k_{-1}) has a scheme with a buffer step of the form $X_1 \rightleftarrows X_3$. A general conclusion is that the kind of scheme of transformations significantly determines the presence of auto-oscillations as a dependence on parameters.

3.4 Catalytic continuous stirred-tank reactor (CSTR)

All of the above kinetic models of catalytic reactions were considered with the assumption of constancy of reactants in the gas phase. Partial pressures of gas-phase substances appear in appropriate kinetic models as parameters. In this section, we will consider models in which these substances are phase variables. Concentrations of these substances are usually observed in specific experiments [1, 2, 114, 152–160].

A nonstationary kinetic model of the catalytic CSTR has the form:

$$\dot{\mathbf{c}} = \mathbf{f}(\mathbf{c}, \mathbf{x}) + v(\mathbf{c}_0 - \mathbf{c}) \,,$$

$$\dot{\mathbf{x}} = \mathbf{g}(\mathbf{c}, \mathbf{x}) \,, \tag{3.101}$$

where $\mathbf{c} = (c_1, \ldots, c_m)$ is a vector of concentrations of substances in the gas phase (observable materials). $\mathbf{x} = (x_1, \ldots, x_n)$ is a vector of concentrations of substances on the catalyst surface (intermediate substances). $\mathbf{c}_0$ is a vector of concentrations at the reactor inlet. v is a flow rate through the reactor of gaseous substances. $\mathbf{f}, \mathbf{g}$ are functions of the kinetic dependencies that are written according to the mechanism of the catalytic reactions. A kinetic subsystem in (3.101) corresponds to the transformations of intermediates on the catalyst surface.

Joint consideration of the systems of equations (3.101) reflects an interaction of the gas phase with a catalyst surface. This interaction is nonlinear. As a formal example, we consider the mechanism of heterogeneous catalytic reactions:

$$1) \; A + Z \rightleftarrows AZ \,,$$

$$2) \; B + AZ \rightleftarrows ABZ \,,$$

$$3) \; A + ABZ \rightarrow Z + A_2B \,, \tag{3.102}$$

where A, B and A_2B are substances in the gas phase and AZ, ABZ are intermediate substances on the catalyst Z. The following kinetic model of type (3.101) corresponds to scheme (3.102):

$$\dot{a} = k_1 az - k_{-1}x - k_3 ay + v(a_0 - a) \,,$$

$$\dot{b} = k_2 bx - k_{-2}y + v(b_0 - b) \,,$$

$$\dot{x} = k_1 az - k_{-1}x - k_2 bz + k_{-2}y \,,$$

$$\dot{y} = k_2 bx - k_{-2}y - k_3 ay \,, \tag{3.103}$$

where a, b are dimensionless concentrations of substances A and B. x and y are concentrations of intermediate substances AZ and ABZ. z is determined as $z = 1 - x - y$. The reaction mechanism (3.102) is linear in the intermediate substances. Therefore, if $a, b = \text{const}$ (for enough high flow rates, which is characteristic for a flow-circulation of laboratory units) there are no critical phenomena in the catalytic subsystem relative to x, y. However, in the full system (3.103) there are nonlinearities of the form az, bx, ay, which can lead, generally speaking, to nonlinear effects in the system.

Above, we considered nonlinear base models of catalytic reactions. Here we will study the specificity of the interaction of the gas phase with a catalytic nonlinear subsystem. For such subsystems, we will use autocatalytic and catalytic triggers and oscillators.

3.4.1 Flow reactor with an autocatalytic trigger

Let a scheme of transformations be

$$1)\ A + Z \rightleftarrows X,$$
$$2)\ X + 2Z \rightarrow 3Z + B, \tag{3.104}$$

where A, B are observable substances and X, Z are intermediate substances. The first step can be interpreted as adsorption. The second step can be considered a reaction on the surface. This step is autocatalytic and trimolecular.

The kinetic model for scheme (3.104) can be written as follows:

$$\dot{x} = k_1 a z - k_{-1} x - k_2 x z^2,$$
$$\dot{a} = -k_1 a z + k_{-1} x + v(a_0 - a), \tag{3.105}$$

where $z = 1 - x$, and x, a are concentrations of the substances X and A. a_0 is a primary concentration of the substance A. v is the rate with which the substance A is supplied. k_i are rate constants of stages, which appear here as parameters.

Equating the right sides of system (3.105) to zero and transforming, we get an equation of steady state:

$$k_1 z(a_0 v - k_2 z^2(1-z)) - v(1-z)(k_{-1} + k_2 z^2) = 0,$$

where $a = (1-z)(k_{-1} + k_2 z^2)/(k_1 z)$.

Parametric dependencies are explicitly expressed from the stationarity equation as inverse dependencies $k_i(z)$:

$$k_1(z) = \frac{v(1-z)(k_{-1} + k_2 z^2)}{z(a_0 v - k_2 z^2(1-z))},$$

$$k_{-1}(z) = \frac{z(a_0 v k_1 - k_2 z(1-z)(v + k_1 z))}{v(1-z)},$$

$$k_2(z) = \frac{v(a_0 k_1 z - k_{-1}(1-z))}{z^2(1-z)(v + k_1 z)},$$

$$a_0(z) = \frac{v(1-z)(k_{-1} + k_2 z^2) + k_1 k_2 z^3(1-z)}{v k_1 z},$$

$$v(z)) = \frac{k_1 k_2 z^3(1-z)}{a_0 k_1 z - (1-z)(k_{-1} + k_2 z^2)}.$$

The corresponding graphs are shown in Figs. 3.54 and 3.55 when parameter values are: $k_1 = 1$, $k_{-1} = 0.01$, $k_2 = 1$, $a_0 = 0.3$, $v = 0.635$. At various values of the second parameter the region of multiplicity of steady states is increased or decreased.

Let us write the elements of the Jacobian matrix for system (3.105):

$$a_{11} = -k_1 a - k_{-1} - k_2 z^2 + 2k_2 z(1-z), \quad a_{12} = k_1 z,$$
$$a_{21} = k_1 a + k_{-1}, \qquad\qquad\qquad a_{22} = -k_1 z - v.$$

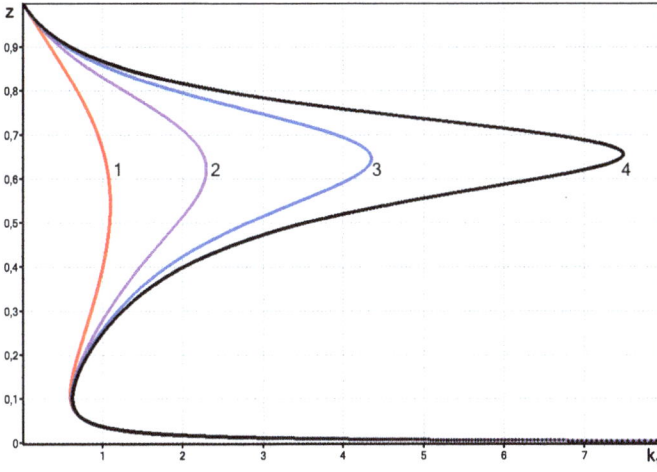

Fig. 3.54: Parametric dependencies $z(k_1)$ at 1) $v = 2$; 2) 0.75; 3) 0.6; 4) 0.55

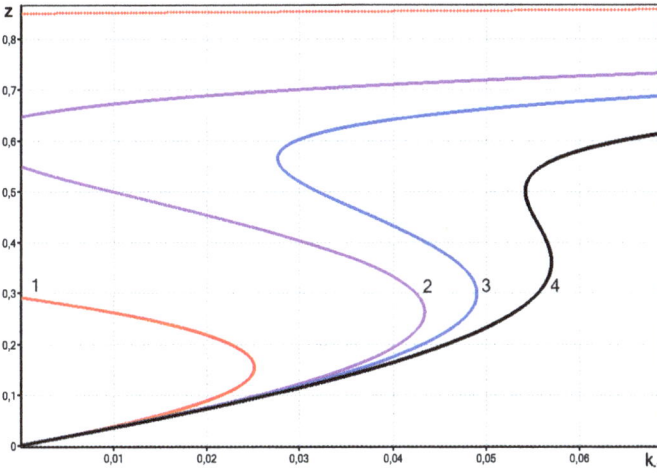

Fig. 3.55: Parametric dependencies $z(k_{-1})$ at 1) $k_2 = 1$; 2) 0.65; 3) 0.6; 4) 0.55

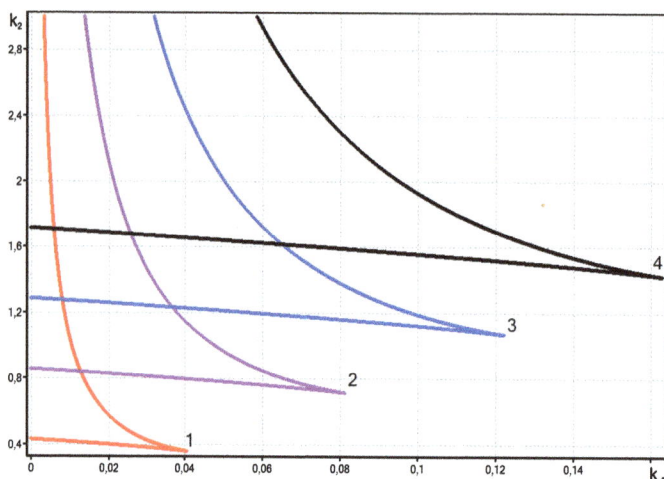

Fig. 3.56: Curves of multiplicity $L_\Delta(k_{-1}, k_2)$ at 1) $a_0 = 0.2$; 2) 0.4; 3) 0.6; 4) 0.8

According to the above scheme for different combinations of parameters we will get the expression for the plots of curves of multiplicity (L_Δ):

$$L_\Delta(k_1, k_{-1}): \quad k_1(z) = \frac{2vk_2z(1-z)^2}{a_0v - 3k_2z^2(1-z)^2},$$
$$k_{-1} = k_{-1}(z, k_{-1}(z)),$$

$$L_\Delta(k_2, k_{-1}): \quad k_2(z) = \frac{a_0vk_1}{z(1-z)^2(2v + 3k_1z)},$$
$$k_{-1} = k_{-1}(z, k_2(z)),$$

$$L_\Delta(v, k_{-1}): \quad v(z) = \frac{3k_1k_2z^2(1-z)^2}{a_0k_1 - 2k_2z(1-z)^2},$$
$$k_{-1} = k_{-1}(z, v(z)),$$

$$L_\Delta(a_0, k_{-1}): \quad a_0(z) = \frac{k_2z(1-z)^2(2v + 3k_1z)}{vk_1},$$
$$k_{-1} = k_{-1}(z, v(z)),$$

Examples of curves L_Δ are shown in Fig. 3.56 and 3.57.
The expressions for curves of neutrality (L_σ) are:

$$\sigma = k_{-1} + k_1z^2 + vz + k_2z^2(2z - 1) = 0. \tag{3.106}$$

$$L_\sigma(k_2, k_{-1}): \quad k_2(z) = \frac{v(a_0k_1 + (1-z)(v + k_1z))}{z(1-z)(2v(1-z) + k_1z)}, \tag{3.107}$$
$$k_{-1} = k_{-1}(z, k_2(z)).$$

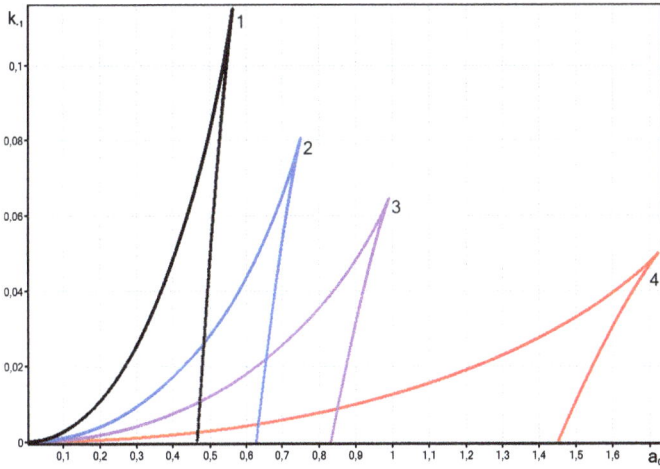

Fig. 3.57: Curves of multiplicity $L_\Delta(a_0, k_{-1})$ at 1) $k_1 = 1$; 2) 0.6; 3) 0.4; 4) 0.2

The analysis of (3.106) and (3.107) shows that σ does not change sign. It allows us to say that the only steady state is always stable. If there are three steady states, two of them are stable and one is unstable.

3.4.2 Flow reactor with a catalytic trigger

Consider the scheme of transformations:

$$1)\ A + Z \rightleftharpoons AZ\,,$$
$$2)\ B_2 + 2Z \rightleftharpoons 2BZ\,,$$
$$3)\ AZ + BZ \to 2Z + AB\,. \tag{3.108}$$

B_2, A and AB are the observable substances. AZ and BZ are the adsorbed substances; Z is the active center of the catalyst surface.

The scheme (3.108) does not contain autocatalytic stages. Therefore for the interpretation of multiplicity of steady states, we require at least three stages and two independent substances. In fact, this scheme is a well-known Langmuir–Hinshelwood mechanism. It is the simplest catalytic trigger which is a system with two stable steady states. The mathematical tools that have been used for this study are elementary. However, we managed to obtain new knowledge, because the study of the kinetic model for this mechanism was carried out with a conscious purpose to search for multiplicity of steady states.

If the concentrations of substances A and B_2 are constant, this scheme is one of the simplest catalytic triggers for which we previously found the region of multiplicity of steady states.

A mathematical model for the scheme of transformations (3.108) can be written:

$$\dot{x} = k_1 az - k_{-1}x - k_3 xy ,$$
$$\dot{y} = 2k_2 b_2 z^2 - 2k_{-2}y^2 - k_3 xy ,$$
$$\dot{a} = -k_1 az + k_{-1}x + v_1(a_0 - a) ,$$
$$\dot{b}_2 = -k_2 b_2 z^2 + k_{-2}y^2 + v_2(b_0 - b_2) . \tag{3.109}$$

Suppose the concentration of a substance A is kept unchanged over time. Then while changing the concentration B_2 over time we get from (3.109):

$$\dot{x} = k_1 az - k_{-1}x - k_3 xy ,$$
$$\dot{y} = 2k_2 b_2 z^2 - 2k_{-2}y^2 - k_3 xy ,$$
$$\dot{b}_2 = -k_2 b_2 z^2 + k_{-2}y^2 + v(b_0 - b_2) ,$$

where: x, y are concentrations of substances on the catalyst surface; a, b_2 are concentrations of substances in the gas phase (assuming that $a \equiv$ const); b_0 is the initial concentration of the substance b_2; v is a flow rate; k_i are rate constants of the stages.

Analysis of mutual arrangement of isoclines and stability of steady states of the subsystem (x, y) allows us to find the region of the existence of a single and unstable steady state. This region guarantees the existence of auto-oscillations. The probability of occurrence of oscillations is higher, where the point of intersection of isoclines is closer to the transition point from the region of existence of a stable steady state to the region of an unstable steady state. One of the possible sets of values of parameters for the existence of auto-oscillations is: $k_1 = 0.3$; $k_{-1} = 0.0005$; $k_2 = 0.1$; $k_{-2} = 0.0005$; $k_3 = 1$; $a = 0.032$; $b_0 = 1$; $v = 0.00065$.

Time dependencies corresponding to this region are shown in Fig. 3.58.

When we increase the value of b_0 or decrease the value of v, the oscillations begin to gradually disappear. The oscillations disappear with an increase in the rate constants of the inverse stages k_{-1}, k_{-2} (Fig. 3.59)

Assuming the constancy of the concentration of the substance b_2, we obtain the system from (3.109):

$$\dot{x} = k_1 az - k_{-1}x - k_3 xy ,$$
$$\dot{y} = 2k_2 b_2 z^2 - 2k_{-2}y^2 - k_3 xy ,$$
$$\dot{a} = -k_1 az + k_{-1}x + v(a_0 - a) . \tag{3.110}$$

As above, for system (3.110) it can be shown that there are oscillations.

The study of system (3.109) by parts allowed us to obtain auto-oscillations for the general situation. In particular, at the following set of parameters: $k_1 = 25.113$; $k_{-1} = 0.03$; $k_2 = 10$; $k_{-2} = 0.02$; $k_3 = 100$; $a_0 = 0.05$; $b_0 = 0.5$; $v_1 = 0.68$; $v_2 = 0.02$ system (3.109) has auto-oscillations (see Figs. 3.60 and 3.61).

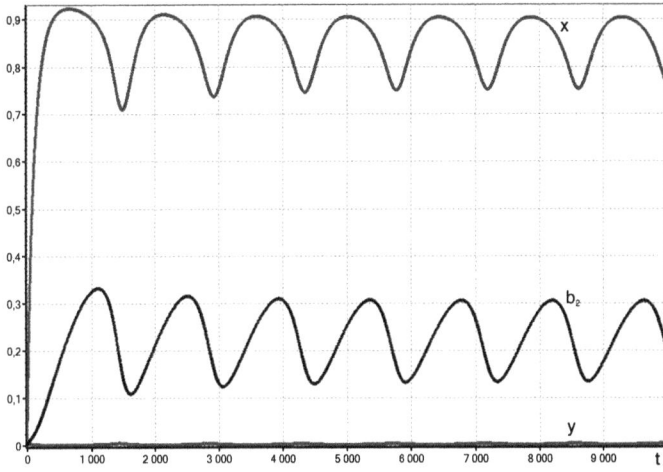

Fig. 3.58: Time dependencies $x(t)$, $y(t)$, $b_2(t)$ at $b_0 = 1$

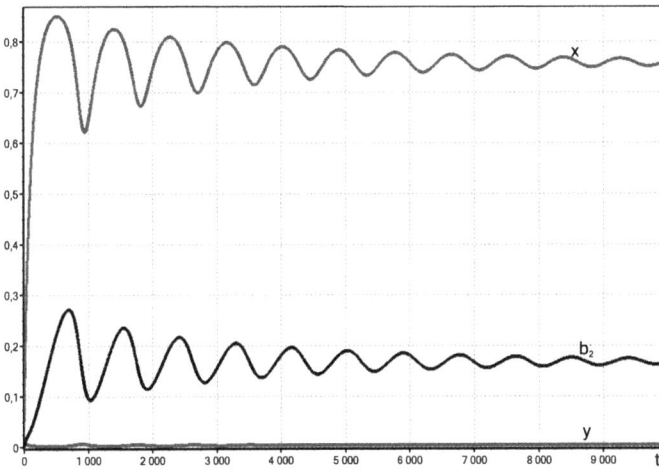

Fig. 3.59: Time dependencies $x(t)$, $y(t)$, $b_2(t)$ at $b_0 = 1.65$

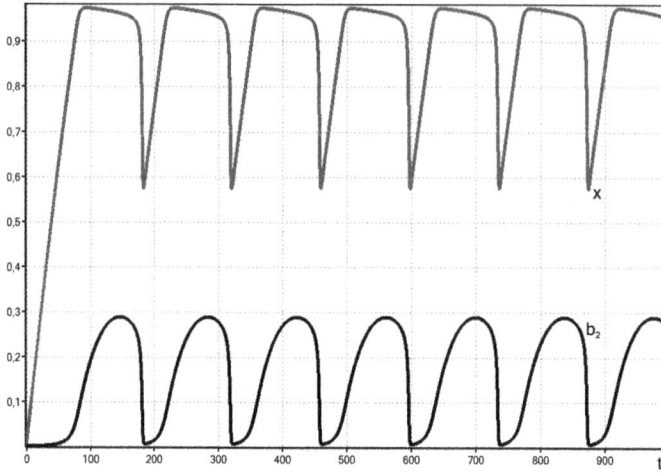

Fig. 3.60: Time dependencies $x(t)$ and $b_2(t)$

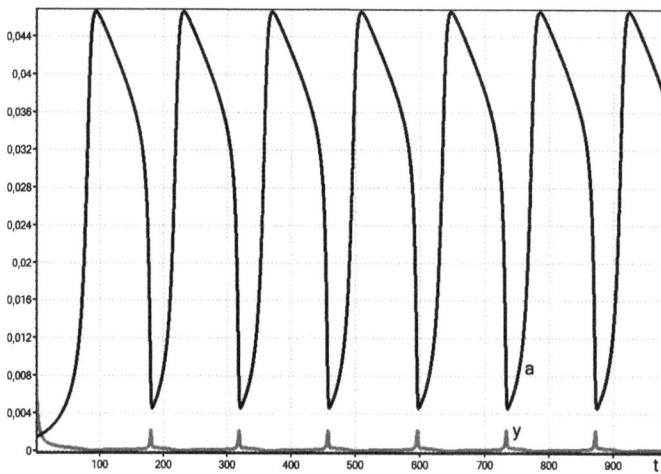

Fig. 3.61: Time dependencies $y(t)$ and $a(t)$

3.4.3 Flow reactor with an autocatalytic oscillator

Consider the mechanism with autocatalysis:

$$1)\ A + Z \rightleftharpoons AZ,$$
$$2)\ AZ + 2Z \rightarrow 3Z + B,$$
$$3)\ B + Z \rightleftharpoons BZ, \tag{3.111}$$

where the step 3) is a buffer step.

The model of the autocatalytic oscillator (3.111) has been studied in previous sections. Two observable substances A and B were added in this scheme. Different impurities can perform as these substances in a real model. We obtain a system of kinetic equations for mechanism (3.111):

$$\dot{z} = -k_1 az + k_{-1} x - k_3 bz + k_{-3} y + k_2 xz^2,$$
$$\dot{y} = k_3 bz - k_{-3} y,$$
$$\dot{a} = -k_1 az + k_{-1} x + v(a_0 - a),$$
$$\dot{b} = k_2 xz^2 - k_3 bz + k_{-3} y + v(b_0 - b). \tag{3.112}$$

The values of coefficients k_i are taken from the previously calculated model of an autocatalytic oscillator. For sufficiently large values of v and with a suitable choice of the variables a_0 and b_0, system (3.112) can experience undamped oscillations.

3.4.4 Flow reactor with a catalytic oscillator

Consider the classic mechanism of Langmuir–Hinshelwood with a buffer step, which above was investigated as the simplest catalytic oscillator:

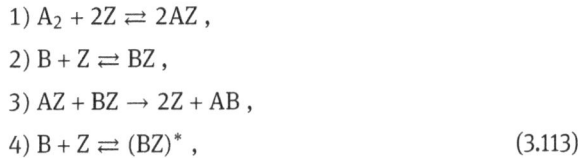

$$1)\ A_2 + 2Z \rightleftharpoons 2AZ,$$
$$2)\ B + Z \rightleftharpoons BZ,$$
$$3)\ AZ + BZ \rightarrow 2Z + AB,$$
$$4)\ B + Z \rightleftharpoons (BZ)^*, \tag{3.113}$$

where A_2, B and AB are substances in the gas phase, and AZ, BZ and $(BZ)^*$ are intermediate substances on the catalyst Z. The following kinetic model of a flow reactor can be written for scheme (3.113):

$$\dot{a} = -k_1 az^2 + k_{-1} x^2 + v(a_0 - a),$$
$$\dot{b} = -k_2 bz + k_{-2} y - k_4 bz + k_{-4} y^* + v(b_0 - b),$$
$$\dot{x} = 2k_1 az^2 - 2k_{-1} x^2 - k_3 xy,$$
$$\dot{y} = k_2 bz - k_{-2} y - k_3 xy,$$
$$\dot{y}^* = k_4 bz - k_{-4} y^*, \tag{3.114}$$

where a and b are concentrations of the substances A_2 and B, x, y and y^* are concentrations of the substances AZ, BZ and $(BZ)^*$ on the catalyst Z, and $z = 1 - x - y - y^*$. The assumption of constancy in the gas phase (a, b = const) in a catalytic subsystem of system (3.114) allows the existence of auto-oscillations. Calculations of equations (3.114) show that for sufficiently large flow rates v and in the selection of appropriate values of a_0 and b_0 there are oscillations in system (3.114).

Note a detailed parametric analysis of models of a flow catalytic reactor needs to be performed. There are several important control parameters in these models, such as v, a_0 and b_0, and the temperature. Variation of these parameters in the existing flow-circulation facilities is possible. Understanding the nonlinear dynamics of catalytic processes, gained through the study of the simplest base models, will allow you to consciously vary the external parameters to control the catalytic processes, with important applications.

For the studied model of the catalytic oscillator from performed calculations it follows that the system is most sensitive to changes of the flow rate of the reactants. This conclusion can be used to solve a control task by a flow catalytic reactor.

Thus, studying mathematical models of flow catalytic systems shows that a flow can be another source of oscillations. The given examples and calculations confirm the fact that interaction on the catalytic surface with the gas phase can lead to critical phenomena. If there is, for example, a multiplicity of steady states in the subsystem (intermediate substances on the catalyst surface), then auto-oscillations can be observed in the system as a whole. Accounting for flow leads to an increase in the number of degrees of freedom and nonlinearity. It provides the possibility of the origin of critical effects.

3.4.5 Kinetic "chaos" induced by noise

It is well known that the behavior of chemical systems in the kinetic region can be quite variable. First, there is a multiplicity of steady states leading to the hysteresis of the stationary reaction rate while changing the parameters. Second, there are auto-oscillations of concentrations of reagents and more complex dynamical regimes, called "chaos" [137–143]. From a mathematical point of view "chaos" in deterministic systems is characterized by the presence of a strange attractor. A strange attractor can occur, for example, as a result of an infinite sequence of period-doubling bifurcations of a limit cycle [136]. A rigorous consideration of this phenomenon is not included in our plans. In this section we give a rather simple mathematical model of the flow catalytic reactor, in which the "chaos" (in our understanding it is a complex aperiodic behavior in time) occurs with small variations of parameters.

An important factor determining the possibility of the presence of complex dynamics is the dimension of the phase space. When the number of degrees of freedom is three or more, the behavior trajectories of a dynamical system can be quite complex.

The local instability of trajectories in phase space in this case leads to the fact that the attracting set can have a complex structure (a strange attractor). Another mechanism of the occurrence of "chaos" is known in physics. It is the interaction of a nonlinear system with an oscillator or a source of multimode oscillations, including white noise. This mechanism is studied by us in [144]. It is shown that kinetic "chaos" for heterogeneous catalytic reactions can be obtained due to the interaction of small fluctuations of the gas phase and nonlinear processes on the catalyst surface in a neighbourhood of critical conditions.

As shown above, the interaction of the gas phase with a catalytic surface is itself a nonlinear process and is fraught with critical phenomena. Usually in the study of a catalytic subsystem it is assumed that the concentrations of substances in the gas phase are constant. It is implicitly assumed that small fluctuations of the gas phase are insignificant. However, as we show, even small fluctuations can be significant if a catalytic subsystem is near critical conditions. For certain ratios of parameters it can lead to "chaos" in the dynamics of the system.

Consider the case when a catalytic subsystem is an autocatalytic oscillator:

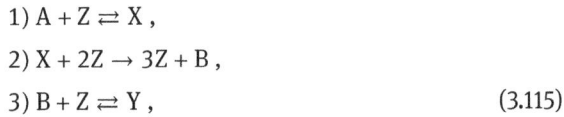

$$1)\ A + Z \rightleftarrows X,$$
$$2)\ X + 2Z \rightarrow 3Z + B,$$
$$3)\ B + Z \rightleftarrows Y, \tag{3.115}$$

where A and B are substances in the gas phase and X and Y are intermediate substances on the surface of catalyst Z. The kinetic model for (3.115), taking into account changes of the substances in the gas phase, is written as

$$\dot{a} = -k_1 a z + k_{-1} z + v(a_0 - a),$$
$$\dot{b} = k_2 x z^2 - k_3 b z + k_{-3} y + v(b_0 - b),$$
$$\dot{x} = k_1 a z - k_{-1} x - k_2 x z^2,$$
$$\dot{y} = k_3 b z - k_{-3} y, \tag{3.116}$$

where $z = 1 - x - y$. When a and b are constant the kinetic subsystem relative to x and y is an oscillator. Its interaction with the gas phase can lead to significant complication of the dynamics of the whole system (3.116). The calculations were performed at values of the parameters k_i where there are oscillations in the catalytic subsystem ($k_1 = 0.11$, $k_{-1} = 0.01$, $k_2 = 1$, $k_3 = 0.0032$, $k_{-3} = 0.002$). The input concentrations of reagents a_0, b_0 and flow rate v were varied. The external impact was considered to have two types: harmonic and white noise. It was assumed that the impact has sufficiently small amplitude. At harmonious impact the flow rate was $v = v_0 \alpha \sin wt$. White noise was simulated by a random number generator $v = v_0(\alpha_1(\alpha_2 - \alpha_1)\text{Random})$.

The harmonic impact leads to the packets of fast oscillations alternating in time [144]. When we apply white noise to the system, the oscillations become more complex and are close to "chaotic" [143]. Such behavior of the solutions of system (3.116) is explained by the fact that the value of the average flow rate v_0 in the calculations was set from a small neighborhood of their critical values. With such values of v_0

for which a steady state becomes unstable without taking into account the pulsations of the v and bifurcation of the birth of a limit cycle is "hard" [69, 70], only the "hard" birth of auto-oscillations (an appearance of a limit cycle with large amplitude at small changing of a parameter) explains a high sensitivity of the dynamics to small external impacts. Strictly speaking, the obtained complex changes of the concentrations of reactants in time cannot be called chaotic. So, the calculation of Lyapunov exponents, estimation of the Fourier power spectrum and the dimension of the attracting set require special mathematical studies. That is not our goal in this book. However, on a physical level of rigor, such complex aperiodic motions are often called chaotic. The input concentrations of the reactants in the gas phase a_0 and b_0 were varied in the calculations in addition to the pulsations v. When we chose a suitable set of their average values, the effect of "chaotization" dynamics was observed.

Thus, in a small neighborhood of critical conditions on the catalyst surface even small fluctuations of the gas phase can lead to the chaotization of the system. This is explained by the high parametric sensitivity of a dynamic system in a neighborhood of the bifurcation values of the parameters. For example, at the bifurcation of birth of a limit cycle the real parts of the eigenvalues are zero, which leads to high sensitivity of the dynamics of the system to external impacts. All this provides a basis for interpreting the occurrence of the kinetic "chaos" induced by noise.

3.5 Two-center mechanisms

In addition to simple auto-oscillations with a clearly defined period and amplitude, in heterogeneous catalytic reactions more complex dynamic behavior can also be observed when the oscillations are irregular. For a description of such complex changes of the rate of catalytic reactions in time, it is necessary to take into account some possible chemical nonideality in the corresponding kinetic model [146, 147]. We think some classical kinetic models, given in previous sections, can be used for the interpretation of complex oscillations, as well as so-called composite mechanisms of reactions occurring on different types of active centers. We will call such mechanisms two-center mechanisms.

3.5.1 Oscillator–trigger model

Consider a mechanism of catalytic reactions occurring on two types of active centers Z_1 and Z_2:

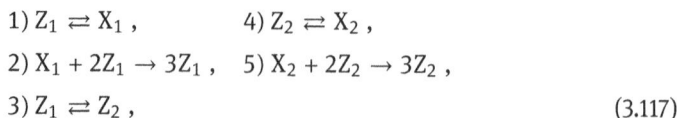

$$
\begin{aligned}
&1)\ Z_1 \rightleftarrows X_1, &&4)\ Z_2 \rightleftarrows X_2, \\
&2)\ X_1 + 2Z_1 \rightarrow 3Z_1, &&5)\ X_2 + 2Z_2 \rightarrow 3Z_2, \\
&3)\ Z_1 \rightleftarrows Z_2,
\end{aligned}
$$

$$(3.117)$$

where X_1 and X_2 are intermediate substances. Stages 1)–3) correspond to the simplest autocatalytic oscillator. Stages 4), 5) satisfy an autocatalytic trigger. They were discussed separately in previous sections. The kinetic model corresponding to the set of stages of (3.117) is

$$\dot{x}_1 = k_1 z_1 - k_{-1} x_1 - k_2 x_1 z_1^2 \,,$$
$$\dot{x}_2 = k_4 z_2 - k_{-4} x_2 - k_5 x_2 z_2^2 \,,$$
$$\dot{z}_2 = k_3 z_1 - k_{-3} z_2 - k_4 z_2 + k_{-4} x_2 + k_5 x_2 z_2^2 \,, \qquad (3.118)$$

where $z_1 = 1 - x_1 - x_2 - z_2$.

The kinetic model (3.118) is a system of three differential equations containing parameters k_i. The nonlinearity of this system is determined by stages 2), 5). The solutions x_1, x_2, z_2 have physical meaning, and they belong to the reaction simplex

$$S_r = \{(x_1, x_2, z_2): \ x_1, x_2, z_2 \ge 0, \ x_1 + x_2 + z_2 \le 1\} \,.$$

The initial conditions are $x_1(0)$, $x_2(0)$, $z_2(0) \in S_r$.

Steady states of system (3.118) are determined from the system of algebraic equations

$$k_1 z_1 - k_{-1} x_1 - k_2 x_1 z_1^2 = 0 \,,$$
$$k_4 z_2 - k_{-4} x_2 - k_5 x_2 z_2^2 = 0 \,,$$
$$k_3 z_1 - k_{-3} z_2 - k_4 z_2 + k_{-4} x_2 + k_5 x_2 z_2^2 = 0 \,. \qquad (3.119)$$

From the second equation of (3.119) we have

$$x_2 = \frac{k_4 z_2}{k_{-4} + k_5 z_2^2} \,.$$

Further, from the third equation (3.119) we can express

$$z_1 = \frac{k_{-3}}{k_3} z_2 \,.$$

Finally using (3.119) we obtain the stationarity equation with one variable z_2:

$$\alpha k_1 z_2 = (k_{-1} + k_2 \alpha^2 z_2^2)\left(1 - z(1 + \alpha) - \frac{k_4 z_2}{k_{-4} + k_5 z_2^2}\right) \,, \qquad (3.120)$$

where $\alpha = k_{-3}/k_3$. The equation (3.120) allows us to explicitly write the inverse parametric dependencies k_1, k_{-1}, k_2, k_4, k_{-4}, k_5 as functions of the variable z_2. For example,

$$k_1(z_2) = (k_{-1} + k_2 \alpha^2 z_2^2)\left(1 - z(1 + \alpha) - k_4 z_2/(k_{-4} + k_5 z_2^2)\right)/(\alpha z_2) \,,$$

$$k_{-1}(z_2) = \frac{k_1 \alpha z_2 - k_2 \alpha^2 z_2^2\left(1 - z(1 + \alpha) - k_4 z_2/(k_{-4} + k_5 z_2^2)\right)}{1 - z(1 + \alpha) - k_4 z_2/(k_{-4} + k_5 z_2^2)} \,.$$

An example of plots of parametric dependencies is presented in Fig. 3.62 with varying values of the parameter k_4 and at the following set of parameters: $k_{-1} = 0.002$, $k_2 = 1$, $k_3 = 0.0032$, $k_{-3} = 0.002$, $k_{-4} = 0.64$, $k_5 = 1$.

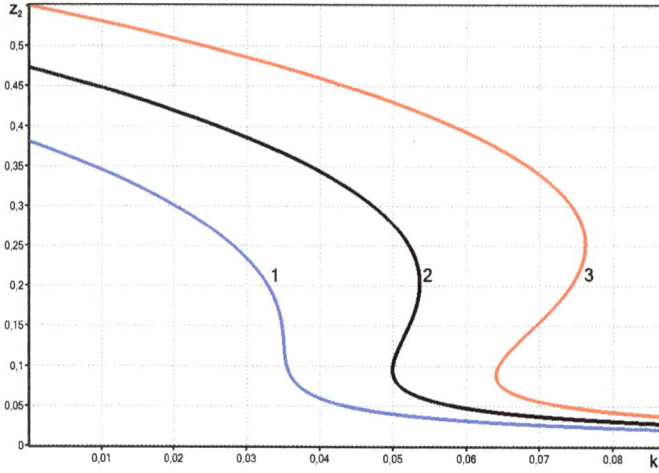

Fig. 3.62: Parametric dependencies $z_2(k_1)$ at 1)$k_4 = 0.3$; 2) 0.2; 3) 0.1

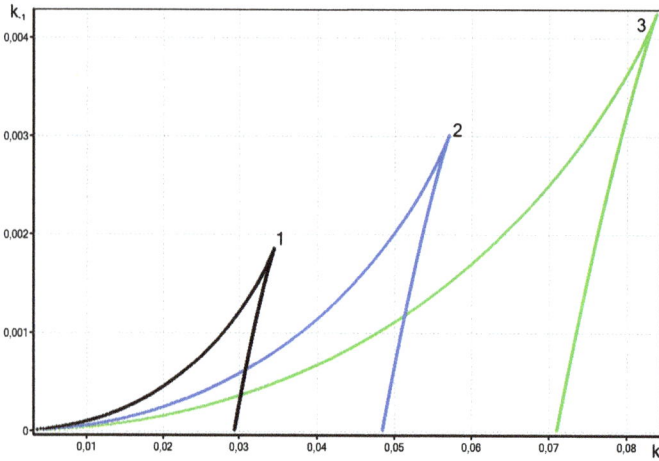

Fig. 3.63: Curves of multiplicity $L_\Delta(k_1, k_{-1})$ at 1)$k_4 = 0.3$; 2) 0.2; 3) 0.1

Knowing the parametric dependencies $k_i(z_2)$ and varying the other parameter as above, it is easy to plot the corresponding bifurcation curves: L_Δ, L_σ and L_ξ, which are determined after calculating the elements of the Jacobian matrix of the right sides of system (3.118).

$$a_{11} = -k_1 - k_{-1} \qquad\qquad a_{21} = 0\,, \qquad\qquad a_{31} = -k_3\,,$$
$$\qquad\quad -k_2 z_2^2 + 2k_2 x_1 z_1\,,$$
$$a_{12} = -k_1 + 2k_2 x_1 z_1\,, \qquad a_{22} = -k_{-4} - k_5 z_2^2\,, \qquad a_{32} = -k_3 + k_5 z_2^2 + k_{-4}\,,$$
$$a_{13} = -k_1 + 2k_2 x_1 z_1\,, \qquad a_{23} = k_4 - 2k_5 x_2 z_2\,, \qquad a_{33} = -k_3 - k_{-3} + 2k_5 x_2 z_2 - k_4\,.$$

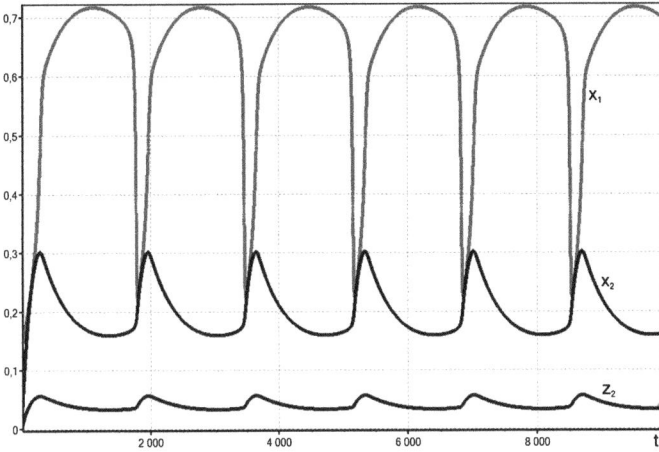

Fig. 3.64: Time dependencies $x_1(t)$, $x_2(t)$, and $z_2(t)$

The coefficients of the characteristic equation determine the bifurcation curves of the number and stability of steady states. For example, the curves of multiplicity in the parameter plane (k_1, k_{-1}) are given in Fig. 3.63, varying the parameter k_4.

Parametric portraits plotted on the basis of the bifurcation curves allow us to allocate the parameter region where oscillations exist, for example: $k_1 = 0.12$, $k_{-1} = 0.007$, $k_2 = 1$, $k_3 = 0.0032$, $k_{-3} = 0.002$, $k_4 = 0.135$, $k_{-4} = 0.64$, $k_5 = 1$ (see Fig. 3.64).

Note if $k_1, k_{-1}, k_2, k_3, k_{-3} \gg k_4, k_{-4}, k_5$, the fast oscillation (x_1, z_2) and the slow trigger x_2 of the subsystems could be identified in system (3.118). In this case, oscillations in the fast subsystem could be alternated with transitions from one branch of the slow subsystem to another. This combination can be characterized by the complex dynamics of the system as a whole. A detailed parametric analysis of model (3.118) remains to be done. In our opinion, it is the main applicant for the title of the basic model of deterministic chaos in the kinetic region.

3.5.2 Oscillator–oscillator model

This model consists of two mechanisms of simple catalytic oscillators which occur in parallel. These reactions are not connected with each other. Let a catalytic reaction proceed on each of the two centers, so we have the following scheme of transformations:

$$\begin{array}{ll} \text{1) } 2Z_1 \rightleftarrows 2X_1 \,, & \text{5) } 2Z_2 \rightleftarrows 2X_2 \,, \\ \text{2) } Z_1 \rightleftarrows Y_1 \,, & \text{6) } Z_2 \rightleftarrows Y_2 \,, \\ \text{3) } X_1 + Y_1 \rightarrow 2Z_1 \,, & \text{7) } X_2 + Y_2 \rightarrow 2Z_2 \,, \\ \text{4) } Z_1 \rightleftarrows S_1 \,, & \text{8) } Z_2 \rightleftarrows S_2 \,, \end{array} \tag{3.121}$$

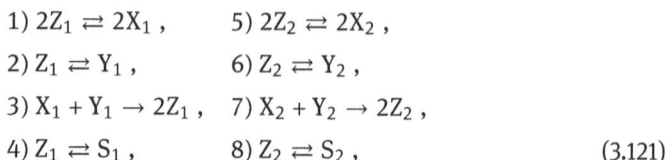

where S_1 and S_2 are substances in the buffer stages.

The following model can be written for scheme (3.121):

$$\begin{array}{ll} \dot{x}_1 = 2k_1 z_1^2 - 2k_{-1} x_1^2 - k_3 x_1 y_1 \,, & \dot{x}_2 = 2k_5 z_2^2 - 2k_{-5} x_2^2 - k_7 x_2 y_2 \,, \\ \dot{y}_1 = k_2 z_1 - k_{-2} y_1 - k_3 x_1 y_1 \,, & \dot{y}_2 = k_6 z_2 - k_{-6} y_2 - k_7 x_2 y_2 \,, \\ \dot{s}_1 = k_4 z_1 - k_{-4} s_1 \,, & \dot{s}_2 = k_8 z_2 - k_{-8} s_2 \,, \\ z_1 = 1 - x_1 - y_1 - s_1 \,, & z_2 = 1 - x_2 - y_2 - s_2 \,. \end{array}$$

If the oscillation periods at each of the centers differ slightly, the behavior of the total reaction rates are $W(t) = k_3 x_1 y_1 k_7 x_2 y_2$ and will be quite complex.

Specificity of this scheme is that there is no interaction reaction of different intermediates, which contain active centers of the same nature. The complex oscillations can be explained by the imposition of oscillations at each active center. When reactions at both centers are independent, stationary and dynamic characteristics of the corresponding kinetic models are simply summed. This situation may apply to a reaction on alloys.

Intermediate steady states implemented if there is the possibility of creating different initial conditions for each center independently of each other. The appearance of intermediate steady states become unlikely if the initial conditions of both centers are not independent. However, in this case, the dependence of the total stationary rate of the reactions on both centers can be more complex than the dependencies of rates at each center.

The total picture becomes even more complex if the reactions proceeding on different centers with different mechanisms. The behavior in time of the observed reaction rate on two centers $W(t) = W_1(t) W_2(t)$ can also be significantly more complex than dependencies $W_1(t)$, $W_2(t)$ separately. Here we give an example of summing the dynamic characteristics which are responsible for auto-oscillatory regimes of the reaction (Fig. 3.65). The set of parameters for plotting time dependencies is: $k_1 = k_5 = 2.5$, $k_{-1} = k_{-5} = 1$, $k_2 = k_6 = 1$, $k_{-2} = k_{-6} = 0.1$, $k_3 = k_7 = 10$, $k_4 = 0.0675$, $k_{-4} = 0.022$, $k_8 = 0.069$, $k_{-8} = 0.023$.

Thus, the observed total dependence $W(t)$ will be complex, although each of the dependencies $W_1(t)$ and $W_2(t)$ is quite simple. In particular, as already noted, the complex oscillatory modes recently discovered in real systems are explained by the fact that the process is proceeding on two centers, where simple oscillations with similar characteristics exist on each of the centers. Here the imitation of a "strange attractor" is possible. On the other hand, if the auto-oscillations of a rate with regular

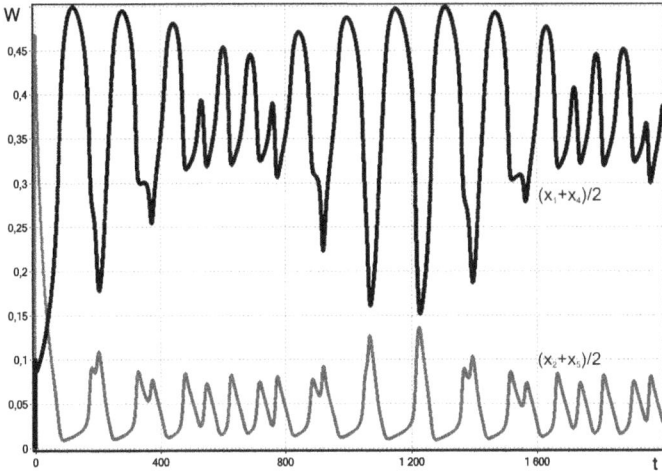

Fig. 3.65: Time dependencies $W(t)$

characteristics are observed, the question arises how and why the synchronization of these oscillations occur on the catalyst containing active centers of a different nature.

3.5.3 Model with a step of interaction of centers $Z_1 \rightleftarrows Z_2$

Consider the case of the connected subsystems

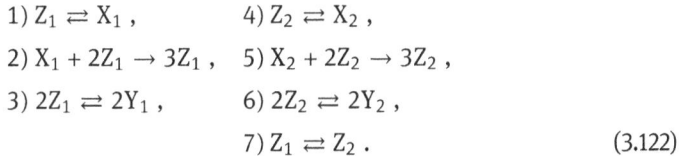

$$
\begin{array}{ll}
1)\ Z_1 \rightleftarrows X_1 , & 4)\ Z_2 \rightleftarrows X_2 , \\
2)\ X_1 + 2Z_1 \rightarrow 3Z_1 , & 5)\ X_2 + 2Z_2 \rightarrow 3Z_2 , \\
3)\ 2Z_1 \rightleftarrows 2Y_1 , & 6)\ 2Z_2 \rightleftarrows 2Y_2 , \\
& 7)\ Z_1 \rightleftarrows Z_2 .
\end{array}
\tag{3.122}
$$

Let us write a kinetic model for mechanism (3.122):

$$
\begin{aligned}
\dot{x}_1 &= k_1 z_1 - k_{-1} x_1 - k_2 x_1 z_1^2 , \\
\dot{y}_1 &= 2k_3 z_1^2 - 2k_{-3} y_1^2 , \\
\dot{x}_2 &= k_4 z_2 - k_{-4} x_2 - k_5 x_2 z_2^2 , \\
\dot{y}_2 &= 2k_6 z_2^2 - 2k_{-6} y_2^2 , \\
\dot{z}_2 &= -k_4 z_2 + k_{-4} x_2 + k_5 x_2 z_2^2 - 2k_6 z_2^2 + 2k_{-6} y_2^2 + k_7 z_1 - k_{-7} z_2 ,
\end{aligned}
$$

where $x_1 + y_1 + z_1 + x_2 + y_2 + z_2 = 1$.

The set of coefficients for the first three stages will be taken from the model of the autocatalytic oscillator found above. The coefficients of the interaction step should be much smaller than the first three. An example of time dependencies that correspond to

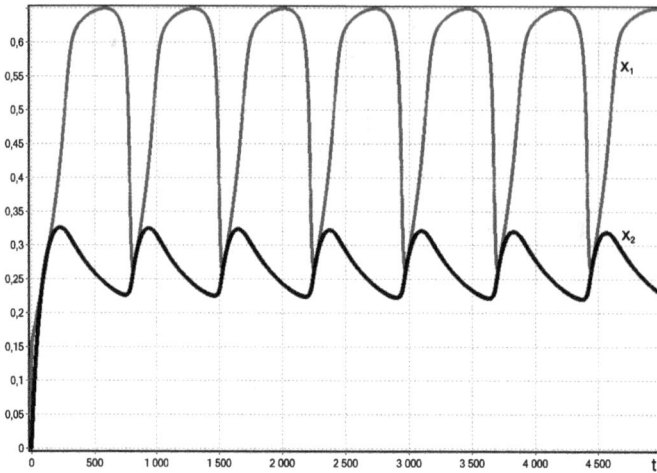

Fig. 3.66: Time dependencies $x_1(t)$ and $x_2(t)$

auto-oscillations is shown in Fig. 3.66 with the following set of parameters: $k_1 = 0.13$, $k_{-1} = 0.01$, $k_2 = k_5 = 1$, $k_3 = k_6 = 0.0032$, $k_{-3} = k_{-6} = 0.002$, $k_4 = 0.008$, $k_{-4} = 0.11$, $k_7 = 10^{-5}$, $k_{-7} = 10^{-4}$.

From the analysis of temporal characteristics, we can make the following conclusions:

1. If both the coefficients of the interaction step are changed while preserving the relations between them, on the center with a smaller interaction coefficient the oscillations do not change.
2. The region of existence of the oscillations is wider in the case where one of the interaction coefficients is equal to zero.
3. The smaller the coefficient, the smaller the period of oscillation.

The amplitudes are almost the same, i.e., the amplitude of oscillation depends on the ratio of the coefficients and not on their values.

It is interesting when the moment of excitation of oscillations occurs on one of the centers even if there is no oscillations on that center or they are the dying oscillations. Picking up the required coefficients of the interaction step, we can induce oscillations on the center.

3.5.4 Model with a diffusion change of interaction centers

Studies of heterogeneous catalytic reactions show that the process can proceed on two types of active centers [113, 118–121, 123–128]. The exchange between centers occurs due to diffusion of the adsorbed substances. If diffusion can be represented as the step of exchange between centers of different nature, then for experimental interpretation it is possible to remain within the framework of the kinetic description. However, the general approaches, developed [1] for the analysis of conditions of occurrence of a multiplicity of steady states and auto-oscillations, are applicable in this case.

Following [93, 94], in this section we will study the conditions of criticality for two-center catalytic mechanisms, including the step of diffusion exchange. They are written in terms of the graph of a complex reaction and in the form corresponding to a feature of a reaction structure, proceeding on two active centers. In particular, it is shown that the exchange between centers can cause auto-oscillations in the system as a whole. If each of the centers works independently, there are no auto-oscillations.

Consider the one-type mechanisms of chemical reactions proceeding on two types of active centers X_0 and Y_0:

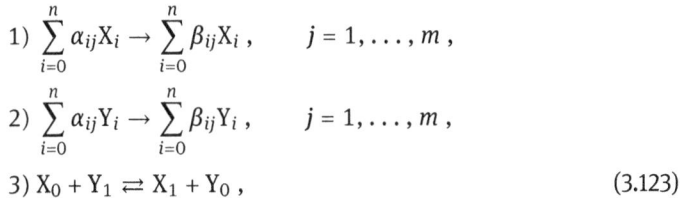

$$1) \sum_{i=0}^{n} \alpha_{ij} X_i \rightarrow \sum_{i=0}^{n} \beta_{ij} X_i , \qquad j = 1, \ldots, m ,$$

$$2) \sum_{i=0}^{n} \alpha_{ij} Y_i \rightarrow \sum_{i=0}^{n} \beta_{ij} Y_i , \qquad j = 1, \ldots, m ,$$

$$3) X_0 + Y_1 \rightleftarrows X_1 + Y_0 , \tag{3.123}$$

where X_1 and Y_1 are the diffusing substances. There is no need to specify that α, β are stoichiometric coefficients, and m_i, b_x, b_y are the coefficients corresponding to the conservation laws of atoms of this type. Here we study a case of only one step of type 3) in system (3.123). Additionally, it is assumed that for each of the active centers, there is one material balance:

$$x_0 + x_1 + \sum_{i=2}^{n} m_i x_i = b_x ,$$

$$y_0 + y_1 + \sum_{i=2}^{n} m_i y_i = b_y , \tag{3.124}$$

where x_i and y_i are the degrees of the covering surface substances for each of the active centers.

In the case where there is no diffusion exchange, i.e., step 3) is absent, the corresponding kinetic model splits into two independent parts. Their conditions of criticality can be studied by plotting and analyzing the characteristic polynomial:

$$P_{n+1}(\lambda) = \lambda^{n+1} + a_1 \lambda^n + \cdots + a_n \lambda = 0 . \tag{3.125}$$

The coefficients a_i are defined as the sum of the contributions of subgraphs of order i of a graph of the first or second step of the reaction. If the reactions are connected by diffusion exchange, the degree of the characteristic polynomial corresponding to the system as a whole becomes equal to $2n + 2$:

$$P_{2n+2}(\lambda) = \lambda^{2n+2} + a_1 \lambda^{2n+1} + \cdots + a_{2n} \lambda^2 = 0 .$$

Here the two youngest coefficients a_{2n+2}, a_{2n+1} are identically equal to zero, which corresponds to the presence of two balances in system (3.124).

It can be shown that the lowest terms of polynomials (3.125) and $P_{2n}(\lambda)$ are not identically equal to zero, but they are related by ratios:

$$a_{2n} = a_n^{(1)} a_n^{(2)} + k_d^+ y_1 a_{n-1,0}^{(1)} a_n^{(2)} + k_d^- y_0 a_{n-1,1}^{(1)} a_n^{(2)} + (k_d^- x_1 a_{n-1,0}^{(2)} + k_d^+ x_0 a_{n-1,1}^{(2)}) a_n^{(1)} ,$$
(3.126)

where $a_n^{(1),(2)}$ is the n-th coefficient of the characteristic polynomial (3.126) corresponding to scheme (3.123); $a_{n-1,i}^{(1),(2)}$ is the sum of contributions from all fragments of scheme (3.123) of order $n-1$ not containing substances with the number i; $k_d^\pm$ are the rate constants of reactions (3.124) in the forward and backward directions.

Note that in the general case the rates of diffusion exchange in the forward and backward directions are not equal: $w_d^+ \neq w_d^-$, and steady states in a fragmented system where step 3) in (3.123) is missing differ from steady states in whole systems. These states are not changed at $w_d^+ = w_d^-$. Therefore, the values of the coefficients $a^{(1)}$, $a^{(2)}$ at the stationary point in (3.126) are not the same as they were in the absence of step 3) in (3.123). However, one can expect that at small k_d^+, k_d^-, they will be close enough.

Similarly to (3.126), the expressions for the coefficients a_{2n-1}, a_{2n-2}, etc. can be written down, but in the general case they are quite bulky and uninformative.

It is an interesting special case of scheme (3.123), when the reaction is not proceeding on the second center, but there is diffusive exchange between centers for substance X_1, i.e., we consider the scheme 1) + 3). Then the ratios (3.126) are simplified.

Some examples are given here. In [94, 131, 132] it is shown for a simple model that diffusion step 3) may serve as a cause of the occurrence of auto-oscillations. For example, for scheme of transformations [94]:

$$1) X_0 \rightarrow X_1 ,$$
$$2) X_1 \rightarrow X_2 ,$$
$$3) 2X_1 + X_2 \rightarrow 3X_0 ,$$
(3.127)

added to step 3) in (3.123), the corresponding kinetic model admits auto-oscillations. However, the scheme:

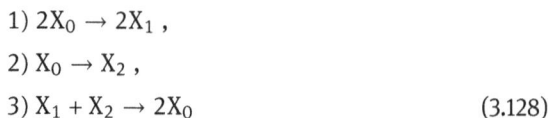

$$1) 2X_0 \rightarrow 2X_1 ,$$
$$2) X_0 \rightarrow X_2 ,$$
$$3) X_1 + X_2 \rightarrow 2X_0$$
(3.128)

together with step 3) in (3.123) cannot describe auto-oscillations [94]. As above, schemes (3.127) and (3.128) correspond to the transformation of intermediates. For example, scheme (3.128) can be interpreted as the scheme of hydrogen oxidation (with the assumption of constant gas phase O_2, H_2 = const) on metal Me:

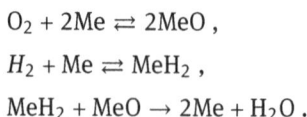

$$O_2 + 2Me \rightleftarrows 2MeO ,$$
$$H_2 + Me \rightleftarrows MeH_2 ,$$
$$MeH_2 + MeO \rightarrow 2Me + H_2O .$$

Consider a scheme consisting of two similar reactions occurring on the two centers:

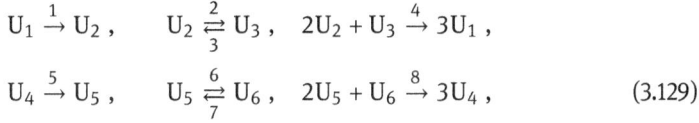

$$U_1 \xrightarrow{1} U_2 , \qquad U_2 \underset{3}{\overset{2}{\rightleftarrows}} U_3 , \qquad 2U_2 + U_3 \xrightarrow{4} 3U_1 ,$$

$$U_4 \xrightarrow{5} U_5 , \qquad U_5 \underset{7}{\overset{6}{\rightleftarrows}} U_6 , \qquad 2U_5 + U_6 \xrightarrow{8} 3U_4 , \qquad (3.129)$$

and connected with a step of diffusion exchange:

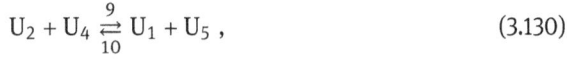

$$U_2 + U_4 \underset{10}{\overset{9}{\rightleftarrows}} U_1 + U_5 , \qquad (3.130)$$

Here the numbers of corresponding reactions are indicated over and under the arrows. According to (3.126) the ratios between the coefficients of the characteristic polynomial are represented in the form

$$a_2 = a_{20}^{(1)} + a_{20}^{(2)} + a_{10}^{(1)} a_{10}^{(2)} + a_2^{(1,2)} .$$

$a_{20}^{(i)}$ are the coefficients of the second order for the scheme on the i-th center without step (10) and $a_{10}^{(i)}$ are the coefficients of the first order. $a_2^{(1,2)}$ are the contributions from the graphs of the second order containing one reaction for each of the centers, taking into account (3.130). We have

$$a_{20}^{(1)} = 3w_4(w_4 P_1^1 + w_3 P_2^1) ,$$
$$a_{20}^2 = 3w_8(w_8 P_1^2 + w_7 P_2^2) ,$$
$$a_{10}^1 = w_4 M_1 + w_3(h_2 + h_3) ,$$
$$a_{10}^2 = w_8 M_2 + w_7(h_5 + h_6) ,$$
$$a_2^{1,2} = (w_9 - w_{10})(w_{10} h_1 h_4 - w_9 h_2 h_4) + w_4 w_{10} M_1(h_4 + h_5) +$$
$$+ w_8 w_3 M_2(h_1 + h_2) + w_9(w_4 P_1^1 + w_3 P_0^1) + w_{10}(w_8 P_1^2 + w_7 P_0^2) ,$$

where $h_i = 1/u_i$, u_i is the concentration of substance U_i in the steady state, and w_j is the rate of the j-th reaction,

$$M_1 = 3h_1 + 4h_2 + h_3 ,$$
$$M_2 = 3h_4 + 4h_5 + h_6 ,$$
$$P_0^1 = h_1 h_2 + h_1 h_3 + h_2 h_3 ,$$
$$P_0^2 = h_4 h_5 + h_4 h_6 + h_5 h_6 ,$$
$$P_1^1 = -h_1 h_2 + h_1 h_3 + h_2 h_3 ,$$
$$P_2^1 = h_1 h_2 + h_1 h_3 + 3h_2 h_3 ,$$
$$P_1^2 = -h_4 h_5 + h_4 h_6 + h_5 h_6 ,$$
$$P_2^2 = h_4 h_5 + h_4 h_6 + 3h_5 h_6 .$$

Further we have the expression

$$a_4 = a_{20}^1 \times a_{20}^2 + a_4^* ,$$

where

$$a_4^* = (w_9 - w_{10})w_9(w_4 + w_3)h_2h_3h_4[w_8(h_5 - h_6) - w_7(h_5 + h_6)]+$$
$$+ (w_{10} - w_9)w_{10}(w_8 + w_7)h_1h_5h_6[w_4(h_2 - h_3) - w_3(h_2 + h_3)]+$$
$$+ 3w_4w_{10}(w_4P_1^1 + w_3P_2^1)(w_8P_1^2 + w_7P_0^2)+$$
$$+ 3w_8w_9(w_4P_1^1 + w_3P_0^1)(w_8P_1^2 + w_7P_2^2).$$

Because $a_4 > 0$ for all h_i, it is enough for the existence of single and unstable steady that $a_2 < 0$. For example, there area auto-oscillations in a kinetic model corresponding to scheme (3.129) + (3.130) with parameters providing inequalities $a_2 < 0$ and $a_4 > 0$. A set of these parameters is: $k_1 = 24$, $k_2 = 200$, $k_3 = 1$, $k_4 = 4 \cdot 10^4$, $k_5 = 0.012$, $k_6 = 0.01$, $k_7 = 0.012$, $k_8 = 0.2 \cdot 10^{-6}$, $k_9 = 25$, $k_{10} = 0.5$. When the initial data are $u_1(0) = 0.6$, $u_2(0) = 10^{-2}$, $u_3(0) = 0.9$, $u_4(0) = 150$, $u_5(0) = 250$, $u_6(0) = 0.37 \cdot 10^{-6}$, the corresponding dynamical system has undamped oscillations in time. Of course, auto-oscillations exist for a range of parameters which are set by the conditions of a change of sign in the coefficients a_i. However, plotting it for different combinations of the varied parameters is already time-consuming enough. We give here only a representative point of this area.

Thus, the coefficients of the characteristic polynomial determining the conditions of criticality for a kinetic model of two-center mechanism can be associated with the corresponding coefficients of its subsystems. In the analysis of specific models, the proposed approach can significantly facilitate a search for conditions for the existence of auto-oscillations in the system as a whole. The examples show that the presence a of diffusion exchange step between centers can become a cause for the appearance of oscillations, despite their absence for each of the subsystems separately.

3.6 Simplest models of CO oxidation on platinum

The oxidation reaction of carbon monoxide on metals, in particular on platinum, is a traditional object of research in heterogeneous catalysis. In his time G.S. Yablonsky had the idea of writing a book "CO oxidation on Pt and the rest of catalysis". Unfortunately, this idea collapsed under the weight of a huge volume of experimental data and the complexity of the task. By the present time, there is a vast number of works on the analysis of such processes. In particular, a large number of articles was published with experimental data evidencing the possibility of complex dynamic behavior of this reaction. We refer only to some works with wide reviews [1, 2, 27, 150].

In this section, we consider only the simplest kinetic models of CO oxidation over Pt, which are fit to our program of plotting and parametric analysis of basic models of chemical kinetics of catalytic reactions with the goal of a description of multiplicity of steady states and auto-oscillations.

A solution to the problem of multiplicity of steady states consists of a selection of the regions in the parameter space with a few steady states (the plotting of bifurcation curves of multiplicity steady states). Here we demonstrate the developed approach of parametric analysis of kinetic models in the plane of the two main parameters – partial pressures of the reactants in the gas phase (CO and O_2) at different temperatures. These bifurcation diagrams are plotted based on the parametric dependencies of steady states, which can be written in an explicit form.

Consider a set of stages

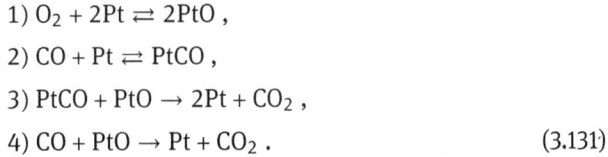

1) $O_2 + 2Pt \rightleftharpoons 2PtO$,

2) $CO + Pt \rightleftharpoons PtCO$,

3) $PtCO + PtO \rightarrow 2Pt + CO_2$,

4) $CO + PtO \rightarrow Pt + CO_2$. (3.131)

Scheme (3.131) is a two-route mechanism of CO oxidation on the platinum surface. The set of stages 1)–3) is the classical adsorption mechanism of Langmuir–Hinshelwood. This mechanism is supplemented by step 4), which reflects the possibility of the shock mechanism of CO oxidation [150]. A justification of mechanism (3.131) is not possible to give in depth here. Our goal is a parametric analysis of kinetic models corresponding to the considered mechanism.

Write a kinetic model for scheme (3.131)

$$\dot{x} = 2k_1 p_O(1 - x - y)^2 - 2k_{-1}x^2 - k_3 xy - k_4 p_{CO}x ,$$
$$\dot{y} = k_2 p_{CO}(1 - x - y) - k_{-2}y - k_3 xy ,$$ (3.132)

where x, y, $1 - x - y = z$ are the dimensionless concentrations of PtO, PtCO, Pt, and accordingly p_O, p_{CO} are partial pressures of O_2, CO in the gas phase, and k_i are the rate constants of the reactions, which have Arrhenius dependence on temperature T. In the dynamic system (3.132) x, y are phase variables, and p_O, p_{CO}, T are parameters.

Steady states for (3.132) are determined at fixed values of the parameters p_O, p_{CO}, T as the solutions of a system of nonlinear algebraic equations:

$$2k_1 p_O(1 - x - y)^2 - 2k_{-1}x^2 - k_3 xy - k_4 p_{CO}x = 0 ,$$
$$k_2 p_{CO}(1 - x - y) - k_{-2}y - k_3 xy = 0 .$$ (3.133)

The solutions of system (3.133) are sought in the reaction simplex

$$S_r = \{(x, y): \ x \geq 0, \ y \geq 0, \ x + y \leq 1\} .$$

From the second equation in (3.133) we get

$$y(x) = \frac{k_2 p_{CO}(1 - x)}{k_2 p_{CO} + k_3 x} .$$ (3.134)

After substituting (3.134) in the first equation (3.133) we obtain the stationarity equation in one variable x:

$$2k_1 p_0 (1 - x - y(x))^2 - 2k_{-1}x^2 - k_3 xy(x) - k_4 p_{CO} x = 0 . \tag{3.135}$$

The equation (3.135) is an algebraic equation of the fourth degree. This equation has one or three real roots on the interval $[0, 1]$. Note that the parameter p_0 is not included in the expression (3.134). So the parametric dependence is obtained from (3.135) in an explicit form:

$$p_0 = \frac{2k_{-1}x^2 + k_3 xy(x) + k_4 p_{CO} x}{2k_1 (1 - x - y(x))^2} . \tag{3.136}$$

Variation of the value of x in the interval $(0, 1)$ allows us to obtain the dependence from (3.136), which is an inverse to the required dependence $x(p_0)$ for fixed p_{CO}, T. Changing the parameters, for example p_{CO}, it is easy to plot the bifurcation curves of multiplicity of steady states from (3.136). Such manipulations were carried out in previous sections for a series of basic models of chemical kinetics.

In the experiment, as a rule, the observed reaction rate is measured, i.e., the rate of formation of carbon dioxide CO_2:

$$W_{CO_2} = k_3 xy + k_4 p_{CO} x . \tag{3.137}$$

The dependencies of the stationary rate W_{CO_2} on parameters are plotted based on the dependencies $x(p_0, p_{CO}, T)$, $y(p_0, p_{CO}, T)$, which are written down according to (3.134), (3.136). In the case of multiplicity of steady states with a variation of parameters, the dependencies of the stationary rate are characterized by hysteresis, which is observed in real experiments [150].

The hysteresis dependencies of the stationary rate (3.137) are implemented under determined limitations on the parameters of model (3.132). For example, multiplicity of steady states is possible if the following inequalities hold:

$$k_{-2} < 0.227 k_3 ,$$
$$8k_1 p_0 > 9k_2 p_{CO}(1 + 3k_2 p_{CO}/k_3)(2/3 + k_4/(3k_2) + k_4 p_{CO}/k_3) .$$

These inequalities show that the presence in the reaction mechanism of (3.131), step 4) leads to a reduction in the region of multiplicity of steady states in the parameter space (p_0, p_{CO}, T). The increase of the rate constants of desorption k_1, k_{-2} leads to the same effect. *Ceteris paribus*, the probability of appearance of a few steady states increases at sufficiently large values of p_0 and small values of p_{CO}, but the value of k_3 must be quite large.

It is easy to show that a single steady state is always stable. If there are three steady states, two of them are stable, and one is unstable. This means that model (3.132) can only describe a multiplicity of steady states. For the interpretation of auto-oscillations, it must be modified.

Specific values of the kinetic parameters $k_i(T) = k_i^0 \exp(-E_i/(RT))$ were defined by V.I. Elokhin [157]:

$$
\begin{array}{ll}
k_1^0 = 0.202 \cdot 10^6 \ c^{-1} \ \text{Torr}^{-1}, & E_1 = 0, \\
k_{-1}^0 = 0.16 \cdot 10^{13} \ c^{-1}, & E_{-1} = 50 \ \text{kcal mol}^{-1}, \\
k_2^0 = 0.45 \cdot 10^6 \ c^{-1} \ \text{Torr}^{-1}, & E_2 = 0, \\
k_{-2}^0 = 10^{13} \ c^{-1}, & E_{-2} = 35.5 \ \text{kcal mol}^{-1}, \\
k_3^0 = 0.4 \cdot 10^4 \ c^{-1}, & E_3 = 11 \ \text{kcal mol}^{-1}, \\
k_4^0 = 0.45 \cdot 10^6 \ c^{-1} \ \text{Torr}^{-1}, & E_4 = 1 \ \text{kcal mol}^{-1},
\end{array}
\tag{3.138}
$$

for the following intervals of parameters (p, T):

$$
300 \ \text{K} \le T \le 800 \ \text{K}, \quad p_O \ge 10^{-8} \ \text{Torr}, \quad p_{CO} \le 10^{-6} \ \text{Torr}. \tag{3.139}
$$

The region (3.139) and the values of the kinetic parameters (3.138) correspond to real experimental conditions. Some results of calculations can be found in [157]. With variation of the temperature, the region of the multiplicity of steady states is moved in the plane (p_{CO}, p_O), disappearing if $T < 370$ K or $T > 500$ K.

Note that the kinetic parameters (3.138) were originally found during the processing of the experiment, in which there were no observed critical effects (multiplicity of steady states in this case). The high parametric sensitivity of the stationary reaction rate in a certain region of pressures p, p_{CO} and temperature T was characteristic. The calculations of the region of the multiplicity of steady states, as it turned out later, have predictive power. The publications [157] were published after [150]. In these works several steady states were found experimentally for the considered reaction. Moreover, the region of parameters (p, T) where the critical effects (hysteresis of stationary reaction rate) were detected turned out very close to the region calculated by us. The corresponding quantitative comparison can be found in [150].

In addition, at the time, we were surprised by the fact that the quite simple and well-known mechanism (3.131) was characterized by rather complex dependencies of the stationary reaction rate. This is partly understandable, because the parameter regions where there is a multiplicity of steady states are quite narrow. It is difficult to hit upon this region accidentally. Only a conscious search leads to these results. In this example we once again can see the productivity of the general scheme: "the experiment (observation) $\rightleftarrows$ understanding".

Kinetic model (3.132) is limited. But in some sense it is a base in constructing more realistic models. However, the procedure of parametric analysis, presented for models of formal mechanisms, has general meaning and can be used by analogy for other specific reactions.

Model (3.132) can be modified, if the dissolution of oxygen in the catalyst volume is taken into account [150]. In this case, the following stages are added to (3.131)

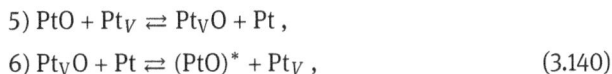

$$
\begin{array}{l}
5) \ \text{PtO} + \text{Pt}_V \rightleftarrows \text{Pt}_V\text{O} + \text{Pt}, \\
6) \ \text{Pt}_V\text{O} + \text{Pt} \rightleftarrows (\text{PtO})^* + \text{Pt}_V,
\end{array}
\tag{3.140}
$$

where Pt_V is the active center in the catalyst volume; Pt_VO is oxygen dissolved in the volume; and $(PtO)^*$ is nonreactive oxygen on the catalyst surface.

A kinetic model corresponding to scheme (3.131)+(3.140) has the form:

$$\dot{x} = 2k_1p_Oz^2 - 2k_{-1}x^2 - k_3xy - k_4p_{CO}x - k_5xz_V + k_{-5}x_Vz,$$
$$\dot{y} = k_2p_{CO}z - k_{-2}y - k_3xy,$$
$$\dot{x}_V = k_5xz_V - k_{-5}x_Vz - k_6x_Vz + k_{-6}xz_V,$$
$$\dot{x}^* = k_6x_Vz - k_{-6}x^*z_V, \tag{3.141}$$

where x_V, x^*, z_V are the concentrations of Pt_VO, PtO^*, Pt_V, accordingly. The values of z, z_V are determined according to two material balances

$$x + y + z + x^* = \text{const},$$
$$x_V + z_V = \text{const}_V, \tag{3.142}$$

where const and const_V correspond to the total numbers of active centers on the catalyst surface and in its volume, respectively.

The model (3.141), (3.142) was proposed in [98–109]. Its full parametric analysis was not carried out. Here we only note that this model is significantly different from (3.132), because there are four independent phase variables in it. This fact significantly expands the diversity of its properties. In particular, the model potentially gives the opportunity to describe auto-oscillations as well as more complex nonstationary regimes. It is sufficient to note that the presence of the buffering substance PtO^*, as considered earlier for catalytic oscillators, can lead to the existence of undamped oscillations in system (3.141). However, the role of the buffer can be to play the step of sharing the surface with a volume of the catalyst (step 5). In the absence of step 6), model (3.141) is simplified:

$$\dot{x} = 2w_1 - w_3 - w_4 - w_5,$$
$$\dot{y} = w_2 - w_3,$$
$$\dot{x}_V = w_5, \tag{3.143}$$

where

$$w_1 = 2k_1p_Oz^2 - 2k_{-1}x^2,$$
$$w_2 = k_2p_{CO}z - k_{-2}y,$$
$$w_3 = k_3xy,$$
$$w_4 = k_4p_{CO}y,$$
$$w_5 = k_5xz_V - k_{-5}x_Vz,$$
$$z = 1 - x - y,$$
$$z_V = \text{const}_V - x_V. \tag{3.144}$$

The kinetic model (3.143), (3.144) for the complex of stages 1)–5) is characterized by three independent phase variables. Our analysis of catalytic oscillators shows that the model can have auto-oscillations. A detailed parametric analysis of the models of (3.141) and (3.143) remains to be done.

Along with model (3.131) in [150] a diffusion model of CO oxidation over Pt was proposed. Instead of stages 5) and 6), we have a diffusion equation of the surface oxygen in the catalyst volume:

$$\frac{\partial x_V}{\partial t} = \frac{D}{L^2} \cdot \frac{\partial^2 x_V}{\partial \xi^2}, \tag{3.145}$$

where ξ is a dimensionless spatial coordinate; $x_V(t, \xi)$ is a dimensionless concentration of diffusing oxygen $Pt_V O$ into the catalyst volume; D is a diffusion coefficient; L is the thickness of the layer in which diffusion occurs. For the partial differential equation, the initial and boundary conditions must be given. Boundary conditions for the catalyst surface ($\xi = 1$) are slightly modified from model (3.132), taking into account the process of the flow of PtO into a volume of the catalyst. Model (3.145)+(3.132) is a model of the microkinetics. Similar models will be discussed in Chapter 4.

Note that in addition to steady states, another important characteristic in chemical dynamics is the relaxation time to a steady state. This task can be included into the scheme of parametric analysis. Knowledge of the dependencies of steady states on the parameters allows us to study in a natural way the roots of the corresponding characteristic equation, which determine the relaxation times to a steady state. Some examples of solutions of this problem were given earlier. However, we do not consider the problem of estimating the relaxation times in more detail here, because it deserves special attention [158]. We note only that the characteristics of the relaxation time of the oxidation reaction CO over Pt are considered in [157]. In particular, when carrying out a series of calculations of model (3.145) with diffusion coefficient

$$D = 10^{-2} \exp\left(\frac{-20,000}{RT}\right) \text{cm}^2 \text{ s}^{-1}$$

it is shown that an influence of diffusion on the unsteady behavior of the reaction can be quite varied. This influence is determined by various factors, for example, the initial composition of the surface and the volume of the catalyst, their steady state, and the relaxations of the kinetic origin [150, 151].

In talking about the diversity of models of the process of CO oxidation on metals, we did not set a goal to consider all diversity of such models in this section. This would lead us too far away from the main topic – parametric analysis of basic models (in a sense, the simplest models). As usual, "real life" is so diverse that when modeling you always have to sacrifice some details.

We complete this section with a brief summary of the results of a detailed parametric study of formal kinetic models, conducted at the time with A.I. Khibnik for a simplified mechanism of catalytic oxidation CO:

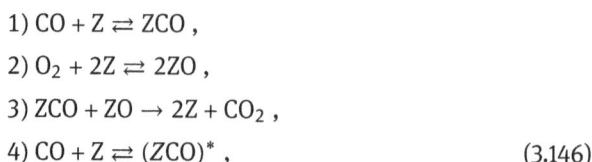

$$1)\ CO + Z \rightleftarrows ZCO\ ,$$
$$2)\ O_2 + 2Z \rightleftarrows 2ZO\ ,$$
$$3)\ ZCO + ZO \rightarrow 2Z + CO_2\ ,$$
$$4)\ CO + Z \rightleftarrows (ZCO)^*\ , \tag{3.146}$$

where $(ZCO)^*$ is a tightly connected form of CO on the catalyst surface Z. With the assumption of constant temperature and of partial pressures of substances in the gas phase, the formal scheme corresponding to mechanism (3.146) is:

$$1)\ Z \rightleftarrows X_1\ ,$$
$$2)\ 2Z \rightleftarrows 2X_2\ ,$$
$$3)\ X_1 + X_2 \rightarrow 2Z\ ,$$
$$4)\ Z \rightleftarrows X_3\ . \tag{3.147}$$

The following kinetic model can be written for scheme (3.147)

$$\dot{x}_1 = k_1 z - k_{-1} x - k_3 x_1 x_2\ ,$$
$$\dot{x}_2 = 2k_2 z^2 - 2k_{-2} x_2^2 - k_3 x_1 x_2\ ,$$
$$\dot{x}_3 = k_4 z - k_{-4} x_3\ ,$$
$$z = 1 - x_1 - x_2 - x_3\ . \tag{3.148}$$

The model (3.148) is similar with precision to the notation discussed earlier in (3.83). The program of full parametric analysis of (3.148), implemented by A.I. Khibnik, includes one, two, and three-parameter local bifurcations of steady states and bifurcations of limit cycles. In particular, the basic bifurcation curves are plotted in the parameter plane (k_1, k_2): curves of multiplicity and neutrality of steady states, Andronov–Hopf bifurcations, bifurcations of a limit cycle and a loop of the separatrices. The study of the mutual location of these curves and their reconstruction while changing the third parameter allowed us to identify 23 types of phase portraits of system (3.148). These typical phase portraits differ in the number and stability of steady states, the presence of a limit cycle, and their mutual location. For details of this rather time-consuming research, we refer the reader to [1]. Here we will just mention that the diversity of possible dynamic behavior of simple kinetic models of the form (3.148) is amazing. On the other hand, our experience of parametric analysis shows that, from a practical point of view, the regions of parameters corresponding to the three steady states and a single unstable steady state (stable limit cycle) are important. Other features of complex dynamics, as rule, are characterized by rather narrow regions of parameters. Therefore, we can conclude that the simple models behave in a relatively simple manner. These basic characteristics (number and stability of steady states) are a goal of our primary parametric analysis. The skeleton of the parametric portrait, plotted on the basis of bifurcation curves L_Δ, L_σ, can be further detailed and supplemented by many other more subtle details.

3.7 Nonideal kinetics

For the equation of chemical kinetics

$$\dot{x} = \sum_{s=1}^{m} \gamma_s w_s(\mathbf{x}, T) , \qquad (3.149)$$

where $\mathbf{x} = (x_1, \ldots, x_n)$ is a vector of the concentrations of reagents; γ_s is the stoichiometric vector with components $\gamma_s = \beta_{is} - \alpha_{is}$, corresponding to the scheme of transformations

$$\sum_{i=1}^{n} \alpha_{is} X_i \rightarrow \sum_{i=1}^{n} \beta_{is} X_i , \quad s = 1, \ldots, m . \qquad (3.150)$$

The rate of a particular step w_s in the simplest case of mass action law has the form

$$w_s(\mathbf{x}, T) = k_s(T) \prod_{i=1}^{n} x_i^{\alpha_{is}} , \quad s = 1, \ldots, m . \qquad (3.151)$$

Marcelin–de Donder kinetics, generalizing the law of mass action in a natural way, defines the reaction rate in the form

$$w_s(\mathbf{x}, T) = w_s^0(T) \exp\left(\sum_{i=1}^{n} \alpha_{is} \mu_i(\mathbf{x}) \right), \qquad (3.152)$$

where with accuracy to a constant multiplier RT the function $\mu_i(\mathbf{x})$ is the chemical potential of the i-th substance X_i in (3.150). For the ideal case $\mu_i = \mu_i^0 + \ln x_i$, we obtain (3.151). For the chemically nonideal system

$$\mu_i(x) = \mu_i^0 + \ln x_i + g_i(\mathbf{x}) ,$$

where a correction for nonideality of $g_i(\mathbf{x})$ in the first approximation can be linear

$$g_i(\mathbf{x}) = \sum_{j=1}^{n} a_{ij} x_j , \quad i = 1, \ldots, n .$$

Corrections of this kind are often used when building kinetic models of nonideal catalytic systems [146, 147, 170].

For closed chemical systems all stages are taken reversible, so

$$w_s(\mathbf{x}, T) = w_s^0(T) \left(\exp\left(\sum \alpha_{is} \mu_i(\mathbf{x}) \right) - \exp\left(\sum \beta_{is} \mu_i(\mathbf{x}) \right) \right) .$$

Natural for this case are the requirements of existence and convexity of thermodynamic functions, which means the symmetry of the matrix corrections on nonideality

$$A_\mu = ((a_{ij})) \qquad (3.153)$$

and positive definiteness of the matrix

$$E\mathbf{x}^{-1} + A_\mu , \qquad (3.154)$$

where E is the identity matrix and $\mathbf{x}^{-1} = (x_1^{-1}, \ldots, x_n^{-1})$ is a vector. These constraints will be called the thermodynamic constraints on the parameters a_{ij}. They have the simplest form if A_μ is a diagonal matrix: the symmetry conditions (3.153) are automatically performed. Conditions for positive definiteness (3.154) mean that $a_{ii} + 1/x_i > 0$, $i = 1, \ldots, n$.

It is possible to show [92] that chemical nonideality, without taking into account thermodynamic constraints, leads to multiplicity of equilibria in a closed system as well as to the possibility of auto-oscillations. However, the correct consideration of constraints to nonideality guarantees the natural dynamic behavior of a closed chemical system, i.e., equilibrium is the only stable state in a reaction polyhedron as a whole.

To any closed system corresponds an open system. In this case the thermodynamic constraints (3.153), (3.154) on the parameters of chemical nonideality, generally speaking, must be fair and for an open system. The absence of detailed equilibrium and the presence of irreversible stages of (3.150) are characteristic for open systems. The Jacobian matrix J_x of system (3.149) for nonideal kinetics (3.152) can be represented in the form

$$(J_x)_{ij} = \sum_{s=1}^{m} \gamma_{si} \sum_{k=1}^{n} w_s \alpha_{sk} \frac{\partial \mu_k}{\partial x} ,$$

i.e.,

$$J_x = J_y \cdot J_\mu ,$$

where the matrix J_y can be interpreted as a matrix of a scheme of transformations (3.150) and the matrix J_μ is responsible for the kinetic nonideality of the elementary stages (3.152).

The following statements are true [92]:
a) if the matrix J_μ is positive definite and J_y is a similar symmetric negative definite matrix, then J_x is stable (also negatively defined);
b) for stability of J_x, negative definiteness of the matrix $J_y J_y^*$ is sufficient (with positive definiteness of J_μ);
c) negative definiteness, referred to in a), b), is sufficient in the reaction polyhedron;
d) if a chemically ideal system corresponding to the scheme (i.e., the matrix J_y) is structurally stable, i.e., the Jacobian matrix J_x is stable under all values of the rate constants of reactions, then a chemically nonideal system with a diagonal matrix J_μ, i.e., $\mu_i = f_i(x_i)$ satisfying thermodynamic constraints, is also structurally stable.

The materiality of the conditions on the matrix J_μ will be shown by a number of formal kinetic models.

Model 1

Consider an open catalytic system (an isomerization reaction):

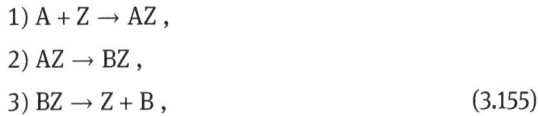

$$1)\ A + Z \rightarrow AZ,$$
$$2)\ AZ \rightarrow BZ,$$
$$3)\ BZ \rightarrow Z + B, \tag{3.155}$$

where A, B are substances in the gas phase; AZ, BZ, Z are intermediate substances and the catalyst, respectively. Let the nonideality of the system be determined by the substance AZ, i.e., the matrix of corrections has the form

$$A_\mu = \begin{pmatrix} a & 0 \\ 0 & 0 \end{pmatrix}. \tag{3.156}$$

The kinetic model can be written for scheme (3.155) as

$$\dot{x} = k_1 z - k_2 x e^{ax},$$
$$\dot{y} = k_2 x e^{ax} - k_3 y, \tag{3.157}$$

where $z = 1 - x - y$; x, y are the dimensionless concentrations of AZ, BZ, respectively. The equations of stationarity for (3.157) can be written as:

$$F(x) = \frac{xe^{ax}}{1 - x} = \frac{k_1 k_3}{k_2(k_1 + k_3)},$$
$$y = k_2 x e^{ax} / k_3.$$

Elementary research shows that if $a \geq -4$, there is one steady state (the function $F(x)$ is monotone, i.e., $F'(x) > 0$), while if $a < -4$ three steady states can exist.

The stability of steady states is determined by the values of the coefficients σ, Δ of the characteristic equation $\lambda^2 + \sigma\lambda + \Delta = 0$, where

$$\Delta = k_1 k_2 F'(x) / x,$$
$$\sigma = \frac{k_1 k_2}{x(k_1 + k_2)} \left((k_1 k_3 + (k_3 - k_1)^2) x / (k_1 k_3) + F'(x) \right).$$

Thus the steady states (3.157) corresponding to $F' < 0$ are unstable, and those corresponding to $F' > 0$ are stable. So if a steady state is single, then it is stable. If three steady states exist, two of them are stable and the third one is unstable. However, thermodynamic constraints on nonideality (3.156) require that $F' > 0$, i.e., the steady state within these constraints is unique and stable.

Model 2

For (3.155), let the nonideality of chemical potentials of the two substances take place, i.e.,

$$A_\mu = \begin{pmatrix} a & 0 \\ 0 & b \end{pmatrix}.$$

The kinetic model has the form

$$\dot{x} = k_1 z - k_2 x e^{ax} ,$$
$$\dot{y} = k_2 x e^{ax} - k_3 y e^{by} . \qquad (3.158)$$

The stationarity equation (3.158) in the previous notation can be written as:

$$y = 1 - x - \alpha x e^{ax} = f(x) , \qquad x = 1 - y + \beta y e^{by} = g(x) ,$$

where $\alpha = k_2/k_1$ and $\beta = k_3/k_1$. The requirement $a, b \geq -4$ is a sufficient condition of monotonicity of the curves $f(x)$, $g(x)$. It is easy to show that if you refuse these demands, up to 9 steady states exist. However, thermodynamic constraints, in this case

$$1 + bx > 0 , \qquad 1 + by > 0 ,$$

provide the monotonicity of the functions $f(x)$, $g(x)$ and thus the uniqueness of steady state. It can be shown, this condition guarantees the stability of the steady state.

Model 3
For scheme (3.155) we consider the case of a matrix of corrections of the form

$$A_\mu = \begin{pmatrix} 0 & a \\ a & 0 \end{pmatrix} . \qquad (3.159)$$

The kinetic model corresponding to (3.155) and (3.159) has the form:

$$\dot{x} = k_1 z - k_2 x e^{ay} ,$$
$$\dot{y} = k_2 x e^{ay} - k_3 y e^{ax} .$$

Steady states are determined from the ratios

$$f(x) = \frac{x}{1-x}(1 + k_2 e^{ax}/k_1)e^{ay} = k_2/k_1 ,$$
$$y = \frac{1-x}{1 + (k_2/k_1)e^{ax}} .$$

The solutions of the equation $f(x) = k_2/k_1$ give all steady states. Thermodynamic constraints on the matrix (3.159) show that $1 - a^2 xy > 0$, i.e., $f' > 0$. Therefore, a steady state is the only one.

Model 4
Consider a nonlinear scheme of transformations

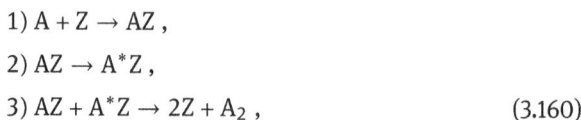

$$1) \ A + Z \rightarrow AZ ,$$
$$2) \ AZ \rightarrow A^* Z ,$$
$$3) \ AZ + A^* Z \rightarrow 2Z + A_2 , \qquad (3.160)$$

where A, A^*, A_2 are substances whose concentrations are constant. For intermediates AZ, A^*Z we have a kinetic model

$$\dot{x} = k_1 z - k_2 x e^{ax} - k_3 xy e^{ax} ,$$
$$\dot{y} = k_2 x e^{ax} - k_3 xy e^{ax} , \qquad (3.161)$$

where x, y are the dimensionless concentrations of AZ, A^*Z, respectively. As in (3.155), the matrix of corrections A_μ has the form (3.156).

It is easy to see that for (3.161) there is an extreme steady state: $x = 0$, $y = 1$, and the internal steady state, for which $y = k_2/k_1$ (at $k_2 < k_1$) and the value of x is determined from the equation

$$f(x) = x e^{ax} (a_1 - x)^{-1} = a_2 ,$$

where $a_1 = 1 - k_2/k_1$ and $a_2 = k_1/(zk_2)$. Implementation of the inequality $a \geq -4/a_1$ is the condition of monotonicity for $f(x)$ on the interval $[0, a_1]$. Thus, if this inequality is true, a steady state is unique. If it is violated, there may exist three steady states.

Note that the specifics of the nonlinear scheme (3.160) resulted in the existence of "forbidden" values of the steady states. Within the framework of thermodynamic constraints $1 + ax > 0$, the value of the derivative of the function $f(x)$ is calculated as:

$$f' = ((1 + ax)(a_1 - x) + x)(a_1 - x)^{-2} e^{ax} ,$$

and this value remains positive in the interval $(0, a_1)$. So, as before, the thermodynamic constraints give the only steady state inside the reaction simplex. A study of stability of steady states allows us to say that the correct account of chemical nonideality preserves not only the number of steady states, but also their type of stability.

Model 5

Examples show, an account of chemical nonideality does not necessarily lead to a qualitative change of the dynamics of a system. It is possible for there to be only quantitative corrections. Generally speaking, this must not be said for open systems. The example of A.N. Ivanova, given below, is a proof of that.

Consider a scheme of transformations

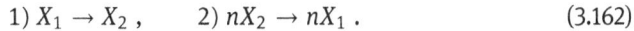

$$1)\ X_1 \rightarrow X_2 , \qquad 2)\ n X_2 \rightarrow n X_1 . \qquad (3.162)$$

Let a matrix of corrections on nonideality have the form

$$A_\mu = \begin{pmatrix} a_{11} & a_{12} \\ a_{12} & a_{22} \end{pmatrix} , \qquad (3.163)$$

i.e., the chemical potentials are set as

$$\mu_1 = \ln x_1 + a_{11} x_1 + a_{12} x_2 ,$$
$$\mu_2 = \ln x_2 + a_{12} x_1 + a_{22} x_2 . \qquad (3.164)$$

A kinetic model can be written for the scheme of transformations (3.162)

$$\dot{x}_1 = -w_1 + nw_2 , \tag{3.165}$$

where in accordance to (3.163), (3.164),

$$w_1 = k_1 e_1^\mu , \qquad w_2 = k_2 e_2^\mu .$$

In the steady state we have $w_1 = nw_2$, where $x_2 = 1 - x_1$. The thermodynamic constraints mean

$$x_1^{-1} + a_{11} > 0 , \quad x_2^{-1} + a_{22} > 0 , \quad (x_1^{-1} + a_{11})(x_2^{-1} + a_{22}) > a_{12}^2 .$$

It can be shown that the parameters of model (3.165) can be chosen in such a way that the steady state is unstable, i.e.,

$$- (x_1^{-1} + a_{11}) + na_{12} + a_{12} - n(x_2^{-1} + a_{22}) > 0 . \tag{3.166}$$

Let $a_{12} = 0.2$, $x_1^{-1} + a_{11} = 0.45$, $x_2^{-1} + a_{22} = 0.1$, then (3.166) means that $n > 2.5$. Let us take $n = 3$. Arbitrarily setting x_1 (where $x_2 = 1 - x_1$), we find a_{11} and a_{22} of the equality $a_{11} = 0.45 - x_1^{-1}$, $a_{22} = 0.1 - x_2^{-1}$. The values of the constants k_1, k_2 are determined by considering the stationarity condition $w_1 = nw_2$.

Thus, the chemical nonideality even at performing of the thermodynamic constraints in an open system can lead to critical effects. In the case of of (3.165) it leads to the instability of a steady state, which was unique and stable in an ideal situation ($a_{ij} = 0$).

Given examples show that the kind of chemical ideality is important. If the matrix J_μ is diagonal, then nonideality, taking into account thermodynamic constraints, does not lead to qualitatively new effects. On the contrary, if J_μ is not diagonal, chemical nonideality can become a cause of critical phenomena in open systems, even when observing thermodynamic constraints. The nondiagonal matrix J_μ actually means accounting for additional interactions, which are not specified in the scheme of transformations. These additional interactions are included in the nondiagonal members of the matrix of derivatives of chemical potentials A_μ.

Note, the critical phenomena of a nonthermal nature in the kinetic region can be purely due to kinetic nonlinearity, i.e., special nonlinearity of a scheme of transformations (for example, the presence of stages of interaction of various intermediate substances). Chemical nonideality allots one more class of systems, which is characterized by the possibility of an interpretation of critical phenomena of a nonthermal nature. The combination of these two factors can lead to quite complex dynamics of a chemical system in the kinetic region.

Nonlinear phenomena due to chemical nonideality can be called *critical phenomena of the second type*. Therefore, when interpreting experimentally observed multiplicity of steady states or auto-oscillations it is important to understand what kind

of critical phenomena take place. M.G. Slinko, et al. [146, 147] introduced nonlinearities of the type e^x, which are interpreted as the influence of the reaction environment
on the catalyst. Such corrections are phenomenological and do not satisfy the thermodynamic constraints. However, as examples given in this section show, with the
existence of thermodynamic constraints in an open system it is possible to describe
critical phenomena in the case of a nondiagonal matrix J_μ or in the case of structurally
unstable schemes of transformations.

Slinko–Chumakov's model

With the description of complex dynamics of oxidation of hydrogen on a metal catalyst
(Nickel) [146, 147], the following scheme of transformations was used

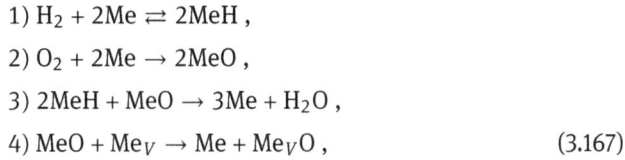

$$1)\ H_2 + 2Me \rightleftarrows 2MeH\,,$$
$$2)\ O_2 + 2Me \rightarrow 2MeO\,,$$
$$3)\ 2MeH + MeO \rightarrow 3Me + H_2O\,,$$
$$4)\ MeO + Me_V \rightarrow Me + Me_V O\,, \qquad (3.167)$$

where Me is a catalyst; MeH, MeO are intermediate substances and Me_V, $Me_V O$ are
the near-surface layer of the catalyst and the oxygen dissolved in it. Stage 4) in mechanism (3.167) corresponds to the exchange of adsorbed oxygen with the volume of the
catalyst.

A kinetic model for scheme (3.167) is

$$\dot{y}_1 = k_1(1 - y_1 - y_2)^2 - k_{-1}y_1^2 - 2k_3y_1^2 y_2\,,$$
$$\dot{y}_2 = k_2(1 - y_1 - y_2)^2 - k_4 y_2 - k_3 y_1^2 y_2\,,$$
$$\dot{y}_3 = \varepsilon(y_2(1 - y_3) - \alpha y_3(1 - y_1 - y_2))\,, \qquad (3.168)$$

where $\dot{y}_i = dy_i/d\tau$, $i = 1, 2, 3$; y_i are reactant concentrations; k_i are the rate constants
of reactions, some of which depend on the reagent concentrations:

$$k_3 = k_3^0 \exp\left(-\mu_3 y_2\right)\,, \qquad k_4 = k_4^0 \exp\left(-\mu_4 y_2 - \mu_5 y_3\right); \qquad (3.169)$$

k_1, k_2, k_{-1}, k_3^0, k_4^0, α, μ_3, μ_4, μ_5, ε are the model parameters. Steady states of
system (3.168) are determined from the conditions of stationarity:

$$k_1(1 - y_1 - y_2)^2 - k_{-1}y_1^2 - 2k_3 y_1^2 y_2 = 0\,,$$
$$k_2(1 - y_1 - y_2)^2 - k_4 y_2 - k_3 y_1^2 y_2 = 0\,,$$
$$y_2(1 - y_3) - \alpha y_3(1 - y_1 - y_2) = 0\,. \qquad (3.170)$$

After elementary transformations of (3.170), we obtain a system of two nonlinear equations with unknowns y_1, y_2:

$$f_1(y_1, y_2) = k_1(1 - y_1 - y_2)^2 - k_{-1}y_1^2 - 2k_3(y_2)y_1^2 y_2 = 0\,,$$
$$f_2(y_1, y_2) = k_2(1 - y_1 - y_2)^2 - k_4(y_1, y_2)y_2 - k_3(y_2)y_1^2 y_2 = 0\,, \qquad (3.171)$$

where $k_3(y_2)$, $k_4(y_1, y_2)$ were determined in (3.169) and

$$y_3 = \frac{y_2}{y_2 + \alpha(1 - y_1 - y_2)} \, . \tag{3.172}$$

As a varying parameter we choose a rate constant k_1 (physically, this means that the partial pressure of hydrogen is changed). Computing the corresponding derivatives for system (3.171), (3.172), we have:

$$\Delta = f'_{1y_1} f'_{2y_2} - f'_{1y_2} f'_{2y-1} \, ,$$
$$\Delta_1 = -f'_{1k_1} f'_{2y_2} \, ,$$
$$\Delta_2 = f'_{1k_1} f'_{2y_1} \, , \tag{3.173}$$

where $f'_{iy_j} = \partial f_i / \partial y_j$, $i, j = 1, 2$:

$$\frac{\partial f_1}{\partial y_1} = -2k_1(1 - y_1 - y_2) - 2k_{-1} - 4k_3(y_2)y_1y_2 \, ,$$

$$\frac{\partial f_1}{\partial y_2} = -2k_1(1 - y_1 - y_2) + 2k_3(y_2)y_1^2(\mu_3 y_2 - 1) \, ,$$

$$\frac{\partial f_1}{\partial k_1} = (1 - y_1 - y_2)^2 \, ,$$

$$\frac{\partial f_2}{\partial y_1} = -2k_2(1 - y_1 - y_2) - 2k_3(y_2)y_1y_2 + \frac{k_4(y_1, y_2)\mu_5 y_2^2 \alpha}{(y_2 + \alpha(1 - y_1 - y_2))^2} \, ,$$

$$\frac{\partial f_2}{\partial y_2} = -2k_2(1 - y_1 - y_2) + k_3(y_2)y_1^2(\mu_3 y_2 - 1) +$$

$$+ k_4(y_1, y_2) \left[y_2 \left(\mu_4 + \frac{\mu_5 \alpha(1 - y_1)}{(y_2 + \alpha(1 - y_1 - y_2))^2} \right) - 1 \right] \, ,$$

$$\frac{\partial f_2}{\partial k_1} = 0 \, .$$

Movement along the curve of parametric dependencies $y_1(k_1)$, $y_2(k_1)$ in the space (y_1, y_2, k_1) is carried out in accordance with a system of two differential equations of the form (3.21), (3.23) or (3.24). In this case, system (3.171) is essentially nonlinear. Because of the presence of the exponents (3.169) it is not reduced to one equation. In contrast to (3.171) system (3.168), discussed earlier, can be reduced to one equation as a result of simple transformations.

The results of calculations for system (3.171) and (3.172) are presented in Figs. 3.67 and 3.68, where the dependencies of steady states are on the parameter k_1: $y_i(k_1)$, $i = 1, 2, 3$.

There is a whole interval of values of k_1, for which system (3.171) has three solutions. This corresponds to the multiplicity of steady states in the dynamic model (3.168). The variation of the second parameter allows us to numerically plot the region of multiplicity of steady states in the plane (k_1, μ_5).

System (3.168) is interesting because it has a complex dynamical regime of the "chaos" type. These regimes often appear when a steady state is singular and unstable

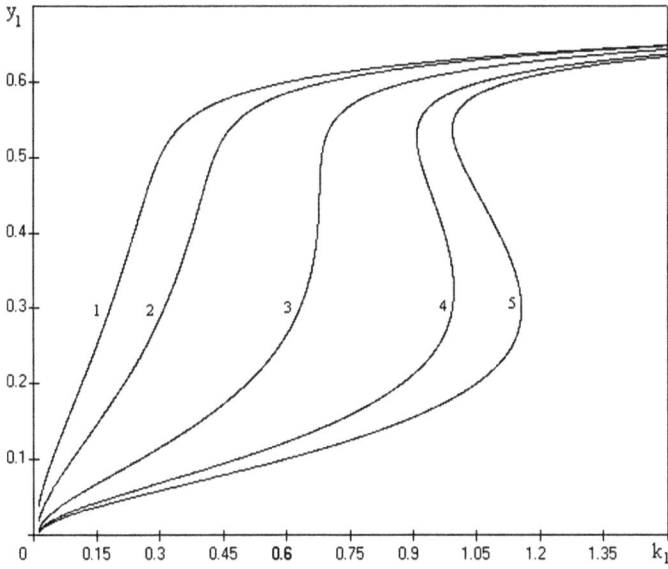

Fig. 3.67: Parametric dependencies $y_1(k_1)$ at $k_{-1} = 0.01$, $k_2 = 15$, $k_3^0 = 100$, $k_4^0 = 2$, $\mu_3 = 30$, $\mu_4 = 12$, $\varepsilon = 0.001$, $\alpha = 10$, 1) $\mu_5 = -10$; 2) $\mu_5 = -8$; 3) $\mu_5 = -6$; 4) $\mu_5 = -5$; 5) $\mu_5 = -4.7$

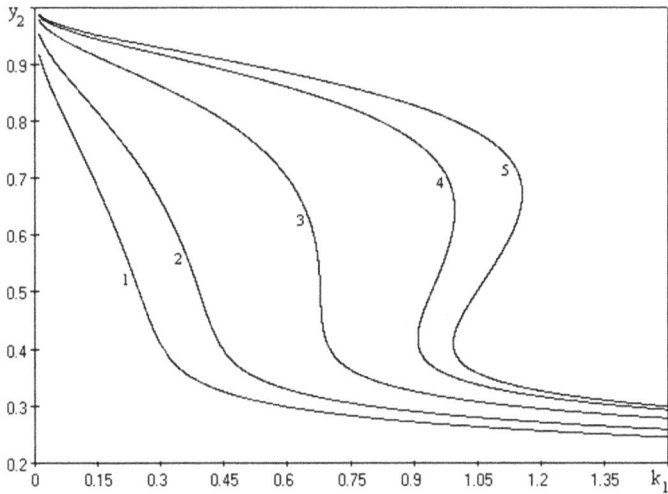

Fig. 3.68: Parametric dependencies $y_2(k_1)$ at the same values as in Fig. 3.68

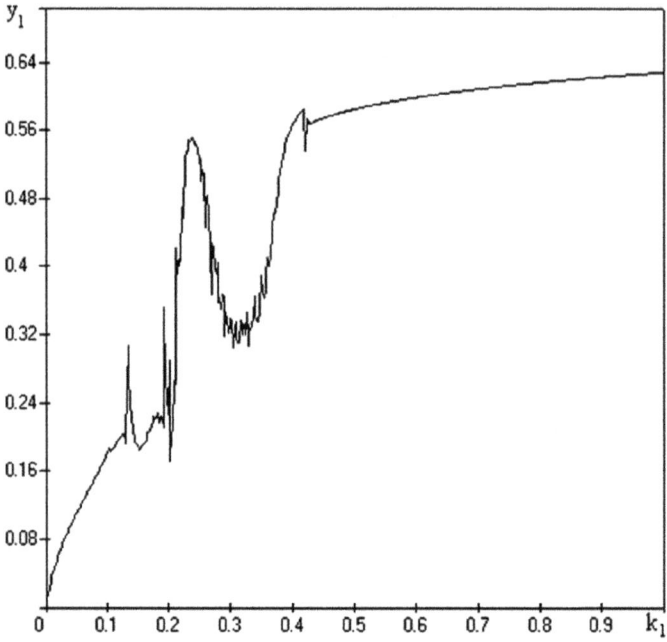

Fig. 3.69: Parametric dependence $y_1(k_1)$ at $k_{-1} = 0.01$, $k_2 = 15$, $k_3^0 = 100$, $k_4^0 = 2$, $\mu_3 = 30$, $\mu_4 = 12$, $\varepsilon = 0.001$, $\alpha = 10$, $\mu_5 = -10$

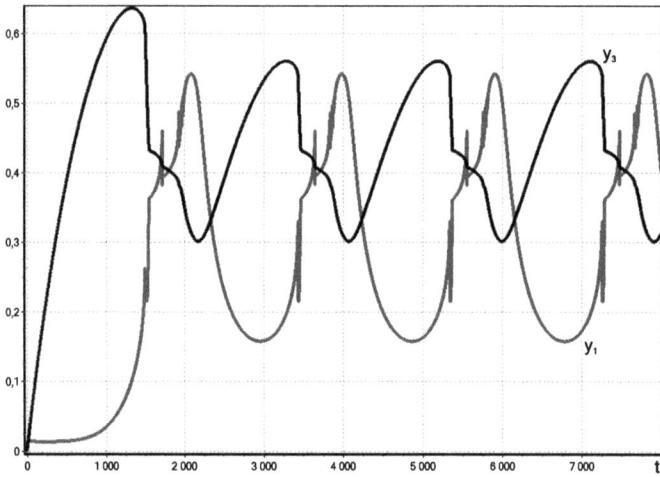

Fig. 3.70: Time dependencies $y_1(t)$ and $y_3(t)$

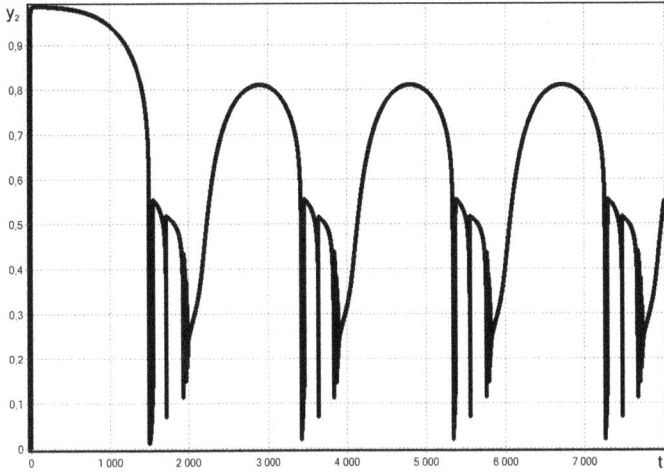

Fig. 3.71: Time dependence $y_2(t)$

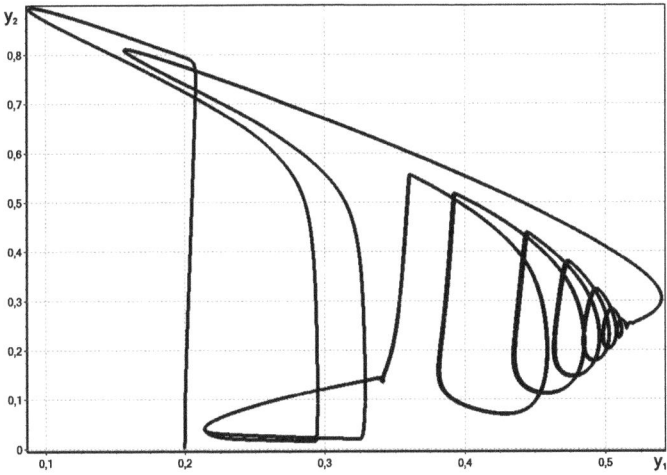

Fig. 3.72: Projection of a phase trajectory of model (3.168) on the plane (y_1, y_2)

and phase trajectories are locally unstable [146, 147]. Parametric dependence, characterized by a singular and unstable steady state, is shown in Fig. 3.69.

Here the oscillations are responsible for the instability of steady states and the existence of the undamped in time solutions $yi(t)$, $i = 1, 2, 3$ of system (3.168). Parametric dependencies in Figs. 3.67–3.69 were obtained by using the proposed variant of implementation of the method of parameter continuation for the system of two nonlinear equations. The complex oscillations are presented in Fig. 3.70 and Fig. 3.71 at the parameter values: $k_1 = 0.2$, $k_{-1} = 0.01$, $k_2 = 15$, $k_3^0 = 100$, $k_4^0 = 2$, $\mu_3 = 30$, $\mu_4 = 12$, $\mu_5 = -10$, $\varepsilon = 0.001$, $\alpha = 10$.

One of the projections of the phase trajectories of corresponding to time dependencies is presented in Fig. 3.72.

3.8 Savchenko's model

There is a huge number of experimental and theoretical studies on the reaction of CO oxidation on platinum. In particular, I.V. Savchenko have proposed a mechanism based on two active centers and a step diffusive exchange between them. He demonstrated the possibility of oscillatory regimes of reaction numerically in the corresponding kinetic model. So, this model can be called Savchenko's model. In works [118–129] the task of parametric analysis of this model was formulated. Here we will give a brief description of the main experimental and theoretical assumptions for analysis of Savchenko's model by published works and carry out the plotting of the parametric dependencies of steady states of this model, including one of its particular cases.

Critical phenomena

In the oxidation reaction of CO on Pt auto-oscillations are observed at pressures $p > 10^{-3}$ Torr. Based on the experimental observation that the surface formation of oxygen occurs at oscillatory conditions, Ladas and others [61, 62] have proposed a mechanism that explains the oscillations by the periodic formation and expenditure of a surface reserve of oxygen. The mechanism of oxidation was described by a system of three differential equations. This mathematical model successfully explains the largest amount of experimental data, in particular the disappearance of the usual (clockwise) hysteresis, which is observed at cyclical variations of p_{CO}. The model, however, could not reproduce the global bifurcation diagram in the parameter area p_{CO}, p_{O_2}, i.e., a change from an aperiodic process at a pressure below $p_{crit} = 10^{-3}$ Torr to oscillatory behavior at pressures above p_{crit}. These changes are associated with the transition of the usual hysteresis (clockwise) at $p < p_{crit}$ into kinetic hysteresis, and oscillations in excess of the critical value of pressure p_{crit}. The disappearance of the hysteresis leads to the typical cross-shaped form of the bifurcation diagram in the parameter area, p_{CO}, p_{O_2}.

As shown in [43], the correct bifurcation diagram is obtained if we take into account the existence of the reflection of the interaction between the CO and oxygen. This result emphasizes the importance of interaction in surface reactions and this gives support for a real model of the subsurface oxidation. The reaction $CO + O_2$ at Pt describes the kinetic oscillations at p_{O_2} above 10^{-3} torr. Based on the reversible formation of surface oxygen and the mechanism of catalytic CO oxidation, R. Imbihl and G. Ertl have proposed a mathematical model that describes the appearance of oscillations and qualitative characteristics of the oscillations [43]. The authors have shown that the surface oxygen model correctly reflects the interaction between the CO and O_2 and reproduces the experimentally defined stability diagram in the parameter area, p_{O_2}, p_{CO}.

Catalytic mechanism with two active centers

At general consideration [114] the adsorption step for simplicity was written as shock. However, the majority of reactions apparently proceeds on the adsorptive mechanism. In this case the competing adsorption of components of A_{ads} and B_{ads} takes place. Authors of the work [43] have made the assumption that in the isolated state one site of a surface of the catalyst is covered with A_{ads}, and the other B_{ads}. Then the "inclusion" of a reaction on the interphase border and a spillover of A_{ads}, B_{ads} has to lead to an essential increase in the rate of reaction. This effect can be shown for many reactions – hydrogenations, oxidations, and synthesis of ammonia [98–123]. In work [43], the CO oxidation reaction was chosen as an example which has been studied in many works in detail. For the oxidation reaction CO on platinum metals it is known that with rather little change of parameters of adsorption, the surface can be covered with O_{ads}, or CO_{ads}. In work [114] it is proposed that the reactions on each of sites M1 and M2 occur according to simple 3-stage schemes, and the sites are connected by a spillover CO (step 4):

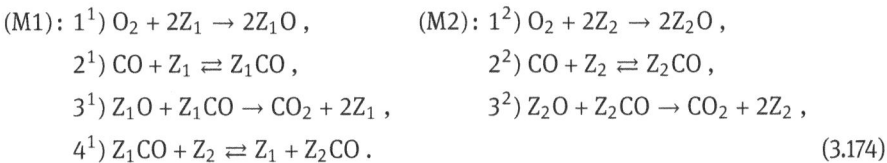

$$(M1): 1^1)\ O_2 + 2Z_1 \rightarrow 2Z_1O, \qquad (M2): 1^2)\ O_2 + 2Z_2 \rightarrow 2Z_2O,$$
$$2^1)\ CO + Z_1 \rightleftarrows Z_1CO, \qquad\qquad 2^2)\ CO + Z_2 \rightleftarrows Z_2CO,$$
$$3^1)\ Z_1O + Z_1CO \rightarrow CO_2 + 2Z_1, \qquad 3^2)\ Z_2O + Z_2CO \rightarrow CO_2 + 2Z_2,$$
$$4^1)\ Z_1CO + Z_2 \rightleftarrows Z_1 + Z_2CO. \tag{3.174}$$

Let us write the kinetic model for scheme (3.174)

$$\dot{x}_1 = k_1^1 z_1^2 - k_3^1 x_1 x_2,$$
$$\dot{x}_2 = k_2^1 z_1 - k_{-2}^1 x_2 - k_3^1 x_1 x_2 + k_4 x_2 z_2 - k_{-4} x_4 z_1,$$
$$\dot{x}_3 = k_1^2 z_2^2 - k_3^2 x_3 x_4,$$
$$\dot{x}_4 = k_2^2 z_2 - k_{-2}^2 x_4 - k_3^2 x_3 x_4 - k_4 x_2 z_2 + k_{-4} x_4 z_1, \tag{3.175}$$

where

$$x_1 = [Z_1O]; \quad x_2 = [Z_1CO]; \quad x_3 = [Z_2O]; \quad x_4 = [Z_2CO]$$

are dimensionless concentrations; Z_1, Z_2 are parts of a free surface. The following rate constants and parameters correspond to the scheme:

$$k_i^1 = 2p_{O_2}s_1(O_2)A(O_2)/N_0 \; ; \quad s_1^2(O_2) = 0.5 \; ;$$
$$k_i^1 = 2p_{CO}s_2(CO)A(CO)/N_0 \; ; \quad s_2^2(CO) = 0.1 \; ;$$
$$k_j^i = k_0 \exp\left(-E_j^i/(RT)\right) , \quad i = 1, 2 \; ; \quad j = -2, 3 \; ;$$
$$A(O_2) = 3.67 \cdot 10^{20} \; ; \quad A(CO) = 3.93 \cdot 10^{20} \; ;$$

$$k_j^i = (10^{-4}/m_i)k_0 \exp\left(-E_j^i/(RT)\right) , \quad i = 1, 2 \; ; \quad j = -4, 4 \; ;$$
$$k_0 = 10^{13} \; ; \quad E_1^i = E_2^i = 0 \; ; \quad E_{-2}^i = 125 \; ; \quad E_3^i = 80 \; ;$$
$$s_1^1(O_2) = 0.3 \; ; \quad s_2^1(CO) = 1 \; ; \quad E_4 = 75 \; ; \quad E_{-4} = E_4 - (E_{-2}^1 - E_{-2}^2) \; .$$

The kinetic model (3.175) contains a considerable number of parameters. First of all, p_{CO}, p_{O_2}, T and N_0 can act as the varied parameters. The plotting of dependencies of stationary rates of reaction from these parameters can be made according to the scheme shown above.

Synchronization of oscillations by surface diffusion (Savchenko's model)

In a heterogeneous catalysis a synchronization of auto-oscillations of reaction rates on the applied particles of a metal is one of the important questions of research of the auto-oscillation mechanism. How do the metal particles, divided by the inert carrier, begin to work in unison? There is an assumption that synchronization happens either through a gas phase or due to diffusion on a surface.

The first is an absolutely natural representation, especially for porous systems, but also for monocrystals the possibility of synchronization of auto-oscillations of CO oxidation rate through a gas phase in the conditions of the isothermal and nongradient reactor [119] is proved.

Surface diffusion of the adsorbed CO molecules can be the reason of synchronization of auto-oscillations on an inhomogeneous surface consisting of sites of two types. As it has been shown [119] in the process of CO oxidation on Pt/Al_2O_3 there is a diffusion of CO_{ads} from the metal to the carrier and back. Therefore, it was represented as expedient to carry out modeling of the CO oxidation reaction on the metal, applied to the inert carrier for the purpose of clarification of a possible role of surface diffusion CO_{ads} in the synchronization of auto-oscillations of the reaction rate.

The model given above corresponds to the scheme of experiments. In [119] V.I. Savchenko has made the assumption that strips of metals M1 and M2 alternate on a surface. Each of the metals has width $\approx 10^3 A$. The strips of metals are divided from each other by a space of the carrier (S) with width $\approx 500 A$. Also it is assumed that [M1] + [S] + [M2] = 1.0. For 1 cm^2 of such a complex surface consisting of $(N_0 = 10^{15})$ centers, the share of boundary atoms is $\theta_L \approx 2.5 \cdot 10^{-3}$.

In work [123] a mathematical model describing an interaction of sites of a surface of two types is formulated. These sites of a surface are interfaced by diffusion CO_{ads}.

The reaction proceeds through a 5-stage mechanism with its own set of constants on particles of each sort M1 and M2. Diffusion of CO_{ads} between particles of metal and the carrier is considered, as well as adsorption-desorption of CO on the carrier.

The reaction scheme is:

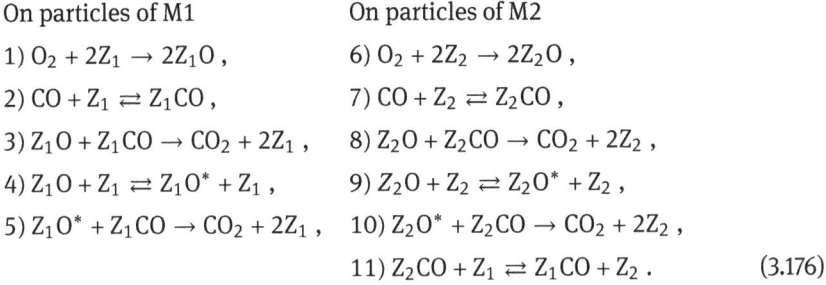

On particles of M1	On particles of M2
1) $O_2 + 2Z_1 \rightarrow 2Z_1O$,	6) $O_2 + 2Z_2 \rightarrow 2Z_2O$,
2) $CO + Z_1 \rightleftarrows Z_1CO$,	7) $CO + Z_2 \rightleftarrows Z_2CO$,
3) $Z_1O + Z_1CO \rightarrow CO_2 + 2Z_1$,	8) $Z_2O + Z_2CO \rightarrow CO_2 + 2Z_2$,
4) $Z_1O + Z_1 \rightleftarrows Z_1O^* + Z_1$,	9) $Z_2O + Z_2 \rightleftarrows Z_2O^* + Z_2$,
5) $Z_1O^* + Z_1CO \rightarrow CO_2 + 2Z_1$,	10) $Z_2O^* + Z_2CO \rightarrow CO_2 + 2Z_2$,
	11) $Z_2CO + Z_1 \rightleftarrows Z_1CO + Z_2$. (3.176)

The system of kinetic equations corresponding to (3.176) has the form:

$$d[Z_1O]/dt = 2w_1 - w_3 - w_4 \,,$$
$$d[Z_1CO]/dt = w_2 - w_3 - w_5 + w_{11} \,,$$
$$d[Z_1O^*]/dt = w_4 - w_5 \,,$$
$$d[Z_2O]/dt = 2w_6 - w_8 - w_9 \,,$$
$$d[Z_2CO]/dt = w_7 - w_8 - w_{10} - w_{11} \,,$$
$$d[Z_2O^*]/dt = w_9 - w_{10} \,,$$

where

$$w_1 = k_1[Z_1]^2 \,,$$
$$w_2 = k_2[Z_1] - k_{-2}[Z_1CO] \,,$$
$$w_3 = k_3[Z_1O][Z_1CO] \,,$$
$$w_4 = k_4[Z_1O][Z_1] - k_{-4}[Z_1O^*][Z_1] \,,$$
$$w_5 = k_5[Z_1O^*][Z_1CO] \,,$$
$$w_6 = k_6[Z_2]^2 \,,$$
$$w_7 = k_7[Z_2] - k_{-7}[Z_2CO] \,,$$
$$w_8 = k_8[Z_2O][Z_2CO] \,,$$
$$w_9 = k_9[Z_2O][Z_2] - k_{-9}[Z_2O^*][Z_2] \,,$$
$$w_{10} = k_{10}[Z_2O^*][Z_2CO] \,,$$
$$w_{11} = k_{11}[Z_2CO][Z_1] - k_{-11}[Z_2][Z_1CO] \,.$$

The reaction rate on M1 is:

$$R(M1) = (w_3 + w_5) \cdot N_0 \,.$$

The reaction rate on $M2$ is:

$$R(M2) = (w_8 + w_{10}) \cdot N_0 \,.$$

Here N_0 is the number of centers for 1 cm^2 of a surface, equal to 10^{15} cm^{-2}. The total reaction rate is: $R_\Sigma = R(M1) \cdot M1 + R(M2) \cdot M2$.

In calculations it was accepted that [M1] = [M2] = 0.4 and [S] = 0.2. The pre-exponential multiplier was considered to equal 10^{13} s^{-1} everywhere.

For the reaction proceeding according to the 5-step scheme on M2, the author has chosen a set of approximate parameters [123] as: E_7 = 112; E_8 = 93; E_9 = 120; E_{-9} = 250; E_{10} = 125 kJ mol^{-1}. Sticking coefficients are S_{O_2} = 0.3 for oxygen and S_{CO} = 1.0 for carbon oxide. The same set of constants was also accepted for M1. The exception was made for a value of E_7, because it was varied while carrying out calculations. For adsorption of CO on the carrier, it was accepted that S_{CO} = 10^{-7}, E_{-1} = 115 kJ mol^{-1}. Calculations were carried out for the conditions: p_{CO} = 10^{-5} mm of mercury, p_{O_2} = $4.0 \cdot 10^{-5}$ mm of mercury, T = 450 K. Though auto-oscillations on M1 and M2 are symmetric, the total rate (R_Σ) changed chaotically. The possibility of quasichaotic behavior of the reaction rate in collaboration of two surface sites which aren't connected among themselves was considered in [119] earlier.

Thus, at a rather high rate of diffusion, even if the properties of particles M1 and M2 significantly differ, surface diffusion of CO provides full synchronization of the oscillations happening on them, and the whole complex surface M1–S–M2 works as a uniform system. At a limited rate of diffusion the synchronization is broken and quasichaotic oscillations are observed [124].

Dynamics of auto-oscillations development.
In [124] a model is considered in which it is assumed that the surface of Pt(110) in CO oxidation consists of two parts. The first part is a source (SN), where the reaction proceeds in a 3-step scheme. The second part is a modified (SM), where the reaction proceeds in a 5-step scheme with two forms of the adsorbed oxygen. In addition, a formation of near-surface β-oxygen on the surface of SM (originally – on the defects of Pt(110)) and its reaction with CO and diffusion of C_{ads} between the surfaces SN and SM is taken into account. The ratio of the sites SN/SM is determined by the concentration of β-oxygen. The calculations describe the dynamics of the development of auto-oscillations: the initial induction period, an appearance and a development of oscillations, an appearance of oscillations with a very large period (tens of minutes), and "packets" of pulses.

At pressures of the reaction mixture $\leq 10^{-2}$ Pa on platinum, auto-oscillations of the reaction rate of CO oxidation are observed [124]. There are two basic approaches to the explanation of auto-oscillations, based on ideas about the influence of the reaction environment on the properties of the metal. The first assumes changes in the structure of the surface induced by adsorption of CO of a phase transition [124]. The second approach takes into account the change of chemical composition. This is a model of an oxidation-reduction introduced by Sales, Turner and Maple (STM model). The STM model suggests the formation of an adsorbed oxide oxygen on the PT surface. In this scheme "oxide" oxygen is understood as a reconstructive adsorbed oxygen (2nd type of adsorption), but not the oxygen of the volume oxide.

Obtained data in [124] allowed us to suggest the following 7-step scheme:

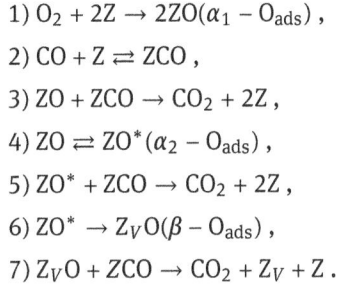

1) $O_2 + 2Z \rightarrow 2ZO(\alpha_1 - O_{ads})$,

2) $CO + Z \rightleftarrows ZCO$,

3) $ZO + ZCO \rightarrow CO_2 + 2Z$,

4) $ZO \rightleftarrows ZO^*(\alpha_2 - O_{ads})$,

5) $ZO^* + ZCO \rightarrow CO_2 + 2Z$,

6) $ZO^* \rightarrow Z_V O(\beta - O_{ads})$,

7) $Z_V O + ZCO \rightarrow CO_2 + Z_V + Z$.

The set of stages 1–5 corresponds to the scheme of oxidation–reduction (STM). Stages 6, 7 describe the formation, i.e., the restoration, of the near-surface β-oxygen. In [124] to describe changes of state all over the surface of Pt(110) in time, dependencies of energy desorption of CO from time $\approx$ (1 J) or of the concentration $\beta - O_{ads}$ were introduced in the system of equations. This formally allowed us to describe the development of auto-oscillations.

Consider in more detail the mechanism proposed by V. Savchenko in [115]. On the original surface of Pt(110) with structure (1×2) the oxygen is adsorbed in the α_1-state, and on the defects (SM, not more that 5%) in the α_1- and α_2-states. The total surface is equal to $SN + SM = 1.0$. On the surface SN the reaction proceeds by the 3-step scheme involving α_1-oxygen (transition $\alpha_1 \rightarrow \alpha_2$ is not possible). On the surface SM, its growth begins with defects (structural violations), and the reaction proceeds according to the 7-step scheme. The first five stages correspond to the STM scheme (α_1- and α_2-oxygen). In addition, the transition α_2-oxygen in β-oxygen and the reaction with CO occurs on sites SM. It is suggested that with the growth of concentration of β-oxygen on sites SM in the reaction process. Once the concentration $[Z_V O]$ reaches some upper critical value θ_β^{VC}, a part of surface SN on the boundary of the islets of modified surface turns into SM. Conversely, if θ_β is below the lower critical concentration (θ_β^{VC}), the edge part SM goes to SN. The share of the boundary atoms is determined by the equation:

$$\theta_L = SM \exp(-\alpha SM),$$

where $\alpha = 3^{-10}$. Diffusion of CO_{ads} between sites SN and SM was taken into account. The activation energy of diffusion of CO_{ads} on the surface of platinum is 35–40 kJ mol^{-1}.

Thus, the scheme of the processes on the surface consisting of the sites SN and SM is represented as follows:

On the sites SM:

1) $O_2 + 2Z \rightarrow 2ZO(\alpha_1 - O_{ads})$,

2) $CO + Z \rightleftarrows ZCO$,

3) $ZO + ZCO \rightarrow CO_2 + 2Z$,

4) $ZO \rightleftarrows ZO^*(\alpha_2 - O_{ads})$,

5) $ZO^* + ZCO \rightarrow CO_2 + 2Z$,

6) $ZO^* \rightarrow Z_V O(\beta - O_{ads})$,

7) $Z_V O + ZCO \rightarrow CO_2 + Z_V + Z$.

8) $ZCO + Z_s \rightleftarrows Z_s CO + Z$,

On the sites SN:

9) $O_2 + 2Z_s \rightarrow 2Z_s O(\alpha_1)$,

10) $CO + Z_s \rightleftarrows Z_s CO$,

11) $ZO + Z_s CO \rightarrow CO_2 + 2Z$.

The kinetic equations reflecting the dynamics of changes of the concentration of adsorbates on the surface were recorded independently for each center and took into account the step of diffusion transition

$$R_{diff} = (k_8 \theta_L [ZCO][Z_s] - k_{-8} \theta_L [ZCO][Z_s])/SM .$$

Only starting at $SM \approx 0.62$, in agreement with the experiment [161, 162], the auto-oscillations of the reaction rate on SM begin to develop. At the same time, due to diffusion of CO the oscillations started on the sites SN. The reaction proceeds in the 3-step scheme there, which itself does not suggest auto-oscillations of rate.

V.I. Savchenko determined that the oscillations on the sites SM and SN occur in antiphase, i.e., the maximum of rate on SM corresponds to a minimum rate on SN. As a result, the oscillation amplitude R_Σ is less than the oscillation amplitude R_{SM}.

3.9 Model of the Belousov–Zhabotinsky reaction

Studies on the Belousov–Zhabotinsky reaction already started more than a half century ago. This quite simple chemical system, which can be implemented even in "home" conditions, demonstrates a large variety of linear and dynamic properties, from simple periodic oscillations to different types of chaos. This system inspired the theoretical school of I. Prigogine on the development of nonlinear thermodynamics of irreversible processes [3, 4]. There is a large volume of literature about this system. Here we will only mention the classical monographs and review papers [1–12]. Note that until now many experimental data on this reaction have not had an explanation. This is referred especially to thin enough transitions between various dynamic modes. The task of constructing detailed kinetic models describing all the variety of properties of a real chemical system are still not solved. Practice shows that this task is unlikely to be resolved.

The purpose of this section is quite modest. Here we will give only one example of a kinetic model of Belousov–Zhabotinsky. This model is responsible for an 11-step scheme of conversion and is investigated in a series of works [167–170]. In the literature there is a significant number of works in which the schemes of transformations are discussed, differing by a number of elementary stages [170]. The record result belongs to [170], where a model was given consisting of 80 stages and includes 27 phase variables. More compact models, including 24 stages, were studied in [170]. However, as researches show, within such bulky models a considerable number of experimental data characterized by complex dynamic behavior cannot be described. A more simple 11-step scheme was considered in [167]. The scheme was more adequate, because it was more convenient for a detailed parametric analysis of the corresponding kinetic model.

The scheme, similar to those given below, was proposed in [170]. It represents a set of the following stages:

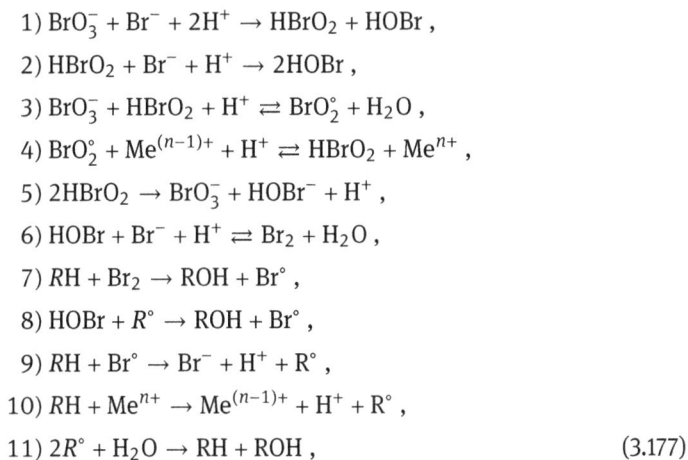

$$1)\ BrO_3^- + Br^- + 2H^+ \rightarrow HBrO_2 + HOBr\,,$$
$$2)\ HBrO_2 + Br^- + H^+ \rightarrow 2HOBr\,,$$
$$3)\ BrO_3^- + HBrO_2 + H^+ \rightleftarrows BrO_2^\circ + H_2O\,,$$
$$4)\ BrO_2^\circ + Me^{(n-1)+} + H^+ \rightleftarrows HBrO_2 + Me^{n+}\,,$$
$$5)\ 2HBrO_2 \rightarrow BrO_3^- + HOBr^- + H^+\,,$$
$$6)\ HOBr + Br^- + H^+ \rightleftarrows Br_2 + H_2O\,,$$
$$7)\ RH + Br_2 \rightarrow ROH + Br^\circ\,,$$
$$8)\ HOBr + R^\circ \rightarrow ROH + Br^\circ\,,$$
$$9)\ RH + Br^\circ \rightarrow Br^- + H^+ + R^\circ\,,$$
$$10)\ RH + Me^{n+} \rightarrow Me^{(n-1)+} + H^+ + R^\circ\,,$$
$$11)\ 2R^\circ + H_2O \rightarrow RH + ROH\,, \qquad (3.177)$$

where RH is responsible malonic acid; Me^{n+} are cerium ions (IV), $Me^{(n-1)+}$ are cerium ions (III); R°, Br°, BrO_2° are radicals. Here we miss the chemical interpretation and justification of mechanism (3.177). In [167] a closed (standing) variant of a scheme is considered, and it is assumed that the concentrations of the reagents BrO_3^-, RH, H^+, H_2O are kept constant. From the balance ratio the concentration of the recovered form of the catalyst $Me^{(n-1)+}$ through oxidization is expressed.

With the given assumptions the following kinetic model corresponds to scheme (3.177):

$$\dot{x}_1 = -w_1 - w_2 - w_6 + w_7 + w_9 ,$$
$$\dot{x}_2 = w_1 - w_2 - w_3 + w_4 - 2w_5 ,$$
$$\dot{x}_3 = w_1 + 2w_2 + w_5 - w_6 - w_8 ,$$
$$\dot{x}_4 = w_3 - w_4 ,$$
$$\dot{x}_5 = w_4 - w_{10} ,$$
$$\dot{x}_6 = w_6 - w_7 ,$$
$$\dot{x}_7 = -w_8 + w_9 + w_{10} - 2w_{11} ,$$
$$\dot{x}_8 = w_8 - w_9 , \tag{3.178}$$

where the expressions for the step rates w_i correspond to the law of mass action

$$w_1 = k_1 B x_1 - k_{-1} x_2 x_3 , \quad w_2 = k_2 x_1 x_2 , \quad w_3 k_3 B x_2 - k_{-3} x_4 ,$$
$$w_4 = k_4 A x_4 - k_{-4} A x_2 , \quad w_5 = k_5 x_2^2 , \quad w_6 = k_6 x_1 x_3 - k_{-6} x_6 ,$$
$$w_7 = k_7 C x_6 , \quad w_8 = k_8 x_3 x_7 , \quad w_9 = k_9 C x_8 ,$$
$$w_{10} = k_{10} C x_5 , \quad w_{11} = k_{11} x_7^2 ,$$
$$A = [Me^{n+,(n-1)+}]_\circ , \quad B = [BrO_3^-]_\circ , \quad C = [RH]_\circ .$$

In (3.178) the current concentrations of reagents are

$$x_1 = [Br^-] , \quad x_2 = [HBrO_2] , \quad x_3 = [HOBr] ,$$
$$x_4 = [BrO_2^\circ] , \quad x_5 = [RH] , \quad x_6 = [Br_2] , \quad x_8 = [Br^+] .$$

Constant concentrations of reagents were accepted as follows:

$$[BrO_3^-]_\circ = 0.08\,mol\,L^{-1} , \quad [Br^-]_\circ = 10^{-5}\,mol\,L^{-1} ,$$
$$[Me^{n+}]_\circ = 5 \cdot 10^{-4}\,mol\,L^{-1} , \quad [Me^{(n-1)+}]_\circ = 0 , \quad [RH]_\circ = 0.2\,mol\,L^{-1} .$$

In the calculation of (3.178) the following set of rate constants was used:

$$k_1 = 2.1 , \quad k_{-1} = 10^4 , \quad k_2 = 3 \cdot 10^6 , \quad k_3 = 42 , \quad k_{-3} = 4.2 \cdot 10^7 ,$$
$$k_4 = 8 \cdot 10^4 , \quad k_{-4} = 8.9 \cdot 10^3 , \quad k_5 = 3 \cdot 10^3 , \quad k_6 = 8 \cdot 10^9 , \quad k_{-6} = 1.1 \cdot 10^2 ,$$
$$k_7 = 4.5 \cdot 10^{-3} , \quad k_8 = 10^5 \div 10^{11} , \quad k_9 = 10^6 , \quad k_{10} = 0.2 , \quad k_{11} = 3.2 \cdot 10^9 .$$

Study of model (3.178) has shown [170] that it has a wide range of complex periodic and aperiodic modes (including quasiperiodic oscillations), which is consistent with available experimental data. Among those modes found in (3.178), there are quasisinusoidal oscillations, quasiperiodic modes, packed oscillations, complex periodic modes, qualitatively diverse chaos, and cascades of period-doubling. When varying parameters the sequence of striping modes is well consistent with [170]. A more thorough investigation of ways of transition to chaos and various variants of complex dynamic modes was carried out in [170].

It turns out that the conditions of stationarity for system (3.178) is reduced to the equation for one variable

$$F(x_7, \boldsymbol{\alpha}) = 0 \, ,$$

where the vector $\boldsymbol{\alpha}$ contains all parameters of a system. That allows us to significantly facilitate a study of steady states of the system and to show that for a given range of parameters there is a single steady state.

When varying the parameter k_8 it is shown that there are two bifurcation points in Andronov–Hopf bifurcation. Calculations allow us to confirm that in (3.178) both sub- and supercritical bifurcations of a limit cycle (hard and soft birth) are implemented. Moreover, the mechanism of the occurrence of chaotic packed oscillations corresponding to the fractal torus in phase space was studied. It was discovered that the birth of a fractal torus occurs as a result of the cascade of doubling bifurcations of resonances on the torus, and its disappearance occurs through the mode of intermittency. Full bifurcation diagrams for a flow system, appropriate to scheme (3.177) were plotted. The flow is carried out according to the original substances.

Thus, model (3.178) corresponds to a significant volume of experimental data. That, according to the authors [169, 170], is because in the extended models (24 stages or more) additional stages form too strong a link between existing cycles in the scheme of reaction. Their interaction in the reaction provides a significant diversity of observed complex periodic and chaotic oscillations. Therefore, we have to admit that the question about the nature of this connection, and in general about the mechanism of occurrence of dual-frequency and chaotic oscillations in the model of the Belousov–Zhabotinsky reaction, remains open.

Bibliography

[1] Bykov VI. Modeling of critical phenomena in chemical kinetics. Moscow, URSS, 2014 (in Russian).

[2] Bykov VI, Tsybenova SB. Nonlinear models of chemical kinetics. Moscow, URSS, 2011 (in Russian).

[3] Glansdorff P, Prigogine I. Thermodynamic theory of stability. Structure, Fluctuation. New York, NY, USA, Wiley, 1971.

[4] Nikolis N, Prigozhin I. Self-organization in nonequilibrium systems: From dissipative structures to orderliness through fluctuations. New York, NY, USA, Wiley, 1977.

[5] Prigozhin I. From existing to occurring: Time and complexity in the physical sciences. Moscow, Nauka, 1985 (in Russian).

[6] Garel D, Garel O. Oscillation chemical reactions. Moscow, Mir, 1986 (in Russian).

[7] Shemyakin FM, Mikhalev PF. Physico-chemical batch processes. Moscow-Leningrad, AS USSR, 1938.

[8] Zhabotinskii AM. Concentration auto-oscillations. Moscow, Nauka, 1974 (in Russian).

[9] Oscillations and traveling waves in chemical systems. Eds. Field RJ, Burger M, New York, NY, USA, Wiley, 1985.

[10]   Field RJ. Experimental and mechanistic characterization of bromate-ion-driven chemical os-
       cillations and traveling waves in closed systems. In: Field RJ, Burger M, eds. Oscillations and
       traveling waves in chemical systems. New York, NY, USA, Wiley, 1985, 55–95.

[11]   Gray P, Scott SK. Isothermal oscillations and relaxation flare in the gas-phase reactions: oxi-
       dation of carbon monoxide and hydrogen. In: Field RJ, Burger M, eds. Oscillations and travel-
       ing waves in chemical systems. New York, NY, USA, Wiley, 1985, 493–528.

[12]   Griffiths JF. Thermokinetic oscillations at homogeneous gas-phase oxidation. In: Field RJ,
       Burger M, eds. Oscillations and traveling waves in chemical systems. New York, NY, USA,
       Wiley, 1985, 529–567.

[13]   Boreskov GK. Catalysis. Siberian branch, Novosibirsk, Nauka, 1971 (in Russian).

[14]   Boreskov GK. On the development of methods for conducting catalytic processes under un-
       steady conditions. Vestnik AN SSSR 1983, 8, 22–30 (in Russian).

[15]   Boreskov GK. The effect of changes in catalyst composition on the kinetics of heterogeneous
       catalysis reactions. Kinet Catal 1972, 13(3), 543–554 (in Russian).

[16]   Boreskov GK, Matros YuSh, Kiselev OV, Bunimovich GA. Realization of a heterogeneous cat-
       alytic process under unsteady conditions. Dokl AN SSSR 1977, 237(1), 160–163 (in Russian).

[17]   Boreskov GK, Yablonskii GS. The evolution of ideas about the regularities of the kinetics of
       reactions in heterogeneous catalysis. J VKHO. 1977, 22(5), 556–561 (in Russian).

[18]   Slinko MG. History of the development of mathematical modeling of catalytic processes and
       reactors. Theor Found Chem Eng 2007, 41(1), 13–29.

[19]   Slinko MG. Kinetics and mechanism of complex catalytic reactions. Kinet Catal 1976, 17(1),
       13–18.

[20]   Slinko MG, Beskov VS, Dubyaga NA. Possibility of existence of several regimes in kinetic
       region of heterogenous catalytic reactions. Dokl AN SSSR 1972, 204(5), 1174–1177.

[21]   Slinko MG, Bykov VI, Yablonskii GS, Akramov TA. Multiplicity of stationary states of heteroge-
       neous catalytic reactions. Dokl AN SSSR 1976, 226(4), 876–879.

[22]   Slinko MG, Slinko MM. Auto-oscillations of the rate of heterogeneous catalytic reactions.
       Kinet Catal 1982, 23(6), 1421–1428.

[23]   Slinko MG, Slinko MM. Self-oscillations of heterogeneous catalytic reaction rates. Russ Chem
       Rev 1980, 49(4), 561–587.

[24]   Slinko MG, Vyatkin YuL, Beskov VS, Ivanov EA. Number and stability of stationary regimes on
       nonporous catalyst grain for a complex reaction. Dokl AN SSSR 1972, 204(6), 1321–1324.

[25]   Slinko MG, Yablonskii GS. Nonstationary and nonequilibrium processes in heterogeneous
       catalysis. In: Krilov OV, Shibanova MD, eds. Problems of kinetics and catalysis-17. Moscow,
       Nauka, 1978, 154–169 (in Russian).

[26]   Slinko MM. Oscillating reactions in heterogeneous catalysis: What new information can be
       obtained about reaction mechanisms? Catal Today 2010, 154(1), 38–45.

[27]   Slinko MM, Jaeger NI. Oscillating heterogeneous catalytic systems. Netherlands, Elsevier
       Science, 1994.

[28]   Slinko MM, Korchak VN, Peskov NV. Mathematical modelling of oscillatory behaviour during
       methane oxidation over Ni catalysts. Appl Catal A 2006, 303(2), 258–267.

[29]   Slinko MM, Ukharskii AA, Peskov NV, Jaeger NI. Chaos and synchronisation in heterogeneous
       catalytic systems: CO oxidation over Pd zeolite catalysts. Catal Today 2001, 70(4), 341–357.

[30]   Slinko MM, Fink T, Loher T, Madden HH, Lombardo SJ, Imbihl R, Ertl G. The NO + H_2 reaction
       on Pt(100): steady state and oscillatory kinetics. Surf Sci 1992, 264(1–2), 157–170.

[31]   Slinko MM, Jaeger NI, Svensson P. Mechanism of the kinetic oscillations in CO oxidation on
       palladium dispersed within a zeolite matrix. J Catal 1989, 118, 349–359.

[32] Slinko MM, Kurkina ES, Liauw MA, Jaeger NI. Mathematical modeling of complex oscillatory phenomena during CO oxidation over Pd zeolite catalysts. J Chem Phys 1999, 111(17), 8105–8114.

[33] Slinko MM, Ukharskii AA, Jaeger NI. Global and nonlocal coupling in oscillating heterogeneous catalytic reactions: CO oxidation on zeolite supported palladium. Phys Chem Chem Phys 2001, 3(6), 1015–1021.

[34] Slinko MM, Ukharskii AA, Peskov NV, Jaeger NI. Chaos and synchronization in heterogeneous catalytic systems: CO oxidation over Pd zeolite catalysts. Catal Today 2001, 70(4), 341–357.

[35] Eigenberger G. Kinetics instabilities in heterogeneously catalyzed reactions. Chem Eng Sci 1978, 33(9), 1263–1268.

[36] Belyaev VD, Slinko MG, Timoshenko VI. Changes in contact potential difference in oscillatory state of heterogeneous catalytic reaction of hydrogen with oxygen on nickel. Kinet Catal 1975, 16(2), 555–563 (in Russian).

[37] Belyaev VD, Slinko MM, Slinko MG, Timoshenko VI. Self-oscillations in the heterogeneous catalytic reaction of hydrogen with oxygen. Dokl AN SSSR 1974, 214(5), 1098–1100 (in Russian).

[38] Belyaev VD, Slinko MM, Timoshenko VI, Slinko MG. Generation of auto-oscillations in the hydrogen oxidation reaction on nickel. Kinet Catal 1973, 14(3), 810–813 (in Russian).

[39] Barelko VV, Beibutian VM, Volodin YuE, Zel'dovich YaB. Thermal waves and nonuniform steady states in a Fe + H_2 system. Chem Eng Sci 1983, 38(11), 1775–1780.

[40] Barelko VV, Volodin YE. About the existence of the branch-chains mechanism for heterogeneous-catalytic reactions. Dokl AN SSSR 1976, 216(5), 1080–1083 (in Russian).

[41] Zhukov SA, Barelko VV, Merzhanov AG. Wave processes on heat generating surfaces in pool boiling. Int J Heat Mass Transfer 1981, 24(1), 47–55.

[42] Zhukov SA, Barelko VV. Nonuniform steady states of the boiling process in the transition region between the nucleate and film regimes. Int J Heat Mass Transfer 1983, 26(8), 1121–1130.

[43] Imbihl R, Ertl G. Oscillatory kinetics in heterogeneous catalysis. Chem Rev 1995, 95(3), 697–794.

[44] Tomilov VN, Zagoruiko AN, Kuznetsov PA. Kinetic investigation and mathematical modeling of oscillation regimes for oxidative dehydrogenation of butene-1. Chem Eng Sci 1999, 54(20), 4359–4364.

[45] Kurdyumov SP. Evolution and self-organization laws in complex systems. Int J Mod Phys C 1990, 1(4), 299–327.

[46] Kurdyumov SP, Kurkina ES, Tel'kovskaya OV. Regimes with sharpening in two-component media. Math Models and Comput Simul 1989, 1(1), 34–50.

[47] Kuretova ED, Kurkina ES. Blowup solutions in a problem for the nonlinear heat equation on a small interval. Comput Math Model 2009, 20(2), 173–191.

[48] Kurkina ES, Averchuk GY. Monte-Carlo simulation of the oscillatory dynamics of a catalytic reaction with lateral interactions. Comput Math Model 2013, 24(4), 526–542.

[49] Kurkina ES, Makarova SM, Slinko MM. Mathematic modeling of oxidation reaction rate self-oscillations over metallic catalysts. Math Models and Comput Simul 1990, 2(1), 14–26.

[50] Kurkina ES, Malykh AV, Makeev AG. Natural waves and chaotic structures in a distributed four-component model of the NO + CO/PT(100) reaction. Comput Math Model 1999, 10(4), 363–378.

[51] Kurkina ES, Peskov NV, Slinko MM, Slinko MG. The origin of chaotic oscillations of CO oxidation rate on Pd-zeolite catalysts. Dokl Akad Nauk 1996, 351(4), 497–501.

[52] Kurkina ES, Semendyaeva NL. Fluctuation-induced transitions and oscillations in catalytic CO oxidation: Monte Carlo simulations. Surf Sci 2004, 558(1), 122–134.

[53] Kurkina ES, Semendyaeva NL. Mathematical modeling of spatial-temporal structures in a heterogeneous catalytic system. Comput Math Model 2012, 23(2), 133–157.

[54] Kurkina ES, Semendyaeva NL. Oscillatory dynamics of co oxidation on platinum-group metal catalyst. Kinet Catal 2005, 46(4), 453–463.

[55] Kurkina ES, Semendyaeva NL, Boronin AI. Mathematical modeling of nitrogen desorption from an iridium surface: a study of the effects of surface structure and subsurface oxygen 1. Kinet Catal 2001, 42(5), 703–717.

[56] Kurkina ES, Peskov NV, Slinko MM. Dynamics of catalytic oscillators locally coupled through the gas phase. Physica D: Nonlinear Phenomena 1998, 118(1), 103–122.

[57] Kurkina ES, Averchuk GY. Monte Carlo simulation of the oscillatory dynamics of a catalytic reaction with lateral interactions. Comput Math Model 2013, 24(4), 526–542.

[58] Kurkina ES, Semendyaeva NL. Fluctuation-induced transitions and oscillations in catalytic CO oxidation: Monte Carlo simulations. Surf Sci 2004, 558(1), 122–134.

[59] Kurkina ES, Semendyaeva NL. Mathematical modeling of spatial-temporal structures in a heterogeneous catalytic system. Comput Math Model 2012, 23(2), 133–157.

[60] Kurkina ES, Tolstunova ED. The general mathematical model of CO oxidation reaction over Pd-zeolite catalyst. Appl Surf Sci 2001, 182(1), 77–90.

[61] Ladas S, Imbihl R, Ertl G. Kinetic oscillations and facetting during the catalytic CO oxidation on Pt(110). Surf Sci 1988, 198(1), 42–68.

[62] Ladas S, Imbihl R, Ertl G. Microfacetting of a Pt(110) surface during catalytic CO oxidation. Surf Sci 1988, 197(1–2), 153–182.

[63] Lashina EA, Kaichev VV, Chumakova NA, Ustyugov VV, Chumakov GA, Bukhtiyarov VI. Mathematical simulation of self-oscillations in methane oxidation on nickel: An isothermal model. Kinet Catal 2012, 53(3), 374–383.

[64] Latkin EI, Elokhin VI, Gorodetskii VV. Monte Carlo model of oscillatory CO oxidation having regard to the change of catalytic properties due to the adsorbate induced Pt(100) structural conversion. J Mol Catal A: Chem 2001, 166(1), 23–30.

[65] Latkin EI, Elokhin VI, Gorodetskii VV. Spiral concentration waves in the Monte Carlo model of CO oxidation over Pd(110) caused by synchronization via CO_{ads} diffusion between separate parts of catalytic surface. Chem Eng J 2003, 91(2–3), 123–131.

[66] Latkin EI, Elokhin VI, Matveev AV, Gorodetskii VV. The role of subsurface oxygen in oscillatory behavior of $CO + O_2$ reaction over Pd metal catalysts: Monte Carlo model. J Mol Catal A: Chem 2000, 158(1), 161–166.

[67] Latkin EI, Sheinin DE, Elokhin VI, Bal'zhinimaev BS. Lattice statistical model for phase transition of active component of catalyst for SO_2 oxidation. React Kinet Catal Lett 1995, 56(1), 169–178.

[68] Lauterbach J, Bonilla G, Pletcher TD. Nonlinear phenomena during CO oxidation in the mbar pressure range: a comparison between $PtSiO_2$ and Pt(100). Chem Eng Sci 1999, 54(20), 4501–4512.

[69] Bautin NN. Behavior of dynamical systems near the boundary of stability. Moscow, Nauka, 1984 (in Russian).

[70] Bautin NN, Leontovich EA. Methods and techniques for qualitative analysis of dynamical systems on the plane. Moscow, Nauka, 1984 (in Russian).

[71] Bykov VI, Tsybenova SB. Parametric analysis of the models of a stirred tank reactor and a tube reactor. Combust Explos Shock Waves. 2001, 37(6), 634–640.

[72] Bykov VI, Tsybenova SB, Kuchkin AG. Modeling of the nitration of amyl in a continuous stirred tank reactor and a tube reactor. Combust Explos Shock Waves. 2002, 38(1), 30–36.

[73] Davidenko DF. On a new method of numerical solution of systems of nonlinear equations. Dokl AN SSSR 1953, 88(4), 601–602.

[74]   Ivanova AN. Conditions for uniqueness of the stationary states of kinetic systems, connected with the structures of their reaction-mechanisms (I, II). Kinet Catal 1979, 20(4), 833–837.

[75]   Ivanova AN, Furman GA, Bykov VI, Yablonskii GS. Catalytic mechanisms with reaction-rate self-oscillations. Dokl AN SSSR 1978, 242(4), 872–875.

[76]   Ivanova AN, Tarnopolskii BL. One approach to the determination of a number of qualitative features in the behavior of kinetic systems, and realization of this approach in a computer (critical conditions, auto-oscillations). Kinet Catal 1979, 20(6), 1271–1277.

[77]   Ivanova AN, Tarnopolskii BL, Furman GA. Autooscillations of reaction-rate in heterogeneous oxidation of hydrogen in ideal-mixing reactor. Kinet Catal 1983, 24(1), 102–109.

[78]   Kholodniok M, Klich A, Kubichek M, Marek M. Methods of analysis of nonlinear dynamical models. Moscow, Mir,1991.

[79]   Fadeev SI, Savchenko VI, Berezin AY. Analysis of auto-oscillations during co oxidation on a nonuniform surface consisting of two types of sites coupled by coads diffusion. React Kinet Catal Lett 1999, 67(1), 155–161.

[80]   Uppal A, Ray WH, Poore AB. On the dynamic behavior of continuous stirred tank reactors. Chem Eng Sci 1974, 29(4), 967–985.

[81]   Uppal A, Ray WH, Poore AB. The classification of the dynamic behavior of continuous stirred tank reactors – influence of reactor residence time. Chem Eng Sci 1976, 31(3), 205–214.

[82]   Vaganov DA, Abramov VG, Samoilenko NG. Finding the regions of existence for oscillating processes in well-stirred reactors. Dokl AN SSSR 1977, 234(3), 640–643.

[83]   Vaganov DA, Samoilenko NG, Abramov VG. Periodic regimes of continuous stirred tank reactors. Chem Eng Sci 1978, 33(8), 1131–1140.

[84]   Novikov EA. Explicit methods for stiff systems. Novosibirsk, Nauka, 1997.

[85]   Bykov VI, Kytmanov AM, Lazman MZ. Elimination methods in polynomial computer algebra. Novosibirsk, Nauka, 1991 (in Russian).

[86]   Bykov VI, Kytmanov AM, Lazman MZ, Passare M. Elimination methods in polynomial computer algebra. Netherlands, Springer Science and Business Media, 2012.

[87]   Bykov VI, Tsybenova SB, Slinko MG. Andronov–Hopf bifurcations in the Aris–Amundson model. Dokl Phys Chem 2001, 378(2), 134–137.

[88]   Bykov VI., Tsybenova SB, Slinko MG. Safe and unsafe boundaries of regions of critical phenomena in the kinetics of exothermic reactions. Dokl Phys Chem 2001, 378(1), 138–140.

[89]   Bykov VI, Gorban AN. Quasithermodynamic characteristic of reactions without the reaction of different substances. J. Phys Chem 1983, 57(12), 2942–2948.

[90]   Bykov VI, Gorban AN. Simplest model of self-oscillations in association reactions. React Kinet Catal Lett 1985, 27(1), 153–155.

[91]   Bykov VI, Gorban AN, Pushkareva TP. Autooscillation model in reactions of the association. J Phys Chem 1985, 59(2), 486–488 (in Russian).

[92]   Bykov VI, Ivanova AN. Chemical nonideality as the reason for critical phenomena. Kinet Catal 1986, 27(1), 63–71.

[93]   Bykov VI, Ivanova AN. Conditions for instability of systems involving complex sets of kinetically controlled catalytic transformations with buffer steps. Dokl Phys Chem 2003, 390(4), 143–146.

[94]   Bykov VI, Ivanova AN. Criticality conditions for the mechanisms of complex catalytic reactions occurring at two types of active sites. Dokl Phys Chem 2003, 393(4), 343–345.

[95]   Bykov VI, Kim VF, Yablonskii GS, Stepanskii YY. Mathematical model for auto-oscillations in the processes of cold-flaming combustion of heptane isooctane mixtures. J Phys Chem 1981, 55(12), 301–3016.

[96]   Bykov VI, Kiselev NV, Tsybenova SB. Macrokinetics of catalytic reactions on surfaces of various geometries. Dokl Chem 2008, 421(1), 161–164.

[97] Andronov AA, Vitt AA, Khaikin CE. The theory of oscillations. Moscow, Nauka, 1981 (in Russian).

[98] Elokhin VI, Bykov VI, Slinko MG, Yablonskii GS. Some aspects of oxidation of carbon monoxide over platinum catalysts. Dokl AN SSSR 1978, 238(3), 615–618.

[99] Elokhin VI, Bykov VI, Slinko MG, Yablonskii GS. Some problems of dynamics of carbon-monoxide oxidation reaction on platinum catalyst. Dokl AN SSSR 1978, 238(3), 615–618.

[100] Elokhin VI, Latkin EI. Statistic lattice model of oscillating and wave phenomena over the catalyst surface during CO oxidation. Dokl Akad Nauk 1995, 344(1), 56–61.

[101] Elokhin VI, Latkin EI, Matveev AV, Gorodetskii VV. Application of statistical lattice models to the analysis of oscillatory and autowave processes in the reaction of carbon monoxide oxidation over platinum and palladium surfaces. Kinet Catal 2003, 44(5), 692–700.

[102] Elokhin VI, Matveev AV, Gorodetskii VV. Self-oscillations and chemical waves in CO oxidation on Pt and Pd: Kinetic Monte Carlo models. Kinet Catal 2009, 50(1), 40–47.

[103] Gorodetskii VV, Elokhin VI, Bakker JW, Nieuwenhuys BE. Field electron and field ion microscopy studies of chemical wave propagation in oscillatory reactions on platinum group metals. Catal Today 2005, 105(2), 183–205.

[104] Elokhin VI, Myshlyavtsev AV, Latkin EI, Resnyanskii ED, Sheinin DE, Bal'zhinimaev BS. Statistical lattice models of physicochemical processes in catalytic reactions: Autooscillations, adsorption on the rough surface, and crystallization. Kinet Catal 1998, 39(2), 246–267.

[105] Elokhin VI, Gorodetskii VV. Atomic scale imaging of oscillations and chemical waves at catalytic surface reactions: experimental and statistical lattice models. Surfactant Sci Ser 2006, 130, 159–189.

[106] Elokhin VI, Latkin EI, Matveev AV, Gorodetskii VV. Application of statistical lattice models to the analysis of oscillatory and autowave processes in the reaction of carbon monoxide oxidation over platinum and palladium surfaces. Kinet Catal 2003, 44(5), 692–700.

[107] Elokhin VI, Myshlyavtsev AV. Catalytic processes: nanoscale simulations. In: Schwarz JA, Contescu CI, Putyera K, eds. Dekker Encyclopedia of Nanoscience and Nanotechnology. New York, NY, USA, Marsel Dekker, 2009, 782–793.

[108] Elokhin VI, Yablonskii GS. Kinetic models of catalytic reactions taking into account the influence of reaction medium. In: Heterogeneous Catalysis: Proc. 7th Int. Symp. Bourgas. 1991, 911–916.

[109] Elokhin VI, Yablonskii GS, Gorban AN. Dynamics of chemical reactions and nonphysical steady states. React Kinet Catal Lett 1980, 15(2), 245–250.

[110] Makeev AG, Nieuwenhuys BE. Mathematical modeling of the NO+ H_2/Pt(100) reaction: "Surface explosion", kinetic oscillations, and chaos. J Chem Phys 1998, 108(9), 3740–3749.

[111] Makeev AG, Slinko MM. Mathematical modeling of the peculiarities of NO decomposition on Rh(111). Surf Sci 1996, 359(1), L467–L472.

[112] Makeev AG, Slinko MM, Janssen NMH, Cobden PD, Nieuwenhuys BE. Kinetic oscillations and hysteresis phenomena in the NO+H_2 reaction on Rh(111) and Rh(533): Experiments and mathematical modeling. J Chem Phys 1996, 105(16), 7210–7222.

[113] Savchenko VI. CO_{ads} spillover and low-temperature activity of heterophase catalysts in CO oxidation. React Kinet Catal Lett 1995, 55(1), 143–151.

[114] Savchenko VI. The chemisorption of oxygen and the oxidation of carbon monoxide on metals. Russ Chem Rev 1986, 55(3), 462–476.

[115] Savchenko VI, Boreskov GK, Kalinkin AV, Salanov AN. State of oxygen on metal surfaces and catalytic activity for the oxidation of carbon monoxide. Kinet Catal 1983, 24(5), 1154–1161.

[116] Savchenko VI, Salanov AN, Bibin VN. Modeling the dynamics of development of reaction rate auto-oscillations in the oxidation of carbon monoxide on platinum. Kinet Catal 1993, 34(1), 187–193.

[117] Bukhtiyarov VI, Boronin AI, Savchenko VI. Two oxygen states and the role of carbon in partial oxidation of ethylene over silver. Surf Sci 1990, 232(1–2), L205-L209.

[118] Savchenko VI. Computer simulation of the transition to chaotic behavior of the oscillations in the rate of CO oxidation on Pt(110). Mendeleev Commun 1991, 1(4), 139–141.

[119] Savchenko VI. Kinetics of reactions on nonuniform surfaces with account of absorbed species migration along the surface. Kinet Catal 1998, 39(1), 124–129.

[120] Savchenko VI, Dadayan KA. On the possible role of adparticles spillover on a heterophase surface in CO using superadditive activity. React Kinet Catal Lett 1995, 55(1), 33–40.

[121] Savchenko VI, Efremova NI. Kinetics of CO oxidation on the surface of heterophase catalysts. Modeling by the Monte Carlo method. React Kinet Catal Lett 1996, 57(1), 49–54.

[122] Savchenko VI, Efremova NI. Kinetics of dissociative adsorption on stepped surfaces of platinum. Kinet Catal 2000, 41(1), 127–131.

[123] Savchenko VI, Ivanov EA, Fadeev SI. Kinetic model of CO oxidation on heterophase surface analyzed regarding CO_{ads} spilover. I. Homotopic method. Effect on temperature and CO pressure. React Kinet Catal Lett 1996, 57(1), 55–60.

[124] Scheintuch M, Schmitz RA. Oscillations in catalytic reactions. Cat Rev – Sci Eng 1977, 15(1), 107–172.

[125] Zyskin AG, Shub FS, Snagovskii YS. Qualitative study of relaxation processes in a 2-step heterogeneous catalytic reaction-accuracy of the quasi-steady-state approximation. Kinet Catal 1990, 31(2), 387–394.

[126] Zyskin AG, Snagovskii YS, Slinko MG. Studies of the dynamic properties of heterogeneous catalytic reactions in a closed adiabatic gradientless system over biographically inhomogeneous catalyst surface. React Kinet Catal Lett 1981, 17(3–4), 263–267.

[127] Zyskin AG, Snagovskii YuS, Slinko MG. Studies of the dynamics properties of heterogeneous catalytic reaction in a closed isothermal gradientless system over biographically inhomogeneous catalyst surface. React Kinet Catal Lett 1981, 17(3–4), 263–267.

[128] Bykov VI, Kytmanov AM, Lazman MZ. Elimination methods in polynomial computer algebra. Dordrecht, Kluwer Academic Publishers, 1998.

[129] Bykov VI, Kytmanov AM. Estimation of the number of steady states for three-stage adsorption mechanisms. I. Parallel scheme. II. Consistent scheme. Kinet Catal 1984, 25(5), 1276–1278.

[130] Bykov VI, Pushkareva TP, Fadeev SI. Parametric analysis of kinetic models. XI. Influence of the number of active sites. React Kinet Catal Lett 1996, 57(1), 133–140.

[131] Bykov VI, Pushkareva TP, Ivanova AN. Parametric analysis of kinetic models. IX. Catalytic triggers with diffusion exchange steps. React Kinet Catal Lett 1995, 56(1), 107–113.

[132] Bykov VI, Pushkareva TP, Ivanova AN. Parametric analysis of kinetic models. X. Catalytic oscillator with a diffusion steps. React Kinet Catal Lett 1995, 56(1), 115–119.

[133] Bykov VI, Pushkareva TP, Savchenko VI. Parametric analysis of kinetic models. XII. Influence of the intensity of diffusion exchange. React Kinet Catal Lett 1996, 57(1), 141–146.

[134] Bykov VI, Trotsenko LS. Parametric analysis of kinetic models. XIII. Catalytic trigger in a flow system. React Kinet Catal Lett 2002, 76(1), 69–74.

[135] Bykov VI, Trotsenko LS. Parametric analysis of kinetic models. XIV. Autocatalytic trigger and oscillator in a flow system. React Kinet Catal Lett 2002, 76(2), 281–286.

[136] Magnitskii NA, Sidorov SV. New methods for chaotic dynamics. Singapore, World Scientific Publ, 2006.

[137] Malinetskii GG, Potapov AB. Nonlinear dynamics and chaos: Basic concepts. Moscow, URSS, 2011 (in Russian).

[138] Malinetskii GG. Mathematical foundations of synergetics: Chaos, structures, computational experiment. Moscow, URSS, 2015 (in Russian).

[139] Malinetskii GG, Potapov AB, Podlazov AV. Nonlinear dynamics. Approaches, results, expectations. Moscow, URSS, 2011 (in Russian).

[140] Malinetskii GG, Shakaeva MS. Analysis of a cellular automaton serving as a model of oscillatory chemical reactions. Sov Phys Dokl 1991, 36(12), 826–829.

[141] Malinetskii GG, Shakaeva MS. Simulation of oscillatory reactions on a surface using cellular automata. Russ J Phys Chem 1995, 69(8), 1380–1384.

[142] Malinetskii GG. The future of applied mathematics. Lectures for young researchers. Moscow, URSS, 2005 (in Russian).

[143] Strogatz SH. Nonlinear dynamics and chaos: with applications to physics, biology, chemistry, and engineering. Boulder, CO, USA, Westview press, 2014.

[144] Bykov VI, Trotsenko LS. Noise-induced kinetic chaos. Russ J Phys Chem 2005, 79(5), 677–680.

[145] Neimark JI, Landa PS. Stochastic and chaotic oscillations (Vol. 77). Berlin, Springer Science + Business Media, 1992.

[146] Chumakov GA, Slinko MG. Kinetic turbulence (chaos) of the hydrogen-oxygen interaction rate on metal-catalysts. Dokl AN SSSR 1982, 266(5), 1194–1198.

[147] Chumakov GA, Slinko MG, Belyaev VD. Complicated behavior of a heterogeneous catalytic reaction rate. Dokl AN USSR 1980, 253(3), 653–658.

[148] Polak LS, Mikhailov AS. Self-organization in nonequilibrium physicochemical systems. Moscow, Nauka, 1983 (in Russian)

[149] Poluektov RA. Dynamical theory of biological populations. Moscow, Nauka, 1974 (in Russian).

[150] Yablonskii GS, Bykov VI, Elokhin VI. The kinetics of model reactions of heterogeneous catalysis. Novosibirsk, Nauka, 1984 (in Russian).

[151] Yablonskii GS, Bykov VI, Gorban AN. Kinetic models of catalytic reactions. Novosibirsk, Nauka, 1983 (in Russian).

[152] Sheplev VS, Slinko MG. Periodic regimes of ideal mixing reactor. Dokl Akad Nauk 1997, 352(6), 781–784.

[153] Sheplev VS, Treskov SA, Volokitin EP. Dynamics of a stirred tank reactor with first-order reaction. Chem Eng Sci 1998, 53(21), 3719–3728.

[154] Vishnevskii AL, Elokhin VI, Kutsovskaya ML. Dynamic model of self-oscillatory evolution in carbon monoxide oxidation over Pt(110). React Kinet Catal Lett 1993, 51(1), 211–217.

[155] Marshneva VI, Boreskov GK. Study of self sustained oscillations in the reaction of carbon monoxide oxidation over platinum applied on silica gel. Kinet Catal 1984, 25(4), 875–883.

[156] Marshneva VI, Boreskov GK, Yablonskii GS, Kim VF. Study of the reaction of carbon monoxide oxidation by oxygen over platinum catalysts. Kinet Catal 1984, 25(3), 662–669.

[157] Yablonskii GS, Bykov VI, Gorban AN, Elokhin VI. Kinetic models of catalytic reactions. Amsterdam, Oxford, New York, Tokyo, Elsevier, 1991.

[158] Gorban AN, Bykov VI, Yablonskii GS. Essays on chemical relaxation. Novosibirsk, Nauka, 1986 (in Russian).

[159] Ladas S, Imbihl R, Ertl G. Kinetic oscillations and facetting during the catalytic CO oxidation on Pt(110). Surf Sci 1988, 198(1), 42–68.

[160] Imbihl R, Ertl G. Oscillatory kinetics in heterogeneous catalysis. Chem Rev 1995, 95(3), 697–794.

[161] Vishnevskii AL, Latkin EI, Elokhin VI. Autowaves on catalyst surface caused by carbon monoxide oxidation kinetics: Imitation model. Surf Rev Lett 1995, 2(4), 459–469.

[162] Vishnevskii AL, Savchenko VI. Self-oscillations of the reaction rate of CO oxidation on Pt(110). React Kinet Catal Lett 1989, 38(1), 167–173.

[163] Elenin GT, Krylov VV, Polezhaev AA, Chernavskii DT. Peculiarities of the formation of contrasting dissipative structures. Dokl AN SSSR 1983, 271(1), 84–88.

[164] Yablonskii GS, Bykov VI, Slinko MG, Kuznetsov YuI. Analysis of steady-state regimes of reaction of carbon-monoxide oxidation on platinum. Dokl AN SSSR 1976, 229(4), 917–919.

[165] Yablonskii GS, Cheresiz VM. Four types of relaxation in chemical kinetics (linear case). React Kinet Catal Lett 1984, 24(1), 49–53.

[166] Yablonskii GS, Bykov VI. Analysis of the structure of the kinetic equation of a complex catalytic reaction (linear single-path mechanism). Theor Exp Chem 1979, 15(1), 29–32.

[167] Karavaev AD, Ryzhkov AB, Noskov OV, Kazakov VP. Generation of a fractal torus in the Belousov–Zhabotinskii reaction model. Dokl Phys Chem 1998, 363(1–3), 367–371).

[168] Noskov OV, Karavaev AD, Spivak SI, Kazakov VP. Modeling of the complex dynamics of the Belousov–Zhabotinskii reaction: the decisive role of fast variables. Kinet Catal 1992, 33(3), 567–574.

[169] Noskov OV, Karavaev AD, Spivak SI, Kazakov VP. Modeling of the complex dynamics of the Belousov–Zhabotinskii reaction: the decisive role of fast variables. Kinet Catal 1992, 33(3), 567–574.

[170] Ryzhkov AB, Noskov OV, Karavaev AD, Kazakov VP. On stationary points and bifurcations of the Belousov–Zhabotinsky reaction. Math Models and Comput Simul 1998, 10(2), 71–78.

[171] Volter BV, Sal'nikov IE. Stability of work regimes of the chemical reactors. Moscow, Khimiya, 1982 (in Russian).

4 Thermokinetic models

Chapter 4 is devoted to the parametric analysis of thermokinetic models such as the Aris–Amundson model, the Zel'dovich–Semenov model, and the Volter–Salnikov model. The main nonlinearity in these models is nonlinearity of the exponential type $\exp(-E/RT)$, which is characterized by the dependence of reaction rate on temperature. The thermokinetic models represent a system of ordinary differential equations in which one of the phase variables is temperature. The parametric portrait for each model is plotted, which is a kind of passport of nonlinear dynamic models. The regions of parameters with a different number and type of stability of steady states are highlighted. In addition, we study the impact of kinetic functions on the number and stability of steady states. The case of catalytic reactions going on in nonisothermal conditions is specially highlighted. The continuous stirred-tank reactor model and the plug flow reactor model with dimensional parameters are considered. In general, the goal of this chapter is to emphasize the specificity of nonisothermal processes in combination with kinetic nonlinearities.

4.1 Continuous stirred-tank reactors (CSTR)

A mathematical model of an exothermic process in a reactor of ideal mixing for one reaction of first order has the form:

$$V\frac{dX}{dt} = -Vk(T)X + q(X^0 - X) ,$$

$$C_p\rho V\frac{dT}{dt} = (-\Delta H)Vk(T)X + q\rho C_p(T_0 - T) + hS(T_x - T) ,$$

$$k(T) = k^0 \exp(-E/(RT)) , \tag{4.1}$$

where V [cm^3] is the volume of the reactor, $k(T) = k^0 \exp(-E/(RT))$ [s^{-1}] is the reaction rate constant, E is the activation energy; R is the universal gas constant; k^0 is the pre-exponent, X and X^0 [mol cm^{-3}] are the current and inlet concentrations of the reagent, respectively, T and T_0 [K] are the current and inlet temperatures of the mixture, q [cm^3 s^{-1}] is the volume flow rate, ρ [mol cm^{-3}] is the density of the mixture, C_p [J/mol^{-1} K^{-1}] is the specific heat of the reactive mixture, S [cm^2] is the area of the heat-exchange surface, h [J (cm^{-2} s^{-1} K^{-1}] is the coefficient of heat transfer through the reactor wall, $(-\Delta H)$ [J mol^{-1}] is the thermal effect of the reaction, t [s] is the astronomical time, and T_w [K] is the temperature of the reactor walls.

The concentration X and the temperature T are the phase variables in system (4.1). The system contains a significant number of thermophysical, kinetic, and geometric parameters. Introducing dimensionless variables and parameters, we can significantly reduce the number of parameters which fully specify the properties of solu-

https://doi.org/10.1515/9783110464948-004

tions of system of equations (4.1). There are several variants for the introduction of dimensionless parameters and variables. Here we consider three main types which are adopted in the theory of combustion, the theoretical foundations of chemical technology, and the theory of polymerization. In the analysis of real processes, sometimes it is necessary to carry out appropriate parametric analysis of model (4.1) without converting it to a dimensionless form.

4.2 Zel'dovich–Semenov model

A mathematical model of a reactor of ideal mixing (4.1) is the traditional object of study in the theory of combustion, and at present it has its own story [1–26]. The modern stage of its parametric analysis was initiated in [1]. To date, 38 different phase portraits of the corresponding dynamical system are found. Further clarification of the variety of the dynamic behavior of this model is made in publications [27]. However, as a rule, parametric portraits are given schematically in one possible plane of dimensionless parameters. Building bifurcation curves, dividing a plane of parameters into areas with different dynamic behaviors of a system, and analysis of their evolution when changing the third parameter allow us to get additional information on the nature of critical phenomena in the considered processes. In this section, we make a parametric analysis of the basic model theory of combustion – a model exothermic reaction carried out in the mode of perfect mixing in a continuous reactor [28]. Here a reaction of the first order is considered in detail. Reactions of an arbitrary n-th order, oxidation reactions, reactions with arbitrary kinetics, etc. will be considered by analogy.

The procedure of parametric analysis includes the study of the number of steady states of an original mathematical model, their stability, the building of dependencies of stationary characteristics from parameters, the determination of curves of local bifurcations (curves of multiplicity and neutrality of the steady states), the construction of parametric and phase portraits of the investigated dynamical system, and the calculation of the time dependencies of its solutions.

It is necessary to note that these models historically preceded the isothermal kinetics models. They were formulated in the framework of combustion theory in the works of N. N. Semenov, B. Ya. Zel'dovich and D. A. Frank-Kamenetsky [29, 30]. The Zel'dovich–Semenov model is the base model, in which the system of dimensionless parameters was suggested by D. A. Frank-Kamenetsky. For this model we investigate several variants of the basic chemical reactions: a first order reaction, a reaction of n-th order, and a combustion reaction considering an oxygen reaction which includes an oxygen reaction with arbitrary kinetic order.

4.2.1 Reaction A → P

Following Frank-Kamenetskii [30], we introduce dimensionless parameters and variables:

$$T^* = \frac{hST_0 + C_p\rho q T_{(0)}}{hS + C_p\rho q} \ , \quad \alpha_* = h + \frac{C_p\rho q}{S} \ , \quad Da = \frac{V}{q}k(T_0) \ ,$$

$$Se = \frac{(-\Delta H)\rho}{\alpha_*(S/V)} \frac{E}{RT^{*2}} k^0 \exp(-E/(RT)) \ , \quad \beta = \frac{(-\Delta H)X^0}{C_p\rho T_0} \ , \quad \gamma = \frac{E}{RT_0} \ ,$$

$$x = \frac{X^0 - X}{X^0} \ , \quad y = \frac{E}{RT^2}(T - T^*) \ , \quad \tau = k^0 t \exp(-E/(RT^*)) \ .$$

Assume that the initial and input conditions for the reactor are the same:

$$X(0) = X^0, \quad q\rho C_p(T_0 - T(0)) = hS(T(0) - T_x) \ .$$

A scheme for the first order reaction is

$$A \rightarrow P \ ,$$

where A is a reacting substance and P is a reaction product. The model of the exothermic process going on in the reactor of ideal mixing is written in dimensionless form:

$$\frac{dx}{d\tau} = f(x)\exp\left(\frac{y}{1 + \beta y}\right) - \frac{x}{Da} = f_1(x, y) \ ,$$

$$\gamma\frac{dy}{d\tau} = f(x)\exp\left(\frac{y}{1 + \beta y}\right) - \frac{y}{Se} = f_2(x, y) \ , \tag{4.2}$$

where $f(x) = 1 - x$ is a kinetic function of the first order, x, y are dimensionless concentration and temperature, and Da, Se, β, γ are dimensionless parameters.

A system of two differential equations of the type (4.2) is called the *Zel'dovich–Semenov model*.

The **steady states** of system (4.2) are the solutions of the equations:

$$\frac{dx}{d\tau} = 0 \ , \quad \gamma\frac{dy}{d\tau} = 0 \ ,$$

or

$$f(x)\exp\left(\frac{y}{1 + \beta y}\right) - \frac{x}{Da} = 0 \ ,$$

$$f(x)\exp\left(\frac{y}{1 + \beta y}\right) - \frac{y}{Se} = 0 \ . \tag{4.3}$$

After the exclusion of $x = (Da/Se)y$ and substitution into the second equation of system (4.3) we obtain the following nonlinear equation

$$(1 - y(Da/Se))\exp\left(\frac{y}{1 + \beta y}\right) = \frac{y}{Se} \ . \tag{4.4}$$

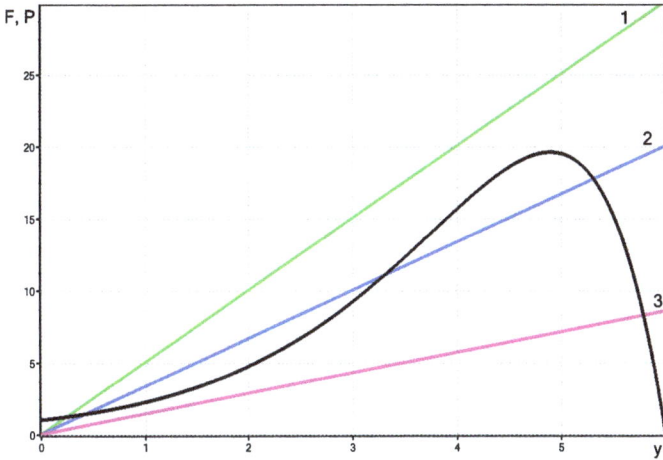

Fig. 4.1: Heat production function ($F(y)$) and heat removal ($P(y)$) at $Da = 0.05$, $\beta = 0.01$, $v = 0.17$, 1) $Se = 0.2$; 2) 0.3; 3) 0.7

Equality (4.4) is the equation of the steady state, where the right side is the heat production function ($F(y)$), and the left side is the function of heat removal ($P(y)$):

$$F(y) = (1 - vy)e(y) ,$$
$$P(y) = y/Se ,$$

where $e(y) = \exp(y/(1 + \beta y))$ and $v = Da/Se$.

Steady states of system (4.2) are studied with the help of Semenov's diagram (Fig. 4.1), where the dependencies of the functions of heat production and heat removal from the temperature of substances in the reactor are shown. The point of intersection of the curves $F(y)$ and $P(y)$ determines the temperature of the steady states. Knowing this temperature we can determine the concentration value for the corresponding steady state from one of the stationarity equations (4.3).

A necessary and sufficient condition for the uniqueness of a steady state is the condition

$$\frac{1}{Se} = P' > F' ,$$

i.e., the rate of heat removal must be greater than the rate of heat production in a steady state (cases 1 and 3 in Fig. 4.1).

The type of stability of the steady state is determined by the roots of the characteristic equation

$$\lambda^2 - \sigma\lambda + \Delta = 0 , \tag{4.5}$$

where σ and Δ are found by using the elements of the Jacobian matrix:

$$a_{11} = \frac{\partial f_1}{\partial x} = -e(y) - \frac{1}{Da} \,,$$

$$a_{12} = \frac{\partial f_1}{\partial y} = \frac{e(y)f(x)}{(1 + \beta y)^2} \,,$$

$$a_{21} = \frac{\partial f_2}{\partial x} = \frac{-e(y)}{y} \,,$$

$$a_{22} = \frac{\partial f_2}{\partial y} = \frac{e(y)f(x)}{(1 + \beta y)^2} - \frac{1}{ySe} \,.$$

The expressions for σ and Δ are

$$\sigma = a_{11} + a_{22} \,,$$

$$\Delta = a_{11}a_{22} - a_{12}a_{21} \,.$$

The roots of the characteristic equation (4.5) are expressed in terms of its coefficients:

$$\lambda_{1,2} = \frac{\sigma}{2} \pm \sqrt{\frac{\sigma^2}{4} - \Delta} \,.$$

Parametric dependencies of the steady states are solutions of the equation

$$G(y) = F(y) - P(y) = 0 \,.$$

The dimensionless parameters Da, Se, β, v are included in the equation of steady state linearly. So it is advisable to take advantage of this and write down parametric dependencies in the form of inverse dependencies $Da = Da(y)$, $Se = Se(y)$, etc. The stationarity equation can be written as:

$$G(y,\ Da,\ Se,\ \beta,\ v) = 0 \,, \tag{4.6}$$

where the dimensionless parameters are selected as arguments.

From the stationarity equation (4.4) after transformations we get the expression for $Se(y)$:

$$Se(y) = \left(Da + \frac{1}{e(y)} \right) y \,. \tag{4.7}$$

Similarly, we get the appropriate parametric dependencies for the other parameters:

$$Da(y) = \frac{Se}{y} - \frac{1}{e(y)} \,,$$

$$\beta(y) = \frac{1}{\ln\,(y/(Se - Day))} - \frac{1}{y} \,, \tag{4.8}$$

$$v = \frac{1}{y} - \frac{1}{e(y)Se} \,.$$

When we plot the dependence $Se(y)$ in accordance with (4.7), we will get the inverse function of the desired parametric dependence $y = y(Se)$. If (4.7) is given in

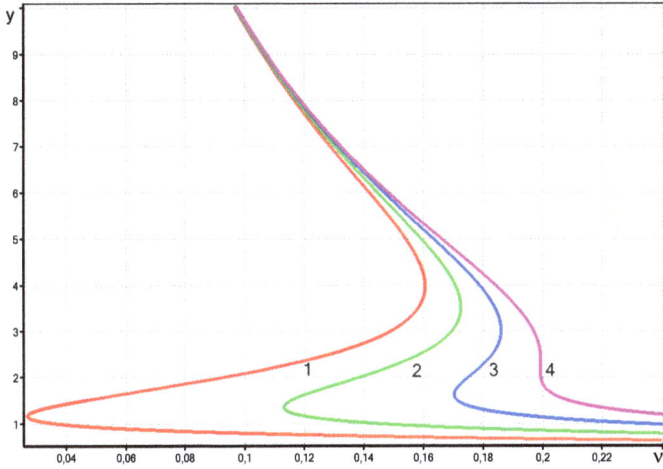

Fig. 4.2: Dependencies of the dimensionless temperature y on parameter v at $Da = 0.05$, $\beta = 0.05$, 1) $Se = 0.4$; 2) 0.45; 3) 0.5; 4) 0.54

graphical or tabular form, the necessary value of y for any fixed value of Se can be found from the corresponding graph or table. An example of the dependence of the stationary temperature from the parameter when changing the second parameter is shown in Fig. 4.2.

Thus, the dependencies of steady states on the parameters (or rather their inverse dependencies) can be obtained without iterative procedures of the solution in the typical case of a nonlinear stationarity equation.

Curves of multiplicity. The curve of multiplicity L_Δ: $\Delta = 0$, is the boundary separating the region of parameters on regions with one and three steady states. Let us select, for example, the plane of parameters (Da, Se) and write down an equation of a curve of multiplicity L_Δ in this plane. For this purpose it is necessary to solve the system

$$f_1(x, y, Da, Se) = 0 \,,$$

$$f_2(x, y, Da, Se) = 0 \,,$$

$$\Delta(y, Da, Se) = 0 \,, \tag{4.9}$$

where Δ is determined in (4.5) by using the elements of the Jacobian matrix. Using (4.6), we reduce system (4.9) to a system of two equations:

$$G(y, Da, Se) = 0 \,,$$

$$\Delta(y, Da, Se) = 0 \,. \tag{4.10}$$

Using the explicit expression $Se(y, Da)$ (4.7) and replacing it in the second equation of (4.10), we obtain the equation of the boundary region of multiplicity of steady states

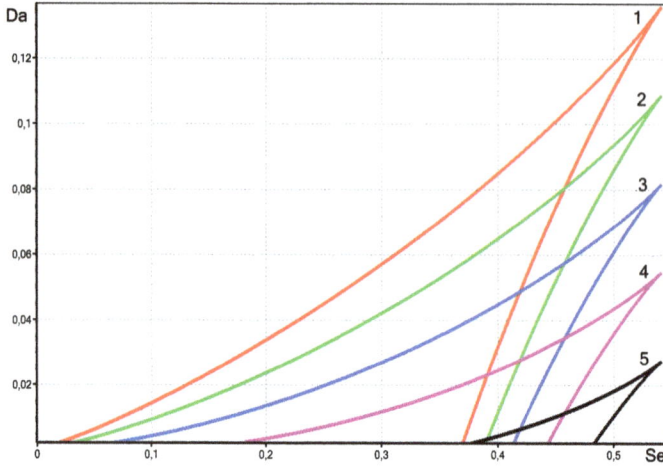

Fig. 4.3: Curves of multiplicity $L_\Delta(Se, Da)$ at $Se = 0.4$, 1) $\beta = 0$; 2) 0.05; 3) 0.1; 4) 0.15; 5) 0.2

in the parameter plane (Da, Se):

$$L_\Delta(Da, Se): \quad Da(y) = \frac{y - (1 + \beta y)^2}{(1 + \beta y)^2 e(y)},$$

$$Se = Se(y, Da(y)). \tag{4.11}$$

Similarly, let us write the parametric form of a curve of multiplicity L_Δ in the other planes of the parameters. So, for the plane (Se, v) we have:

$$L_\Delta(Se, v): \quad Se(y) = \frac{y^2}{(1 + \beta y)^2 e(y)},$$

$$v = v(y, Se(y)). \tag{4.12}$$

For planes of parameters involving y a curve of multiplicity is represented by two parallel lines, between which three steady states exist. Examples of curves of multiplicity are shown in Fig. 4.3 with variation of the third parameter.

In the case when it is difficult to obtain expressions for curves of multiplicity in any of the planes of parameters in an explicit form, we propose a procedure to graphical plotting for L_Δ. The procedure consists of plotting a series of parametric dependencies on one of parameters at variation of the second parameter. Note that these dependencies can be plotted for all parameters (4.7)–(4.8). The interval of multiplicity of steady states is located between the turning points of the parametric curve (the maximum and minimum values of the curves (4.7)–(4.8)). When varying the second parameter the bounds of this interval limit the region of multiplicity of steady states on the appropriate plane. An example implementation of the described graphic procedure is presented in Fig. 4.4.

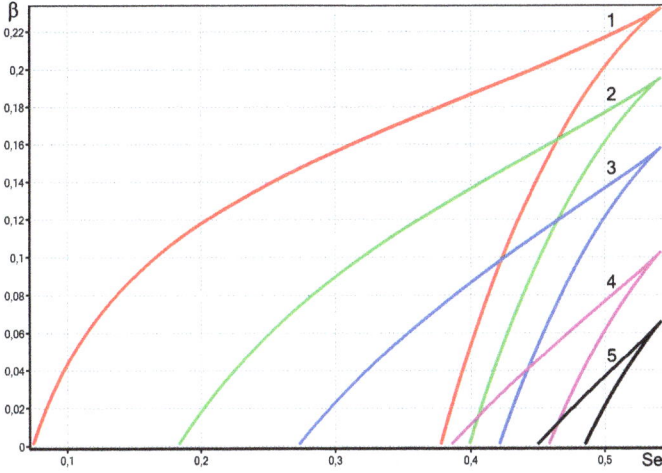

Fig. 4.4: Curves of multiplicity $L_\Delta(Se, Da)$ at 1) $\beta = 0$; 2) 0.05; 3) 0.1; 4) 0.15; 5) 0.2

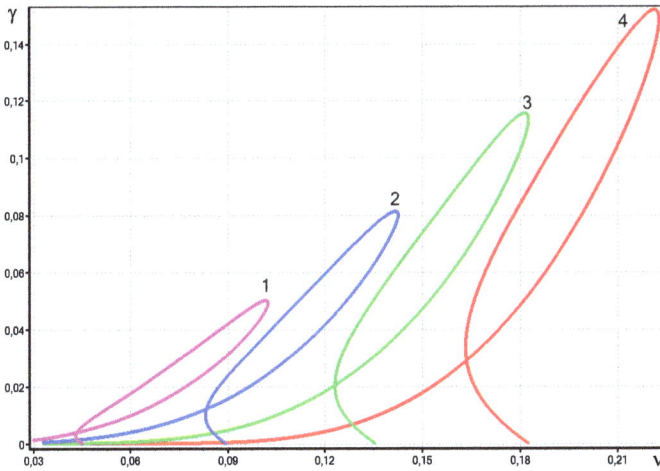

Fig. 4.5: Curves of neutrality $L_\sigma(v, \gamma)$ for $Se = 0.45$, 1) $\beta = 0$; 2) 0.04; 3) 0.08; 4) 0.12

Curves of neutrality. The curve of neutrality $L_\sigma(\sigma = 0)$ determines the type of stability of the steady state. For the plotting of L_σ it is necessary to solve the following system of equations:

$$G(y, Da, Se) = 0 \,,$$
$$\sigma(y, Da, Se) = 0 \,. \tag{4.13}$$

When we replace an explicit expression of the dependence $Da(y, Se)$ of (4.8) in the second equation of (4.13), we will get the equation of the curve of neutrality explicitly, similarly to L_Δ.

$$L_\sigma(Se, v): \qquad Se(y) = \frac{Da(y - (1 + \beta y)^2)}{y(1 + \beta y)^2(Dae(y) + 1)},$$

$$v = v(y, Se(y)),$$

$$L_\sigma(Da, Se): \qquad Da(y) = (a(y) \pm \sqrt{a^2(y) - 1})/e(y),$$

$$Se = Se(y, Da(y)),$$

$$L_\sigma(y, Da): \qquad y(y, Da) = \frac{Dae(y)(y - (1 + \beta y)^2)}{y(1 + \beta y)^2(Dae(y) + 1)^2},$$

$$Da = Da(y),$$

$$L_\sigma(y, Se): \qquad y(y, Se) = \frac{(e(y)Se - y)(y - (1 + \beta y)^2)}{(e(y)Se(1 + \beta y))^2},$$

$$Se = Se(y),$$

$$L_\sigma(y, \beta): \qquad y(y, \beta) = \frac{Dae(y)(y - (1 + \beta y)^2)}{y(1 + \beta y)^2(Dae(y) + 1)^2},$$

$$\beta = \beta(y),$$

$$L_\sigma(y, v): \qquad y(y, v) = \frac{ve(y)Se(y - (1 + \beta y)^2)}{y(1 + \beta y)^2(e(y)vSe + 1)^2},$$

$$v = v(y), \tag{4.14}$$

where

$$a(y) = (1/(2y(1 + \beta y)^2) - (1/(2yy)) - 1).$$

If it is impossible to obtain the expressions for curves of neutrality in any plane of parameters in an explicit form, as above we can offer a procedure of graphical plotting for L_σ. Similarly, the series of parametric dependencies are built on one of the parameters while varying a second parameter. The points corresponding to the change of sign σ are plotted on a parametric plane. Example curves of neutrality on explicit expressions (4.14) are shown in Fig. 4.5.

Parametric portraits. The mutual arrangement of curves of multiplicity (L_Δ) and neutrality (L_σ) specifies the parametric portrait of a system. It defines different regions of parameters for which the number and type of stability of steady states differ. In our case, there are six such regions (see Fig. 4.6). The use of this specific feature of the task allows us to plot the corresponding bifurcation curves without special computational difficulties. Moreover, the explicit form of these dependencies, (4.11)-(4.12) and (4.14) gives us the opportunity to study the influence of third parameters on the appearance of curves L_Δ and L_σ (see Fig. 4.3–4.5).

Phase portraits. For the dynamic behavior of a system, it is convenient to investigate the phase portraits. In accordance with the parametric portrait of our system, we can distinguish six types of phase portraits.

Region 1 is characterized by a single stable steady state (low and high temperature). Regions 2 and 6 correspond to unstable steady states, which provide the exis-

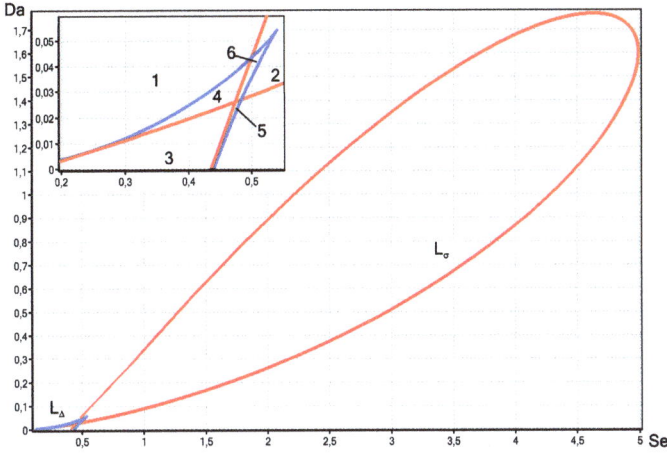

Fig. 4.6: Parametric portrait in (Se, Da) at $\beta = 0.15$, $\gamma = 0.01$

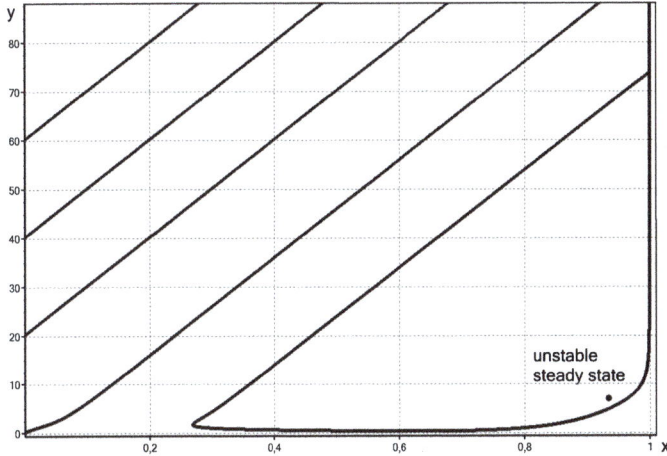

Fig. 4.7: Phase portrait in (x, y) at $Da = 0.08$, $Se = 0.6$, $\beta = 0.05$, $\gamma = 0.01$

tence of auto-oscillations in the system. These regions differ in that there is one steady state in region 2, but there are three steady states in the region 6. In regions 3, 4, and 5 there are also three steady states. Region 3 is characterized by one unstable and two stable low and high temperature steady states (Fig. 4.6). Regions 4 and 5 correspond to one stable and two unstable steady states. In Fig. 4.7 the phase portrait for an unstable steady state is shown.

Time dependencies. Numerical integration of the original dynamical system (4.2) allows us to obtain time dependencies $x(t)$ and $y(t)$ at different initial conditions and parameters. An example of the time dependence is shown in Fig. 4.8 for the case of one unstable steady state.

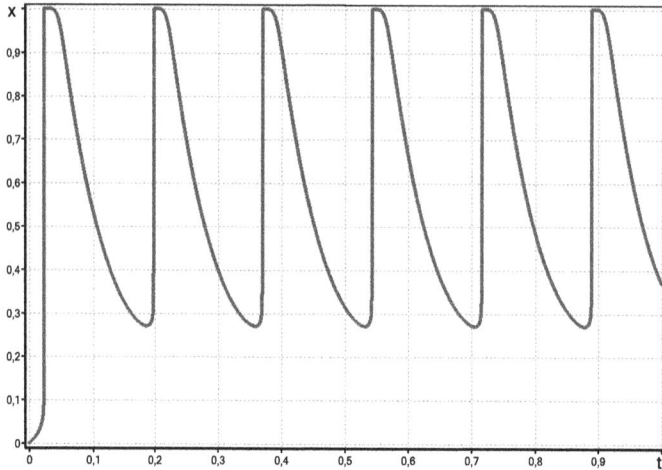

Fig. 4.8: Time dependence $x(t)$ at $Da = 0.08$, $Se = 0.6$, $\beta = 0.05$, $\gamma = 0.01$

Regions of technologically safe modes. Analysis of time dependencies and phase portraits illustrates that the stabilization of solutions to the low-temperature steady state may occur with significant dynamic cast (a sharp temperature increase at a specific point in time). This can be undesirable from a technological point of view. For other initial data and parameters of the system the stabilization of a steady state may have a monotonic character. On the corresponding parametric portrait this region can be highlighted and presented as the region of technologically safe modes.

4.2.2 The oxidation reaction A + O$_2$ → P

We now complicate the reaction scheme considered above. If earlier it was assumed that oxygen is in excess, now we let oxygen be one of the original substances. The reaction scheme is as follows

$$A + O_2 \rightarrow P ,$$

where A is the original substance, P is the combustion product and O_2 is an oxidation.
A mathematical model of the CSTR can be written in the form:

$$V\frac{dX_A}{dt} = -Vk(T)X_A X_{O_2} + q(X_A^0 - X_A) ,$$

$$V\frac{dX_{O_2}}{dt} = -Vk(T)X_A X_{O_2} + q(X_{O_2}^0 - X_{O_2}) , \tag{4.15}$$

$$C_p \rho V \frac{dT}{dt} = (-\Delta H_p)Vk(T)X_A X_{O_2} + q\rho C_p(T_0 - T) + hS(T_x - T) ,$$

where $k(T) = k^0 \exp(-E/(RT))$.

From the assumption of equality of input and output conditions:

$$X_A(0) - X_A^0 = X_{O_2}(0) - X_{O_2}^0 ,$$

$$q\rho C_p(T_0 - T(0)) = hS(T(0) - T_x) ,$$

model (4.15) is written in dimensionless form:

$$\frac{dx}{dt} = f(x)(1 - \alpha x)e(y) - \frac{x}{Da} = f_1(x, y) ,$$

$$y\frac{dy}{dt} = f(x)(1 - \alpha x)e(y) - \frac{y}{Se} = f_2(x, y) , \qquad (4.16)$$

where $f(x) = 1 - x$, $e(y) = \exp(y/(1 + \beta y))$, α is a dimensionless parameter characterizing the ratio of fuel/oxygen. Other dimensionless parameters are defined similarly to the basic model of Zel'dovich–Semenov.

The procedure of parametric analysis of a model consisting of two equations is described in detail for the previous reaction. Here we only write down the main results for (4.16) immediately in its final form.

Steady states are determined from the system

$$f(x)(1 - \alpha x)e(y) - \frac{x}{Da} = 0 ,$$

$$f(x)(1 - \alpha x)e(y) - \frac{y}{Se} = 0 . \qquad (4.17)$$

Expressing x and substituting into the second equation of system (4.17), we get an equation of steady state in the form of Semenov's diagram:

$$\left(1 - y\left(\frac{Da}{Se}\right)\right)\left(1 - \alpha y\left(\frac{Da}{Se}\right)\right)e(y) = \frac{y}{Se} . \qquad (4.18)$$

Here the functions of heat production and heat removal are determined as:

$$F(y) = (1 - vy)(1 - \alpha vy)e(y) ,$$

$$P(y) = \frac{y}{Se} .$$

As before the parameter v is $v = Da/Se$.

Types of stability of steady states. A study of stability of steady states is carried out according to the standard scheme. The elements of the Jacobian matrix for (4.16) have the following form:

$$a_{11} = \frac{\partial f_1}{\partial x} = (2\alpha x - \alpha - 1)e(y) - \frac{1}{Da} ,$$

$$a_{12} = \frac{\partial f_1}{\partial y} = \frac{(1 - x)(1 - \alpha x)e(y)}{(1 + \beta y)^2} ,$$

$$a_{21} = \frac{\partial f_2}{\partial x} = \frac{(2\alpha x - \alpha - 1)e(y)}{y} ,$$

$$a_{22} = \frac{\partial f_2}{\partial y} = \frac{(1 - x)(1 - \alpha x)e(y)}{y(1 + \beta y)^2} - \frac{1}{ySe} .$$

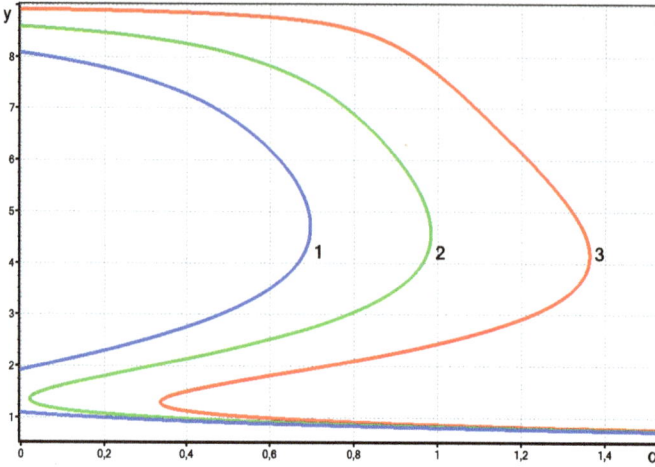

Fig. 4.9: Dependencies of the stationary temperature $y(\alpha)$ at $Da = 0.05$, $Se = 0.45$, 1) $\beta = 0.02$; 2) 0.05; 3) 0.07

Parametric dependencies. To plot a parametric dependence we rewrite equation (4.18) as follows:

$$G(y, \xi) = F(y, \xi) - P(y, \xi),$$

where $\xi = (\alpha, \beta, Da, Se)$ is a vector of parameters.

A linear dependence of $F(y)$ and $P(y)$ in (4.18) from the parameters allows us to obtain, as before, the explicit form of the dependencies of steady states from various parameters (or rather their inverse dependencies):

$$Da(y) = \frac{Se}{2y}\left(1 + \frac{1}{\alpha}\right) - \sqrt{\left(\frac{Se}{2y}(1 + 1/\alpha)\right)^2 - \frac{Se}{\alpha y}\left(\frac{Se}{y} - \frac{1}{e(y)}\right)},$$

$$Se(y) = \frac{Day}{2}(1 + \alpha) + \frac{y}{2e(y)} + \sqrt{\left(\frac{Day}{2}(1 + \alpha) + \frac{y}{2e(y)}\right)^2 - \alpha(Day)^2},$$

$$\beta(y) = \frac{1}{\ln\left(Sey/((Se - Day)(Se - \alpha Day))\right)} - \frac{1}{y},$$

$$v(y) = \frac{1}{2y}(1 + 1/\alpha) - \sqrt{\left(\frac{1}{2y}(1 + 1/\alpha)\right)^2 - \frac{1}{\alpha y}\left(\frac{1}{y} - \frac{1}{e(y)Se}\right)},$$

$$\alpha(y) = \frac{Se}{Da}\left(\frac{1}{y} - \frac{1}{e(y)(Se - Day)}\right). \tag{4.19}$$

The explicit form of the obtained dependencies allows us to analyze the influence of other parameters on the stationary characteristics. In particular, the dependence of the stationary temperature on the parameter α is presented in Fig. 4.9.

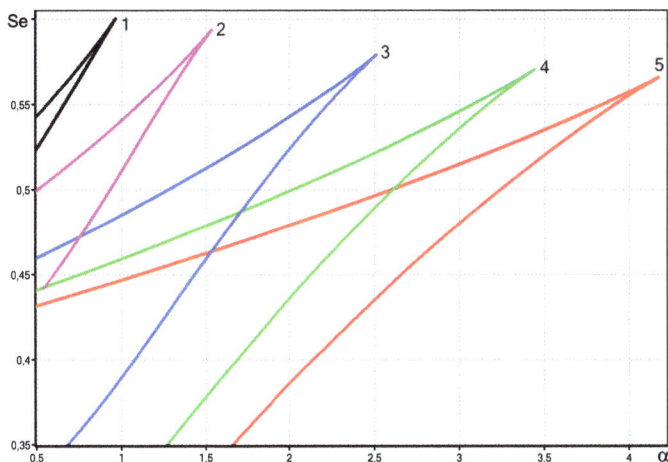

Fig. 4.10: Curves of multiplicity $L_\Delta(\alpha, Se)$ at $\beta = 0.05$, 1) $Da = 0.025$; 2) 0.03; 3) 0.04; 4) 0.06; 5) 0.08

Curves of multiplicity. The curve L_Δ is obtained from the system:

$$G(y, \xi) = 0 \,,$$
$$\Delta(y, \xi) = 0 \,, \tag{4.20}$$

where Δ, as before, is determined by coefficients of the Jacobian matrix. System (4.20) is represented as:

$$L_\Delta: \quad (1 - vy)(1 - \alpha vy)e(y) - \frac{y}{Se} = 0 \,,$$

$$(1 - vy)(1 - \alpha vy)\frac{e(y)}{(1 + \beta y)^2} + (2\alpha vy - \alpha - 1)ve(y) - \frac{1}{Se} = 0 \,.$$

Using an explicit expression of the parametric dependencies (4.19), we write the equation of the curve of multiplicity of steady states in the plane, for example, (Se, α):

$$L_\Delta(Se, \alpha): \quad Se(y) = Day + \frac{e(y)}{2bc} + \sqrt{\left(Day + \frac{e(y)}{2bc}\right)^2 - Da(Day^2 + c(1 + y/b))} \,,$$

$$\alpha = \alpha(y, Se) \,, \tag{4.21}$$

where $b = (1 + \beta y)^2$, $c = y^2/e(y)$. In other planes of parameters the explicit formulas of the type (4.21) become too bulky, so we do not give them here. An example of plotting multiplicity curves (4.21) while varying the third parameter is given in Fig. 4.10.

Direct calculations of bulky expressions can be avoided by applying the above graphic procedure of plotting multiplicity curves by known parametric dependencies.

Neutrality curves. To plot the neutrality curve it is necessary to solve the following system of equations:

$$G(y, \xi) = 0 \,,$$
$$\sigma(y, \xi) = 0 \,,$$

which for the studied model (4.16) takes the form:

$$L_\sigma: \quad (1 - vy)(1 - \alpha vy)e(y) - \frac{y}{Se} = 0 \,,$$

$$(2\alpha vy - \alpha - 1)e(y) + (1 - vy)(1 - \alpha vy)\frac{e(y)}{y(1 + \beta y)^2} - \frac{1}{Da} - \frac{1}{ySe} = 0 \,.$$

Substituting $\alpha(y)$ from the first equation of the system into the second one, we obtain the expressions for the curves of neutrality corresponding to different combinations of dimensionless parameters:

$$L_\sigma(y, Da): \quad Da(y) = \frac{Se}{2y}(1 + 1/\alpha) - \sqrt{\left(\frac{Se}{2y}(1 + 1/\alpha)\right)^2 - \frac{Se}{\alpha y}\left(\frac{Se}{y} - \frac{1}{e(y)}\right)} \,,$$

$$y = y(y) \,,$$

$$L_\sigma(y, Se): \quad Se(y) = \frac{Day}{2}(1 + \alpha) + \frac{y}{2e(y)} + \sqrt{\left(\frac{Day}{2}(1 + \alpha) + \frac{y}{2e(y)}\right)^2 - \alpha(Day)^2} \,,$$

$$y = y(y) \,,$$

$$L_\sigma(y, \beta): \quad \beta(y) = \frac{1}{\ln\left(Sey/((Se - Day)(Se - \alpha Day))\right)} - \frac{1}{y} \,,$$

$$y = y(y) \,,$$

$$L_\sigma(y, \alpha): \quad \alpha(y) = \frac{Se}{Da}\left(\frac{1}{y} - \frac{1}{e(y)(Se - Day)}\right)$$

$$y = y(y) \,,$$

$$\tag{4.22}$$

where

$$y(y) = \frac{Day(Se - Day)(y - A)}{A\, Se(Day^2(Dae(y) + 1) + e(y)Se(Se - 2Day))} \,, \quad A = (1 + \beta y)^2 \,.$$

By varying the parameter α corresponding to a share of oxygen content in the reaction mixture, it is possible to observe changes in the neutrality curve (see Fig. 4.11). As for the multiplicity curve, in other planes of parameters the explicit formulas of the type (4.22) become too bulky, so we do not show the results here.

Parametric portraits. Typical parametric portraits (the curves L_Δ, L_σ) are shown in Fig. 4.12. The curve L_Δ (the region inside the dotted vertical lines) bounds the region of the three steady states. The curve L_σ determines the change of stability of steady state. It is easy to highlight the region with the only unstable steady state, which specifies the presence of auto-oscillations in system (4.16).

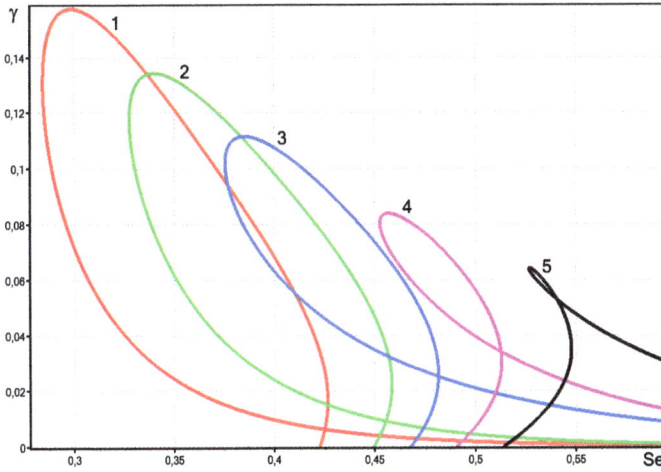

Fig. 4.11: Curves of neutrality $L_\sigma(Se, \gamma)$ at $\beta = 0.01$, $Da = 0.05$, 1) $\alpha = 0$; 2) 0.6; 3) 1; 4) 1.5; 5) 2

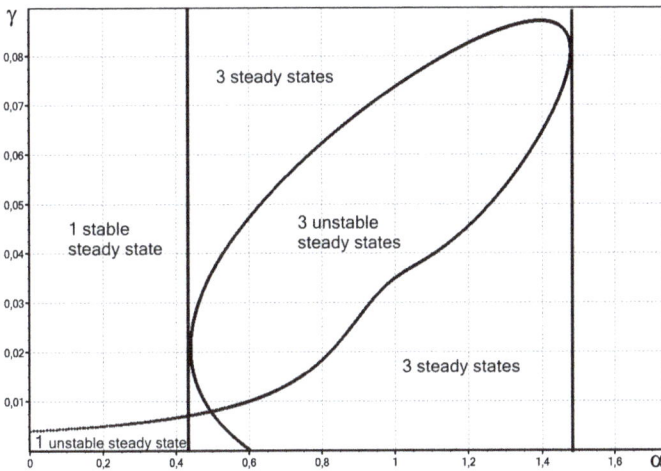

Fig. 4.12: Parametric portrait (α, γ) at $\beta = 0.01$, $Da = 0.05$, $Se = 0.45$

4.2.3 Reaction nA → P

The Zel'dovich–Semenov model for a reaction of the n-th order has the form:

$$\frac{dx}{dt} = (1 - x)^n e(y) - \frac{x}{Da} = f_1(x, y) ,$$

$$\gamma \frac{dy}{dt} = (1 - x)^n e(y) - \frac{y}{Se} = f_2(x, y) . \tag{4.23}$$

Steady states are determined from the system

$$(1 - x)^n e(y) - \frac{x}{Da} = 0 ,$$

$$(1 - x)^n e(y) - \frac{y}{Se} = 0 ,$$

which can be simplified to the one equation after transformations:

$$(1 - vy)^n e(y) - \frac{y}{Se} = 0 , \qquad (4.24)$$

where $v = Da/Se$. The functions of heat production and heat removal are defined as

$$F(y) = (1 - vy)^n e(y) ,$$

$$P(y) = \frac{y}{Se} .$$

Parametric dependencies. We write down the dependencies of steady states from the parameters of system from the equations (4.24):

$$Da(y) = \frac{Se}{y} \left(1 - \left(\frac{y}{e(y)Se} \right)^{1/n} \right) ,$$

$$v(y) = \frac{1}{y} \left(1 - \left(\frac{y}{e(y)Se} \right)^{1/n} \right) ,$$

$$\beta(y) = \frac{1}{\ln (y/(Se(1 - vy)^n))} - \frac{1}{y} . \qquad (4.25)$$

Assuming that $v = \text{const}$, $Se(y)$ can be expressed as:

$$Se(y) = y/((1 - vy)^n e(y)) .$$

Examples of plotting the parametric dependencies of the stationary temperature on the parameters with variation of the reaction order n are shown in Fig. 4.13.

The elements of the Jacobian matrix for system (4.23) have the form:

$$a_{11} = -n(1 - x)^{n-1} e(y) - \frac{1}{Da} , \qquad a_{12} = \frac{(1 - x)^n e(y)}{(1 + \beta y)^2} ,$$

$$a_{21} = \frac{-n(1 - x)^{n-1} e(y)}{y} , \qquad a_{22} = \frac{(1 - x)^n e(y)}{y(1 + \beta y)^2} - \frac{1}{ySe} .$$

Curves of multiplicity can be sought, as above, from the stationarity equation and the condition of equality to zero of Δ.

$$L_\Delta : \qquad (1 - vy)^n e(y) - \frac{y}{Se} = 0 ;$$

$$\frac{ne(y)(1 - vy)^{n-1}}{ySe} - \frac{e(y)(1 - vy)^n}{yDa(1 + \beta y)^2} + \frac{1}{yDaSe} = 0 .$$

The expressions for the multiplicity curves in different planes of parameters can be written in explicit form:

$$L_\Delta(Da, Se): \qquad Da = Da(y, Se(y)),$$

$$Se(y) = \frac{y}{e(y)}\left(1 - \frac{1}{n} + \frac{y}{n(1 + \beta y)^2}\right)^n,$$

$$L_\Delta(Se, v): \qquad v = v(y, Se(y)),$$

$$Se(y) = \frac{y}{e(y)}\left(1 - \frac{1}{n} + \frac{y}{n(1 + \beta y)^2}\right)^n. \qquad (4.26)$$

Examples of plotting the dependencies (4.26) while varying the third parameter are given in Fig. 4.14. Here variation of the parameter n is shown.

Curves of neutrality can be expressed using the stationarity equation and the condition of equality to zero of σ.

$$L_\sigma: \qquad (1 - vy)^n e(y) - \frac{y}{Se} = 0;$$

$$- n(1 - vy)^{n-1} e(y) + \frac{(1 - vy)^n e(y)}{y(1 + \beta y)^2} - \frac{1}{Da} - \frac{1}{ySe} = 0. \qquad (4.27)$$

Substituting $Se(y)$ (4.25) from the first equation (when $v = \mathrm{const}$) into the second one lets us express $y(y)$:

$$L_\sigma(y, Se): \qquad Se(y) = \frac{y}{(1 - vy)^n e(y)}, \qquad y(y) = y(y, Se(y)),$$

$$L_\sigma(y, Da): \qquad Da(y) = \frac{Se}{y}\left(1 - \frac{y}{e(y)Se}\right)^{\frac{1}{n}}, \qquad y(y) = y(y, Da(y)),$$

$$L_\sigma(y, \beta): \qquad \beta(y) = \frac{1}{\ln\left(y/(Se(1 - vy)^n)\right)} - \frac{1}{y}, \qquad y(y) = y(y, \beta(y)),$$

$$L_\sigma(y, v): \qquad v(y) = \frac{1}{y}\left(1 - \left(\frac{y}{e(y)Se}\right)^{\frac{1}{n}}\right), \qquad y(y) = y(y, v(y)), \qquad (4.28)$$

where

$$y(y) = \frac{y/(1 + \beta y)^2 - 1}{\frac{y}{(1 - y/(Se(y)))^{1/n}} + \frac{ny}{(y/(Se(y)))^{1/n}}}.$$

Substituting $Da(y, Se)$ into the second equation of system (4.27) we receive $Se(y)$:

$$L_\sigma(Da, Se): \qquad Da = Da(y, Se),$$

$$Se(y) = \frac{y(2b)^n}{e(y)\left(b + n - 1 \pm \sqrt{(b + n - 1)^2 - 4bn}\right)^n},$$

where

$$b = (1/y)\left(1/(1 + \beta y)^2 - 1/y\right).$$

Examples of plotting some dependencies (4.28) are shown in Fig. 4.15 while varying the third parameter n.

Parametric portraits. An example of a parametric portrait for the parameter plane (y, β) is given in Fig. 4.16.

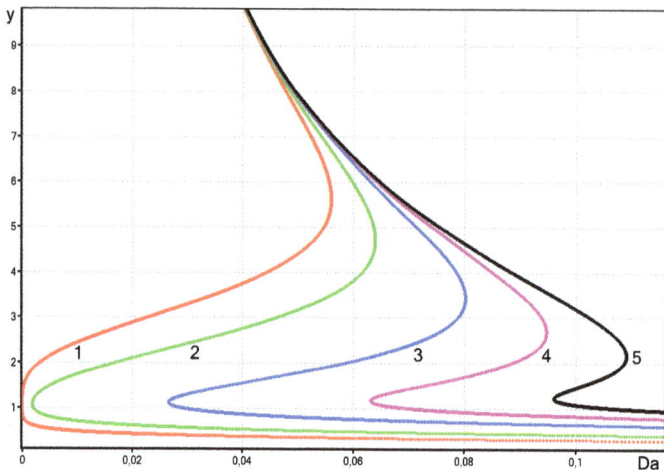

Fig. 4.13: Dependencies of dimensionless temperature from Da at $Se = 0.4$, $\beta = 0.01$, 1) $n = 0.3$; 2) 0.5; 3) 1; 4) 1.5; 5) 2

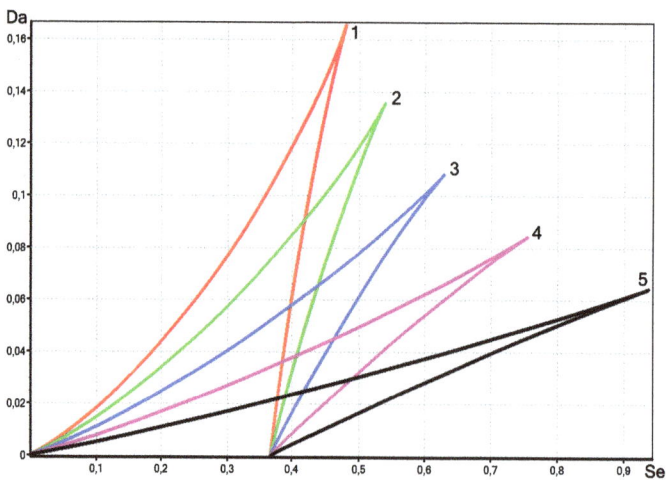

Fig. 4.14: Curves of multiplicity $L_\Delta(Se, Da)$ at $\beta = 0$, 1) $n = 0.5$; 2) 1; 3) 2; 4) 4; 5) 8

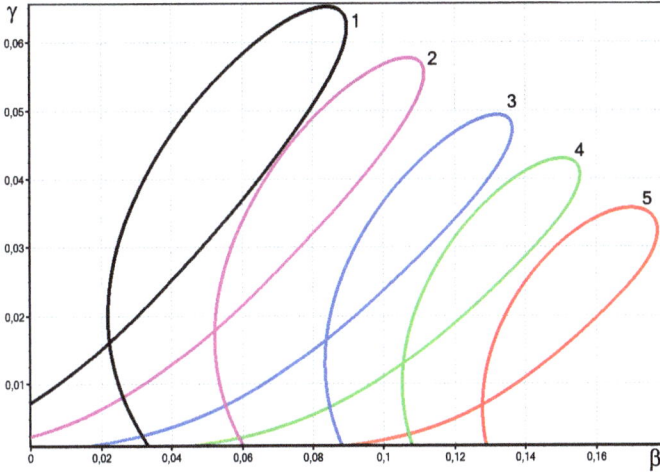

Fig. 4.15: Curves of neutrality $L_\sigma(\beta, \gamma)$ at $Da = 0.05$, $Se = 0.45$, 1) $n = 0.3$; 2) 0.5; 3) 0.7; 4) 1; 5) 1.3

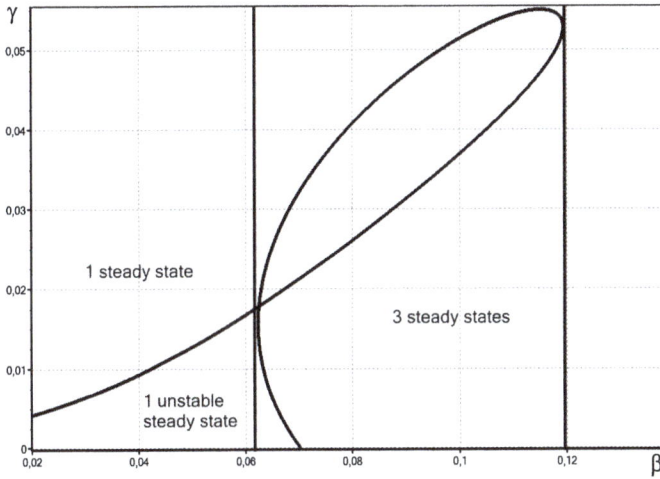

Fig. 4.16: Parametric portrait (β, γ) at $Da = 0.05$, $Se = 0.45$, $n = 0.9$

4.2.4 Reaction A → P with arbitrary kinetics

A dimensionless model for arbitrary kinetics is as follows:

$$\frac{dx}{dt} = f(x)e(y) - \frac{x}{Da} = f_1(x, y),$$

$$\gamma\frac{dy}{dt} = f(x)e(y) - \frac{y}{Se} = f_2(x, y), \tag{4.29}$$

where the kinetic function $f(x)$ can be rather arbitrary. For example, for an oxidation reaction

$$nA + mO_2 \rightarrow P$$

this function is

$$f(x) = (1 - x)^n(1 - \alpha x)^m$$

Steady states are determined by the system

$$f(x)e(y) - \frac{x}{Da} = 0,$$

$$f(x)e(y) - \frac{y}{Se} = 0.$$

The stationarity equation has the form:

$$f(x)e(y) - \frac{y}{Se} = 0. \tag{4.30}$$

The functions of heat production and heat removal are:

$$F(y) = f(x)e(y),$$

$$P(y) = \frac{y}{Se}.$$

Coefficients of the characteristic equation. As above, $\sigma = a_{11} + a_{22}$, $\Delta = a_{11}a_{22} - a_{12}a_{21}$, where

$$a_{11} = f'(x)e(y) - \frac{1}{Da}, \qquad a_{12} = \frac{e(y)}{(1 + \beta y)^2}f(x),$$

$$a_{21} = \frac{e(y)}{\gamma}f'(x), \qquad a_{22} = \frac{e(y)}{\gamma(1 + \beta y)^2}f(x) - \frac{1}{\gamma Se}.$$

Parametric dependencies. For an arbitrary kinetic function of the form $f(x)$ we express the dependence of the stationary temperature on the parameter β from equation (4.30):

$$\beta(y) = \frac{1}{\ln(y/(Sef(x)))} - \frac{1}{y}.$$

Examples of plotting the dependencies of stationary temperature on parameter β for the oxidation reaction of the n-th order are shown in Fig. 4.17.

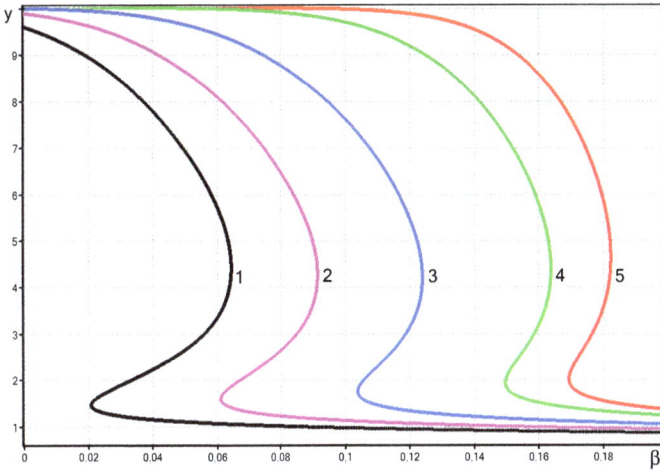

Fig. 4.17: Dependencies of stationary temperature on parameter β at $Da = 0.05$, $Se = 0.5$, $\alpha = 0.2$,
1) $n = 0.3$; 2) 0.5; 3) 1; 4) 1.5; 5) 2

Curves of multiplicity for different parameter planes can be expressed

$$L_\Delta: \qquad f(x)e(y) - \frac{y}{Se} = 0 ;$$

$$\frac{1}{\gamma DaSe} - \frac{e(y)f'(x)}{\gamma Se} - \frac{e(y)f(x)}{Day(1 + \beta y)^2} = 0 .$$

In the plane (y, β) the curve of multiplicity is, as before, two parallel lines for any $f(x)$.
    Curves of neutrality. To plot the neutrality curve it is necessary to solve system
of equations:

$$L_\sigma: \qquad f(x)e(y) - \frac{y}{Se} = 0 ;$$

$$f'(x)e(y) + \frac{f(x)e(y)}{y(1 + \beta y)^2} - \frac{1}{Da} - \frac{1}{\gamma Se} = 0 . \qquad (4.31)$$

Let us substitute $\beta(y)$ into the second equation of system (4.31) and express $y(y)$:

$$L_\sigma: \qquad \beta(y) = \frac{1}{\ln\left(y/(Sef(x))\right)} - \frac{1}{y} ,$$

$$y(y, \beta) = \frac{Da\left(e(y)Sef(x) - (1 + \beta y)^2\right)}{Se(1 + \beta y)^2(1 - Dae(y)f'(x))} .$$

Examples of calculations are given in Fig. 4.18 for a kinetic function of the form $f(x) = (1 - x)^n(1 - \alpha x)$.
    Thus, consistent application of the procedure of parametric analysis to the classical model of the Zel'dovich–Semenov allowed us to write down explicit expressions

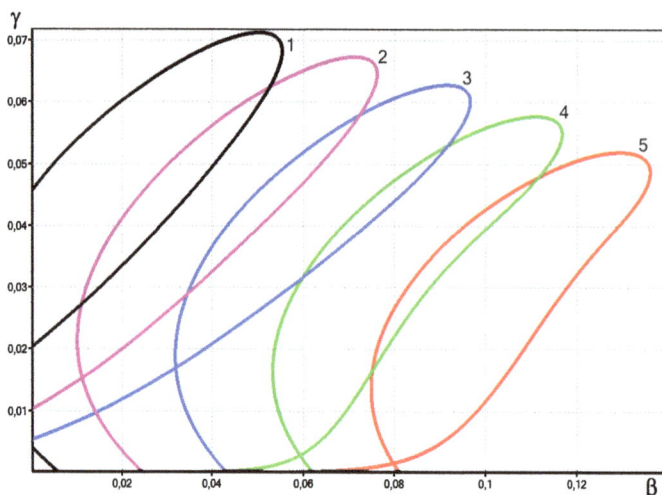

Fig. 4.18: Curves of neutrality $L_\sigma(\beta, \gamma)$ at $Da = 0.05$, $Se = 0.45$, $n = 0.2$, 1) $\alpha = 1.4$; 2) 1.2; 3) 1; 4) 0.8; 5) 0.6

for curves of local bifurcations of the steady states for various combinations of dimensionless parameters and investigate the influence of other parameters on them. The obtained results give a rather complete description of the features of one of the basic models of the theory of thermal explosion. They have methodical value and are also useful from a practical point of view. The presence of explicit expressions for the bifurcation curves in the plane of dimensionless parameters allows us to plot these curves in the planes of the dimensional parameters, which correspond to the specific geometry and thermophysical characteristics of real exothermic processes occurring in the CSTR.

Comparative analysis of the plotted parametric portraits for different kinetic functions shows that the form of the kinetics significantly influences the critical conditions of combustion, multiplicity of steady states, and the existence of auto-oscillations. The obtained results give a quantitative estimate of this influence.

4.2.5 Semenov diagram as a stability criterion

The Semenov diagram is widely used in engineering practice for estimation of ignition conditions. In 1978 in Alma-Ata on the symposium on combustion, academician Ya.B. Zel'dovich said with enthusiasm in his bright speech that the analysis of the mutual arrangement of the curves of heat production and heat removal in the simplest case of a reaction of the zero order allows us to make conclusions about the stability of the only steady state. If there are three such states, then the one in the middle is unstable and the extreme steady states are stable. However, in the general case (when there are two or more degrees of freedom in the system, for example, a temperature and a concentration of reactants) this simple criterion does not work [31].

Semenov's diagram for (4.29) as the equality of curves of heat production (F) and heat removal (P) for a single reaction with arbitrary kinetic function $f(x)$ has the form:

$$f(vy)e(y) = \frac{y}{Se} , \qquad (4.32)$$

where $e(y) = \exp(y/(1 + \beta y))$, $v = Da/Se$.

The ratio (4.32) specifies steady states of the original dynamical system (4.29), but it is not enough for the analysis of their stability in the typical case. Studies of the model show that this model can have a single unstable steady state, or all three steady states can be unstable. There are other options.

The stability of steady states for (4.29) is determined by the roots of the characteristic equation

$$\lambda^2 - \sigma\lambda + \Delta = 0 ,$$

where $\sigma = a_{11} + a_{22}$, $\Delta = a_{11}a_{22} - a_{12}a_{21}$,

$$a_{11} = f_x'(x)e(y) - \frac{1}{Da} , \qquad a_{12} = \frac{e(y)}{(1 + \beta y)^2}f(x) ,$$

$$a_{21} = \frac{e(y)}{\gamma}f_x'(x) , \qquad a_{22} = \frac{e(y)}{\gamma(1 + \beta y)^2}f(x) - \frac{1}{\gamma Se} .$$

It is easy to show that taking into account the conditions of stationarity (4.32)

$$\Delta = a_{22}(F_y' - P_y') , \qquad (4.33)$$

$$\sigma = -\frac{1}{Da} - \Delta Da + \left(1 - \frac{Da}{\gamma Se}\right)f_x'(vy)e(y) , \qquad (4.34)$$

where F_y', P_y' are the derivatives of the functions of heat production and removal, respectively.

The expression (4.33) is a consequence of the more general case. Let the original dynamic model have the form:

$$\dot{x}_1 = f_1(x_1, x_2) , \qquad (4.35)$$

$$\dot{x}_2 = f_2(x_1, x_2) . \qquad (4.36)$$

The steady states for (4.35), (4.36) are determined from the system:

$$f_1(x_1, x_2) = 0 , \qquad (4.37)$$

$$f_2(x_1, x_2) = 0 . \qquad (4.38)$$

If the variable x_2 is expressed from the second equation through x_1, i.e., $x_2 = x_2(x_1)$, we have from the identity $f_2(x_1, x_2(x_1)) \equiv 0$:

$$x_{21}' = -\frac{f_{21}'}{f_{22}'} , \qquad (4.39)$$

where the sign $'$ means differentiation by a variable, the number of which is determined by the second digit of the index. After substituting (4.39) into the first equation (4.37) we obtain the analogue of the Semenov's diagram (4.32), i.e.,

$$f_1(x_1, x_2(x_1)) = 0 .\tag{4.40}$$

Differentiation of (4.40), taking into account (4.39), gives:

$$\Delta = f_{22}' \frac{df_1}{dx_1} ,\tag{4.41}$$

where $\Delta = f_{11}'f_{22}' - f_{12}'f_{21}'$ on the solutions (4.37), (4.38).

Thus, the graphic definition of the steady states from the Semenov diagram (4.32) gives a sufficient condition for instability: if in the steady state $F_y' > P_y'$, then at $a_{22} < 0$ (it usually holds for kinetics with positive local order) we have $\Delta < 0$, which determines the instability of the steady state. When $\Delta > 0$ the stability is determined by the sign of σ.

The expression (4.34) specifies the relationship between the characteristic of the Semenov diagram (a value of Δ in the steady state) and the value of σ. So, when $\Delta \geq 0$ the sign of σ is defined (when $f_x' < 0$) by the ratio of the parameters Da, Se, y. If

$$Da \leq ySe ,\tag{4.42}$$

and $\Delta > 0$ than $\sigma < 0$, i.e., a steady state is stable. The inequality (4.42) is a sufficient condition that the Semenov diagram can be used as a stability criterion of a steady state. The condition (4.42) can be weakened considerably, if we introduce into consideration a stationary value of y. When $\Delta = 0$ directly from (4.34) we have

$$e(y) < \frac{1}{-f_x'(vy)} Da \left(\frac{Da}{ySe} - 1 \right) .\tag{4.43}$$

For the reaction of the first order ($f(x) = 1 - x$) the expression (4.43) becomes easier:

$$e(y) < \frac{1}{Da \left(\frac{Da}{ySe} - 1 \right)} .\tag{4.44}$$

If there are three steady states ($y_* < y^0 < y^*$), from the Semenov diagram and (4.33) it follows that the middle steady state y^0 is unstable. The extreme steady states y_*, y^* are stable under the execution of (4.42) or (4.44). Note that (4.44) can be rewritten in the form

$$y^* < g(Da, Se, y, \beta) ,\tag{4.45}$$

where the right side of the inequality depends only on the system parameters. Inequality (4.45) shows that the Semenov diagram can be used as a stability criterion with higher probability for relatively small values of a stationary temperature.

The inequalities (4.44)–(4.45) are only sufficient conditions. The exact boundaries of a multiplicity of steady states and their stability are defined by the corresponding bifurcation curves L_Δ and L_σ (the curves of multiplicity and neutrality of steady states).

As shown above, they can be written out explicitly for model (4.2) in almost all combinations of planes of the dimensionless parameters. For example, for the reaction $A + O_2 \rightarrow P$ (the kinetic function $f(x) = (1 - x)(1 - \alpha x)$, where α is a ratio of fuel/oxygen) we have the formulas (4.21), (4.22) for L_Δ and L_σ. Similar expressions can be written for arbitrary kinetic functions $f(x)$.

A detailed parametric analysis of model (4.2) shows that it can have over 38 different phase portraits [6, 23, 24, 27]. However, our goal is not to enumerate all the variety of the Zel'dovich–Semenov model properties.

The main conclusion is that the answer to the question "Can the Semenov diagram be accepted as a stability criterion for the steady states?" is positive, but when executing inequalities, for example, (4.42) or (4.43)–(4.45). With appropriate condition, which, of course, must be previously checked, the Semenov diagram gives the engineer a clear stability criterion of a steady state. If there is one steady state then it is stable. If there are three steady states, the middle steady state is stable and the extreme steady states are unstable. If inequalities of the type (4.42)–(4.45) are not executed, then there is no guarantee that the steady states for which $F' < P'$ are stable. The presence of a region with a single and unstable steady state is another proof.

The task of graphical determination of the stability of steady states by the Semenov diagram can be formulated for complex kinetics. In this case, a heat balance equation similar to (4.1) is written as:

$$\dot{T} = W(T, \mathbf{x}) + \alpha(T_0 - T) , \qquad (4.46)$$

where T is a temperature, $\mathbf{x}$ is a vector of reagent concentrations, $W(T, \mathbf{x})$ is a function of heat production, and $\alpha(T_0 - T)$ characterizes heat removal. Similarly to (4.1), the kinetic subsystem is specified in accordance with a complex mechanism of reaction and mass transfer conditions

$$\dot{\mathbf{x}} = \sum_{s=1}^{m} \gamma_s w_s(T, \mathbf{x}) + v(\mathbf{x}^0 - \mathbf{x}) , \qquad (4.47)$$

where γ_s is a stoichiometric vector and w_s is a rate of the s-th reaction. The term $v(\mathbf{x}^0 - \mathbf{x})$ characterizes the exchange of substances with the environment.

If we can express $\mathbf{x} = \mathbf{x}(T)$ from the stationarity conditions, then we will obtain the usual Semenov diagram from (4.47):

$$W(T, \mathbf{x}(T)) = \alpha(T - T_0) . \qquad (4.48)$$

The establishment of a relationship between the characteristics (4.48) and the conditions for uniqueness and stability of a steady state is the subject of further consideration.

The mathematical model of Zel'dovich–Semenov is the basic model of combustion theory. Like a drop of water, it contains a reflection of the large variety of dynamic properties of the whole "ocean" of thermokinetic models. Naturally, the variety of properties of systems of type (4.46)–(4.47) is much wider than the properties of (4.2), (4.29). It is enough to say that the considered models are characterized by a dimension of phase space which is equal to two. In the study of models with three-dimensional phase space a fundamentally new property of the dynamics appears. That is deterministic chaos. Moreover, when moving to distributed systems dissipative structures and nonstationary spatial structures arise. However, knowledge of the main nonlinear and nonstationary properties of the base models of the theory of combustion is necessary in more complex situations as well. It seems to us that the obtained results can serve as a first approximation in the analysis of more complex models of specific processes. Despite the fact that parametric analysis is far from exhaustive, the plotted parametric portraits can be useful from the general methodical point of view and from the point of view of applications. As our experience shows, knowledge of the critical conditions in the specifics of dimensionless parameters allows us to simply estimate the values of specific thermophysical and kinetic parameters of the real process, at which a multiplicity of steady states, hysteresis, and auto-oscillation regimes can arise.

4.3 Aris–Amundson model

The mathematical model of Aris–Amundson is one of the basic models of the theoretical bases of chemical technology [5, 9–15]. In the literature models of this type are called CSTR (continuous stirred-tank reactor) models. Given the significant contribution to the theory of chemical reactors of R. Aris and N. Amundson, we propose calling the corresponding base model the *Aris–Amundson model*.

The models which will be considered in this section are in a sense similar to the Zel'dovich–Semenov model. Their parametric analysis is carried out according to the scheme shown above. Therefore we mainly represent here the results of parametric analysis: dependencies of steady states on parameters, the basic bifurcation curves, parametric portraits, typical phase portraits, etc. The plotted parametric portraits are not only illustrations of the implementation of the general approach of a parametric analysis but also the specific results of the studied models. They contain the main quantitative characteristics of the conditions of occurrence of critical phenomena in those processes for which a given class of models is adequate.

4.3.1 Reaction A → P

Following R. Aris and N. Amundson, we will do another nondimensionalization of the CSTR (4.1) and will carry out the parametric analysis in accordance with the above scheme.

Model and parameters. For one chemical reaction of the first order: A → P, we introduce the following dimensionless parameters and variables [2]:

$$Da = \frac{V}{q}k(T_0), \quad \gamma = \frac{E}{RT_0}, \quad \beta = \frac{(-\Delta H)X^0}{\rho C_p T_0}, \quad y^* = \frac{T_x}{T_0},$$

$$x = \frac{X^0 - X}{X^0}, \quad s = 1 + \frac{hS}{q\rho C_p}, \quad \tau = \frac{q}{V}t, \quad y = \frac{T}{T_0}.$$

Let us write system (4.1) in a dimensionless form:

$$\frac{dx}{d\tau} = f(y)(1 - x) - x = f_1(x, y),$$

$$\frac{dy}{d\tau} = \beta f(y)(1 - x) - s(y - 1) = f_2(x, y), \qquad (4.49)$$

where $f(y) = Da \exp(y/(1 - 1/y))$. For model (4.1) it is usually assumed that the initial and input conditions for the reactor are the same:

$$X(0) = X^0, \qquad q\rho C_p(T_0 - T(0)) = hS(T(0) - T_x).$$

When $T_x = T_0$ for the dimensionless model (4.49) we have $y^* = 1$ and therefore $y(0) = 1$. In a system of two differential equations (4.49) the variables x, y are the phase variables, while β, Da, γ, s are the parameters of the system. Note that in the theory of combustion D. Frank-Kamenetsky [30] suggested another system of introduction of dimensionless values. B. Voltaire [32–34] introduced his own system of nondimensionalization in the theory of polymerization reactors.

The use of modern mathematical software tools allows us to move deeply into the analysis of the whole variety of nonlinear dynamic properties of model (4.49). Describing the state of the problem of parametric analysis of nonlinear dynamic models of the type (4.49), we highlight two directions. In works [23–25] a rather complex technique of mathematical analysis of the Lyapunov coefficients, global bifurcations and other characteristics of a model are considered for studying it in depth. These techniques can be referred to as the first direction. In this case, corresponding bifurcation curves are often shown schematically, because the restructuring of phase portraits occurs when changing parameters with five to seven significant digits. Such narrow regions of changes of parameters are of interest only in mathematical terms. They are not significant from an engineering point of view. The second direction of works on parametric analysis can be characterized as a study "in breadth" [35–54]. In these works the most important phase portraits are highlighted, and the basic bifurcation curves in different planes of parameters are plotted. All technology of parametric analysis can be represented in the form of an information system, which aims to create and store a database of models and the results of their parametric analysis.

As above, the goal of our study is to find the region of parameters of model (4.49) where there are auto-oscillations. Further, we need to investigate the effect of different parameters on the characteristics of these auto-oscillations and to identify the structure of the partition of the space of parameters β, Da, γ, s in the regions that correspond to qualitatively different types of dynamic behaviour and to specify the phase portrait for each region.

Steady states of system (4.49) are the solutions of the stationarity system

$$f_1(x, y) = 0 ,$$

$$f_2(x, y) = 0 . \tag{4.50}$$

After the exclusion of x the system can be written in the form

$$F(y) = P(y) , \tag{4.51}$$

where, as above, $F(y)$ has the meaning of the heat production function, and $P(y)$ is the heat removal function in the stationary conditions:

$$F(y) = \frac{\beta f(y)}{1 + f(y)} ,$$

$$P(y) = s(y - 1) .$$

We denote

$$F(\infty) = F_\infty = \frac{\beta f(\infty)}{1 + f(\infty)} , \qquad f(\infty) = Da \exp(y) .$$

The solutions of (4.49) will be considered in the polyhedron of reaction

$$D = \{x, y: \ 0 \le x \le 1, \quad 1 \le y \le y^\infty\} ,$$

where $y^\infty = 1 + (1/s)F_\infty$ at $y^* = 1$. If $(x(0), y(0)) \in D$, then for all $\tau > 0$ the solutions of system (4.49) are $(x(\tau), y(\tau)) \in D$. This guarantees the existence in D of at least one steady state. We find the steady-state value of y by solving (4.51). The value of x in the steady state is found according to (4.50) with the formula

$$x = \frac{s}{\beta}(y - 1) ,$$

or the same

$$x = \frac{f(y)}{1 + f(y)} .$$

The system of equations (4.49) allows one or more steady states, which can be graphically defined on the van Heerden diagram (the analogue of Semenov diagram in the theory of combustion) [2]. Figure 4.19 presents the corresponding graphs for the the the one low-temperature and high-temperature states (lines 1 and 3) and for three steady states (line 2).

A necessary and sufficient condition for the existence of the singular steady state has the form:

$$s = P' > F' = \beta \frac{f'(y)}{(1 + f(y))^2} ,$$

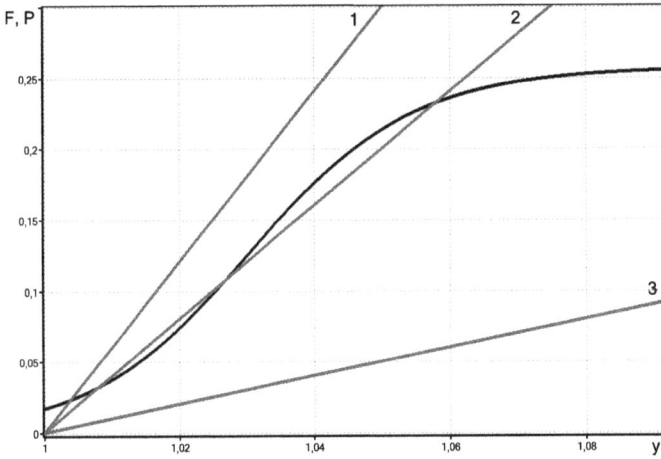

Fig. 4.19: The van Heerden diagram at $Da = 0.07$, $\beta = 0.25705$, $\gamma = 89$ and 1) $s = 1$; 2) $s = 4$; 3) $s = 6$

i.e., if the rate of heat removal in the steady state is more than the rate of heat production, this provides the existence of the singular steady state.

Stability conditions of steady states. Let us repeat that the type of stability of steady states is determined by the roots of the characteristic equation:

$$\lambda^2 - \sigma\lambda + \Delta = 0 , \tag{4.52}$$

where the coefficients σ and Δ are the elements of the Jacobian matrix of the right sides of system (4.49) in the steady state. Let us write down these matrix elements for (4.49):

$$a_{11} = \frac{\partial f_1}{\partial x} = -1 - f(y) ,$$

$$a_{12} = \frac{\partial f_1}{\partial y} = \frac{yf(y)(1-x)}{y^2} ,$$

$$a_{21} = \frac{\partial f_2}{\partial x} = -\beta f(y) ,$$

$$a_{22} = \frac{\partial f_2}{\partial y} = \frac{\beta yf(y)(1-x)}{y^2} - s .$$

The steady state being studied is unstable, if at least one of the inequalities, $\sigma < 0$ or $\Delta > 0$, does not perform. When $\Delta < 0$ the stationary point is a saddle (unstable point). When $\Delta > 0$ and $\sigma > 0$ the stationary point is an unstable node or unstable focus. If the system parameters are such that the steady state is singular and unstable, then there are auto-oscillations in the region D.

The solutions of the characteristic equation (4.52) have the form

$$\lambda_{1,2} = \frac{\sigma}{2}\left(1 \pm \sqrt{1 - \frac{4\Delta}{\sigma^2}}\right) .$$

If $\sigma^2 > 4\Delta$, then the roots of the characteristic equation are real numbers. If $\sigma^2 < 4\Delta$, the roots are complex numbers. Subject to the conditions of stationarity (4.51) we have

$$\sigma = \frac{\beta y f(y)}{y^2(1 + f(y))} - f(y) - s - 1 , \tag{4.53}$$

$$\Delta = s(1 + f(y)) - \frac{\beta y f(y)}{y^2(1 + f(y))} . \tag{4.54}$$

Parametric dependencies of steady states are the solutions of equation (4.51) or the equation

$$G(y) = F(y) - P(y) = 0 . \tag{4.55}$$

Let us write down equation (4.55) in the form

$$G(y, Da) = 0 , \tag{4.56}$$

i.e., along with y let us highlight the parameter Da. In the case where it is necessary to plot parametric dependencies, for example, $x(Da)$ or $y(Da)$, it is advisable to use the specifics of this task. Since the parameter Da, included in (4.56), is linear, then for $G(y, Da, s, \beta, y)$ we will receive

$$\frac{\beta f(y)}{1 + f(y)} - s(y - 1) = 0 . \tag{4.57}$$

Let us express the parametric dependencies from (4.57):

$$s(y) = \frac{\beta f(y)}{(1 + f(y))(y - 1)} ,$$

$$\beta(y) = \frac{s(y - 1)(1 + f(y))}{f(y)} ,$$

$$Da(y) = \frac{s(y - 1)}{e(y)(\beta - s(y - 1))} ,$$

$$y(y) = \frac{y}{y - 1} \ln \frac{s(y - 1)}{Da(\beta - s(y - 1))} . \tag{4.58}$$

When we plot the dependence $Da(y)$ in accordance with (4.58), we will get the inverse function of the desired parametric dependence $y(Da)$. Knowing $y(Da)$, we can plot $x(Da)$, because $x = f(y)/(1 + f(y))$.

Thus, we write down the expressions needed to plot the dependencies for stationary values x on the parameters:

$$(x, s): \quad s = s(y),$$

$$x(y, s(y)) = \frac{s(y)}{\beta}(y - 1);$$

$$(x, \beta): \quad \beta = \beta(y),$$

$$x(y, \beta(y)) = \frac{s}{\beta(y)}(y - 1);$$

$$(x, Da): \quad Da = Da(y),$$

$$x(y, Da(y)) = \frac{f(y)}{1 + f(y)};$$

$$(x, y): \quad y = y(y),$$

$$x(y, y(y)) = \frac{f(y)}{1 + f(y)}. \tag{4.59}$$

The dependence of steady states on the parameters Da, β, s, y can be plotted without any iterative procedure of solving nonlinear equations (4.51). An example of the plotting of the obtained parametric dependencies and analyzing the impact of other parameters on them is shown in Fig. 4.20–4.21.

The use of explicit dependencies of steady states on the parameters (4.58) and (4.59) also allows us to plot the dependencies of the eigenvalues of the Jacobian matrix in the steady state on the parameters. An example of such dependence is shown in Fig. 4.22. Such parametric dependencies graphically illustrate the change of types of stability of steady states when varying the parameters. The transition of eigenvalues from negative to positive characterizes the loss of stability of the steady state.

If such a transition is realized through purely imaginary eigenvalues, it corresponds to the Andronov–Hopf bifurcation. In Fig. 4.22 the real roots are indicated by red and green lines. Complex roots have real parts (blue lines) and imaginary parts (black lines). In the case of the existence of three steady states, the three pairs of dependencies λ on the parameters appear. The merging and disappearance of branches of the corresponding curves mean the bifurcation of steady states.

Multiplicity curves define the regions of parameters with one or more steady states. Most obvious is the presentation of the results in the form of multiplicity curves on the plane, i.e., in the space of any two parameters. Let us select two of them, for example Da and s, considering the other parameters to be fixed. Further, we plot a curve of multiplicity of steady states in the parameter plane (Da, s). This requires the

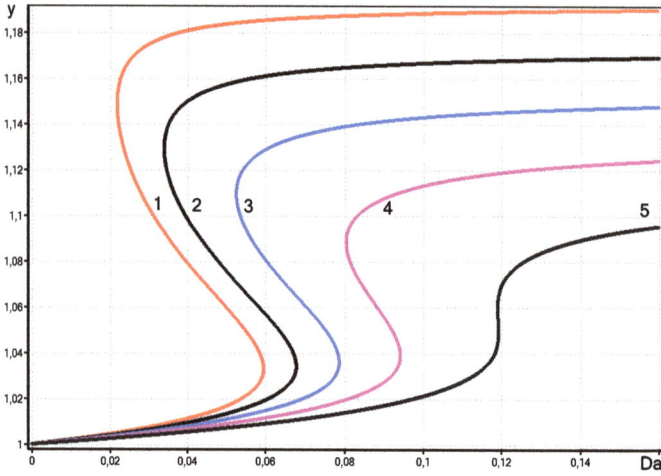

Fig. 4.20: Dependencies of the temperature on the parameter Da at $s = 2.6$, $\gamma = 39$, 1) $\beta = 0.5$; 2) 0.45; 3) 0.4; 4) 0.35; 5) 0.3

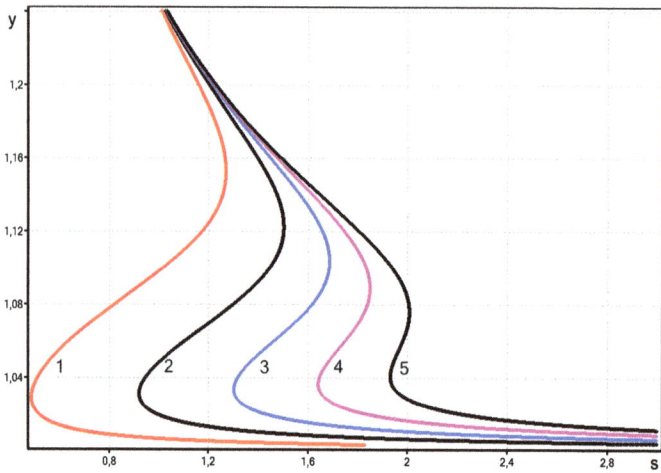

Fig. 4.21: Dependencies of the temperature on the parameter s at $\beta = 0.25$, $\gamma = 39$, 1) $Da = 0.02$; 2) 0.04; 3) 0.06; 4) 0.08; 5) 0.1

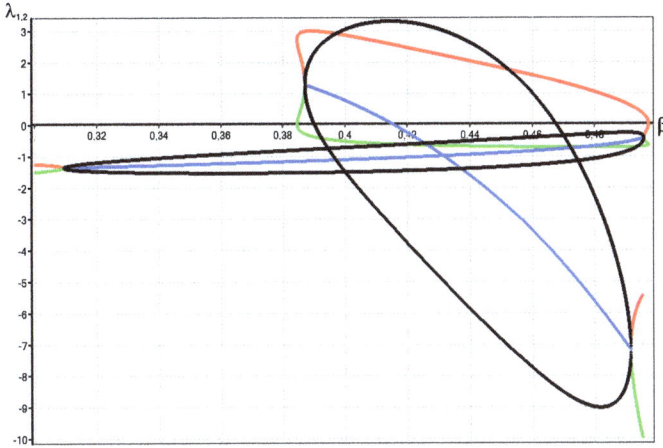

Fig. 4.22: Dependence of the eigenvalues on the parameter β at $Da = 0.06$, $s = 2.6$, $\gamma = 39$

solution of a system of nonlinear equations:

$$f_1(x, y, Da, s) = 0 \,,$$

$$f_2(x, y, Da, s) = 0 \,,$$

$$\Delta(x, y, Da, s) = 0 \,, \tag{4.60}$$

where Δ is defined according to (4.53). After standard transformations for L_Δ we will receive

$$\frac{\beta f(y)}{1 + f(y)} - s(y - 1) = 0 \,,$$

$$s(1 + f(y)) - \frac{\beta y f(y)}{y^2(1 + f(y))} = 0 \,. \tag{4.61}$$

Now, using the explicit expression $s(y, Da)$ obtained from (4.58) and substituting it into the second equation of (4.61), we write down the equation of the boundary of the region of multiplicity of steady states. L_Δ in the parametric plane (Da, s) has the form:

$$L_\Delta(y, Da, s(y, Da)) = 0 \,. \tag{4.62}$$

Thus, solving equation (4.62) explicitly relative to Da for every fixed s, we find the corresponding value $Da(y)$. Finally, plotting the border of the region of multiplicity of steady states is reduced to calculation of the function $L_\Delta(Da, s)$, given in parametric form:

$$L_\Delta(Da, s): \quad Da(y) = \left(\frac{y(y - 1)}{y^2} - 1 \right) \frac{1}{e(y)} \,,$$

$$s = s(y, Da(y)) \,. \tag{4.63}$$

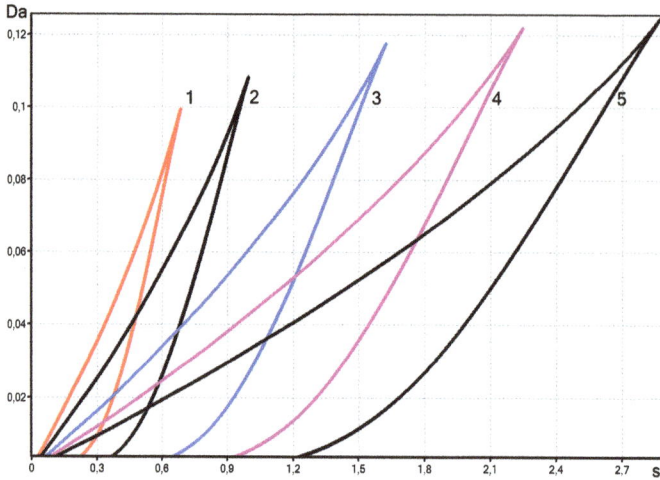

Fig. 4.23: Curves of multiplicity $L_\Delta(s, Da)$ at $\beta = 0.25$, 1) $\gamma = 15$; 2) 20; 3) 30; 4) 40; 5) 50

Similarly, the equations of curves of multiplicity of steady states in the other parametric planes (Da, β), (Da, γ), $(Da, s/\beta)$ are written down at $(y \in (1, y_\infty))$:

$$L_\Delta(Da, \beta): \quad Da(y) = \left(\frac{y(y-1)}{y^2} - 1\right)\frac{1}{e(y)},$$

$$\beta = \beta(y, Da(y));$$

$$L_\Delta(Da, s/\beta): \quad Da(y, s/\beta(y)) = \frac{(y-1)}{e(y)(\beta/s - (y-1))},$$

$$\frac{s}{\beta}(y) = \frac{y(y-1) - y^2)}{y(y-1)^2};$$

$$L_\Delta(Da, \gamma): \quad Da = Da(y, \gamma(y)),$$

$$\gamma(y) = \frac{\beta y^2}{(y-1)(\beta - s(y-1))}. \tag{4.64}$$

Two lines, emerging from the origin, form the border of the region of multiplicity of steady states in the plane (β, s). To plot them it is sufficient to find two points on the plane (β, s) where $\Delta(y, \beta, s) = 0$. This can be done using the graph of the curve of multiplicity of steady states, for example, in the plane $(Da, s/\beta)$. Examples of plotted curves of multiplicity in one of the planes of the dimensionless parameters in accordance with (4.63)–(4.64) are given in Fig. 4.23–4.24.

In the case when it is difficult to obtain the expressions for curves of multiplicity in any of the planes of parameters in an explicit form, it is possible to offer the procedure of graphical plotting L_Δ. It consists of plotting a series of parametric dependencies for one of the parameters while varying a second parameter. Note that these dependencies can be plotted for all parameters. It is obvious that the interval of multiplicity of steady states is located between the turning points of the parametric curve. When varying the

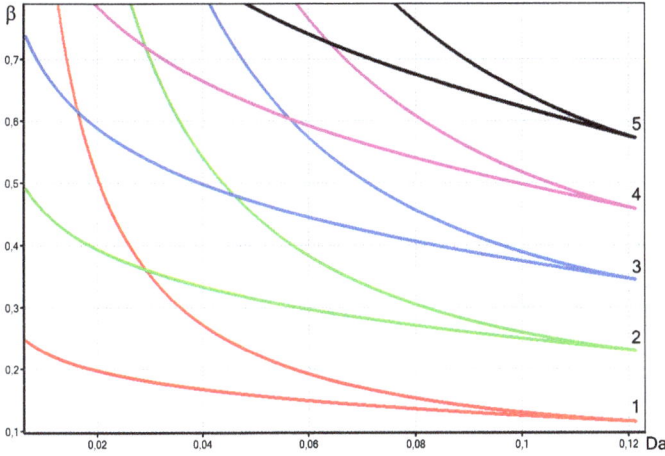

Fig. 4.24: Curves of multiplicity $L_\Delta(Da, \beta)$ at $\gamma = 39$, 1) $s = 1$; 2) 2; 3) 3; 4) 4; 5) 5

second parameter the bounds of this interval determine the region of multiplicity of steady states on the corresponding plane.

Curves of neutrality. The curve L_σ ($\sigma = 0$) corresponds to the bifurcation values of the parameters, i.e., it allows us to analyze the change of stability of a steady state. The curve L_σ is determined by the system of equations:

$$\frac{\beta f(y)}{1 + f(y)} - s(y - 1) = 0,$$

$$\frac{\beta y f(y)}{y^2(1 + f(y))} - f(y) - s - 1 = 0.$$

Finally we obtain the equation of the curve of neutrality for different combinations of dimensionless parameters in explicit form:

$L_\sigma(Da, s):$ $Da = Da(y, s(y))$,

$$s(y) = \frac{\beta}{2(y-1)} \pm \sqrt{\left(\frac{\beta}{2(y-1)}\right)^2 - \frac{\beta}{(y(y-1)/y^2 - 1)(y-1)}} ;$$

$L_\sigma(Da, \beta):$ $Da(y) = \left(\frac{sy(y-1)}{y^2} - s - 1\right)\frac{1}{e(y)}$,

$\beta = \beta(y, Da(y))$;

$L_\sigma(Da, y):$ $Da = Da(y, y(y))$,

$$y(y) = \frac{(1 + s + s(y-1)/(\beta - s(y-1)))y^2}{s(y-1)} ; \tag{4.65}$$

Examples of dependencies of type (4.65) while varying the third parameter are shown in Fig. 4.25–4.26.

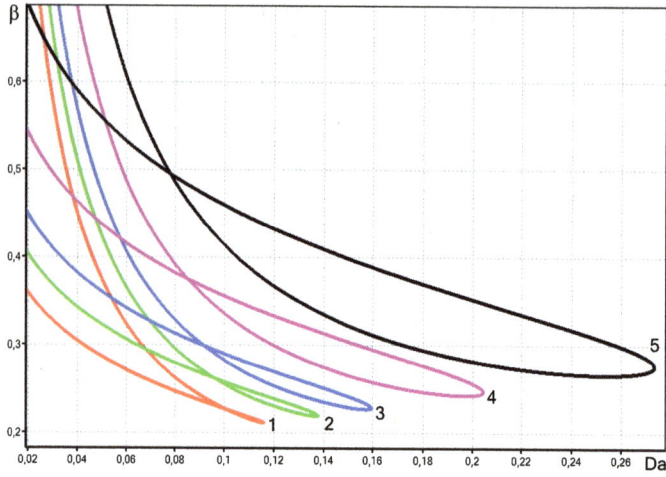

Fig. 4.25: Curves of neutrality $L_\sigma(Da, \beta)$ at $\gamma = 39$, 1) $s = 1.8$; 2) 2; 3) 2.2; 4) 2.6; 5) 3.2

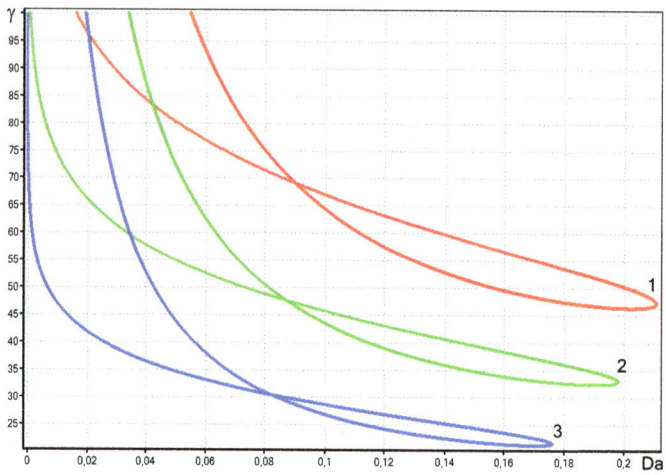

Fig. 4.26: Curves of neutrality $L_\sigma(Da, \gamma)$ at $s = 2.6$, 1) $\beta = 0.2$; 2) 0.3; 3) 0.5

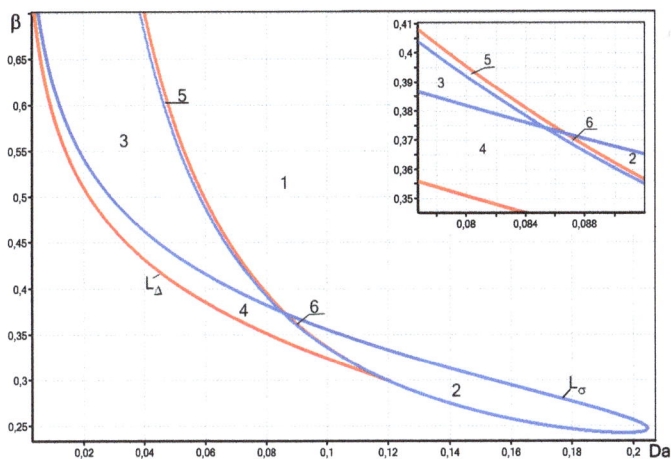

Fig. 4.27: Parametric portrait on the plane (Da, β) at $s = 2.6$, $\gamma = 39$

Parametric portrait. The mutual position of the curves of multiplicity (L_Δ) and neutrality (L_σ) specifies the parametric portrait of system (4.49). It defines the different regions of parameters, which differ by number and type of stability of steady states. In our case, there are six regions in the parametric portrait (see for example, Fig. 4.27).

Curve of the Andronov–Hopf bifurcation. The Andronov–Hopf bifurcation curve gives important information about the behavior of a dynamical system, which corresponds to the existence of purely imaginary roots of the characteristic equation (4.52) [55–57]. Bifurcation values of a parameter are determined by the conditions:

$$\operatorname{Re}\lambda = 0 , \qquad \sigma^2 < 4\Delta .$$

The Andronov–Hopf bifurcation curves limit the regions of parameters. "Hard" or "soft" bifurcations occur at varying values of the parameters. "Soft" bifurcations determine dangerous and safe boundaries of critical phenomena in the kinetics of exothermal reactions [53, 54].

Phase portraits. In accordance with the parametric portrait of system (4.49), six types of phase portraits can be distinguished. A typical phase portrait for the case of a singular and unstable steady state, where there are auto-oscillations, is presented in Fig. 4.28.

Time dependencies. The integration of the original dynamical system (4.49) allows us to obtain time dependence $x(t)$ and $y(t)$ for different initial conditions and parameters. An example of time dependence of the dimensionless temperature is given in Fig. 4.29 for a single unstable steady state.

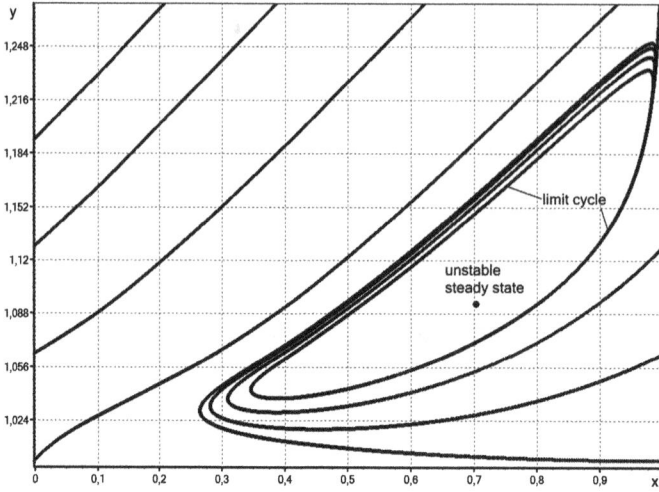

Fig. 4.28: Phase portrait for region 2 at $Da = 0.14$, $\beta = 0.4$, $s = 3$, $\gamma = 33$, $x_{st} = 0.704249$, $y_{st} = 1.093899$

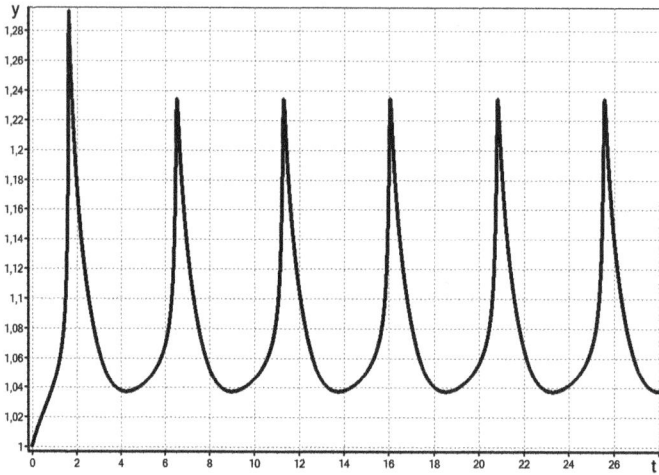

Fig. 4.29: Time dependence of temperature at $Da = 0.14$, $\beta = 0.4$, $s = 3$, $\gamma = 33$

4.3.2 Reaction of the *n*-th order

For the reaction of the n-th order $nA \rightarrow P$ dimensionless model of a CSTR has the form

$$\frac{dx}{d\tau} = f(y)(1 - x)^n - x = f_1(x, y),$$

$$\frac{dy}{d\tau} = \beta f(y)(1 - x)^n - s(y - 1) = f_2(x, y),$$ (4.66)

where n is the reaction order. Let us carry out the parametric analysis for more complex (compared to (4.49)) model (4.66).

Steady states. The system of stationarity for (4.66) has the form

$$f(y)(1 - x)^n - x = f_1(x, y),$$

$$\beta f(y)(1 - x)^n - s(y - 1) = f_2(x, y).$$

From here, similarly to (4.51), we obtain the equation for determining steady states:

$$F(y) = \beta f(y)\left(1 - \frac{s}{\beta}(y - 1)\right)^n,$$

$$P(y) = s(y - 1).$$

Parametric dependencies. As above, the parametric dependencies can be obtained as:

$$Da(y) = \frac{s(y - 1)}{\beta \exp\left(y(1 - 1/y)\right)(1 - (s/\beta)(y - 1))^n},$$

$$y(y) = \frac{y}{y - 1} \ln \frac{s(y - 1)}{Da\beta(1 - (s/\beta)(y - 1))^n}.$$

The expressions for the parameters s and β could not be obtained explicitly. To plot them we can use a modified method of continuation on a parameter, proposed in [5, 35].

The elements of the Jacobian matrix have the form

$$a_{11} = -nf(y)(1 - x)^{n-1} - 1, \quad a_{12} = \frac{y}{y^2}f(y)(1 - x)^n,$$

$$a_{21} = -n\beta f(y)(1 - x)^{n-1}, \quad a_{22} = \frac{y}{y^2}\beta f(y)(1 - x)^n - s.$$

The coefficients of the characteristic equation are:

$$\sigma = \frac{y}{y^2}\beta f(y)(1 - x)^n - nf(y)(1 - x)^{n-1} - s - 1,$$

$$\Delta = s(1 + nf(y)(1 - x)^{n-1}) - \frac{y}{y^2}\beta f(y)(1 - x)^n.$$

The curve of multiplicity L_Δ is determined by the system

$$\beta f(y)\left(1 - \frac{s}{\beta}(y - 1)\right)^n - s(y - 1) = 0,$$

$$s(1 + nf(y)(1 - (s/\beta)(y - 1))^{n-1}) - \frac{y}{y^2}\beta f(y)(1 - (s/\beta)(y - 1))^n = 0.$$

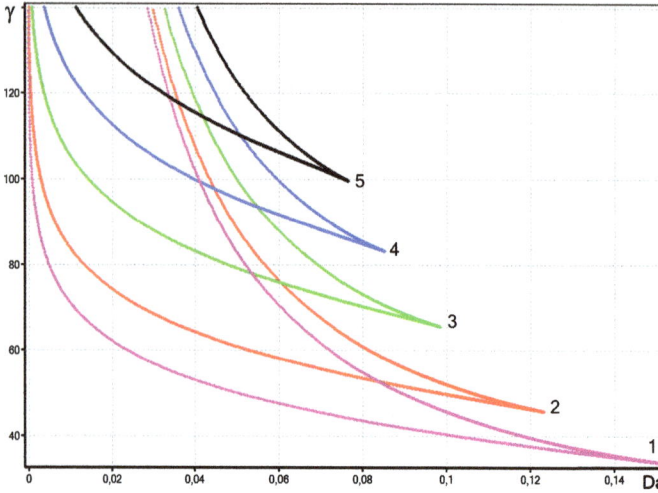

Fig. 4.30: Curves of multiplicity $L_\Delta(Da, \gamma)$ at $\beta = 0.25$, $s = 2.6$, 1) $n = 0.2$; 2) 1; 3) 2; 4) 3; 5) 4

From these equations, after elementary transformations, we obtain for $L_\Delta(Da, \gamma)$:

$$L_\Delta(Da, \gamma): \quad \gamma(y) = \frac{y^2}{y - 1} + \frac{nsy^2}{\beta - s(y - 1)} ,$$

$$Da(y, \gamma(y)) = \frac{s(y - 1)}{\beta \exp\left(\gamma(1 - 1/y)\right)(1 - (s/\beta)(y - 1))^n} . \tag{4.67}$$

An example of a plot of the curve of multiplicity (4.67) at varying n is shown in Fig. 4.30. To plot L_Δ in other planes of parameters, either the above described graphical procedure can be used or numerical methods of continuation on a parameter [5, 35, 48].

The curve of neutrality L_σ is determined by the system

$$\beta f(y)\left(1 - \frac{s}{\beta}(y - 1)\right)^n - s(y - 1) = 0 ,$$

$$\frac{y}{y^2}\beta f(y)(1 - (s/\beta)(y - 1))^n - nf(y)(1 - (s/\beta)(y - 1))^{n-1} - s - 1 = 0 ,$$

i.e., for $L_\sigma(Da, \gamma)$ we obtain

$$L_\sigma(Da, \gamma): \quad \gamma(y) = \frac{y^2(s + 1)}{s(y - 1)} + \frac{ny^2}{\beta - s(y - 1)} ,$$

$$Da(y, \gamma(y)) = \frac{s(y - 1)}{\beta \exp\left(\gamma(1 - 1/y)\right)(1 - (s/\beta)(y - 1))^n} .$$

Plotting the curve of neutrality L_σ for other planes of parameters requires the use of standard computational procedures of continuation on a parameter.

Parametric and phase portraits, and also time dependencies, can be plotted by analogy with a first order reaction.

4.3.3 The oxidation reaction

For the oxidation reaction $A + O_2 \rightarrow P$ a dimensionless model of a CSTR has the form

$$\frac{dx}{d\tau} = f(y)g(x) - x = f_1(x, y) ,$$

$$\frac{dy}{d\tau} = \beta f(y)g(x) - s(y - 1) = f_2(x, y) , \qquad (4.68)$$

where $f(y) = Da \exp(y(1 - 1/y))$ is a temperature dependence; $g(x) = (1 - x)(1 - \alpha x)$ is a kinetic dependence; α is the dimensionless parameter that characterizes the ratio of total amount of reaction mixture to the amount of oxygen.

Steady states. The system of stationarity for model (4.68) has the form

$$f(y)g(x) - x = 0 ,$$

$$\beta f(y)g(x) - s(y - 1) = 0 .$$

After excluding x, we obtain the equation for determination of steady states

$$f(y)(1 - (s/\beta)(y - 1))(\beta - \alpha s(y - 1)) = s(y - 1) . \qquad (4.69)$$

Parametric dependencies are expressed from equation (4.69):

$$Da(y) = \frac{\beta s(y - 1)}{e(y)(\beta - s(y - 1))(\beta - \alpha s(y - 1))} ,$$

$$\beta(y) = s(y - 1)\left(\frac{1}{2}\left(1 + \alpha + \frac{1}{f(y)}\right) + \sqrt{\frac{1}{4}\left(1 + \alpha + \frac{1}{f(y)}\right)^2 - \alpha} \right) ,$$

$$\gamma(y) = \frac{y}{y - 1} \ln \frac{\beta s(y - 1)}{Da(\beta - s(y - 1))(\beta - \alpha s(y - 1))} ,$$

$$\alpha(y) = \beta\left(\frac{1}{s(y - 1)} - \frac{1}{f(y)(\beta - s(y - 1))} \right) ,$$

$$s(y) = \frac{\beta}{y - 1}\left(\frac{2 + \alpha f(y)}{2\alpha f(y)} \pm \sqrt{\left(\frac{2 + \alpha f(y)}{2\alpha f(y)}\right)^2 - \frac{1}{\alpha}} \right) . \qquad (4.70)$$

An example of plotting the dependencies of steady states on the parameters on the basis of formulas (4.70) is shown in Fig. 4.31.

Curves of multiplicity L_Δ are determined from the system

$$f(y)(1 - (s/\beta)(y - 1))(\beta - \alpha s(y - 1)) - s(y - 1) = 0 ,$$

$$\frac{y}{y^2}f(y)(1 - (s/\beta)(y - 1))(\beta - \alpha s(y - 1)) + \frac{f(y)}{\beta}(2\alpha s^2(y - 1) - s\beta(1 + \alpha)) - s .$$

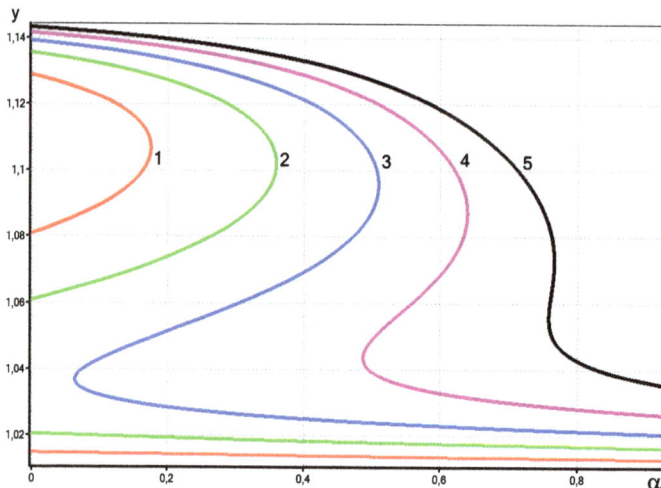

Fig. 4.31: Dependencies of temperature on parameter α at $\beta = 0.4$, $s = 2.6$, $\gamma = 39$, 1) $Da = 0.06$; 2) 0.07; 3) 0.08; 4) 0.09; 5) 0.1

As above, we can obtain the explicit expressions for the multiplicity curves in different planes of the dimensionless parameters:

$$L_\Delta(Da, \beta): \quad \beta(y) = \frac{sD}{2}(A - 1/B) \pm \sqrt{\left(\frac{sD}{2}(A - 1/B)\right)^2 - \alpha s^2 A(A - 2/B)},$$

$$Da = Da(y) \, ;$$

$$L_\Delta(Da, \alpha): \quad \alpha(y) = \frac{\beta(BC + \dot{s})}{s((y - 1)(BC + 2s) - \beta)} \, ,$$

$$Da = Da(y) \, ;$$

$$L_\Delta(Da, \gamma): \quad \gamma(y) = \frac{y^2}{y - 1} - \frac{2\alpha s^2 y^2(y - 1) - \beta s D y^2}{CE} \, ,$$

$$Da = Da(y) \, , \tag{4.71}$$

where $A = y - 1$, $B = 1/A - y/y^2$, $C = \beta - sA$, $D = 1 + \alpha$, $E = \beta - \alpha sA$.

One of the results of the calculations of L_Δ by the formulas (4.71) for various combinations of the dimensionless parameters is shown in Fig. 4.32.

Curves of neutrality L_σ are determined from the system

$$f(y)(1 - (s/\beta)(y - 1))(\beta - \alpha s(y - 1)) - s(y - 1) = 0 \, ,$$

$$\frac{\beta y}{y^2} f(y)(1 - (s/\beta)(y - 1))(1 - \alpha(s/\beta)(y - 1)) - f(y)(1 + \alpha(1 - (s/\beta)(y - 1))) - s - 1 = 0 \, .$$

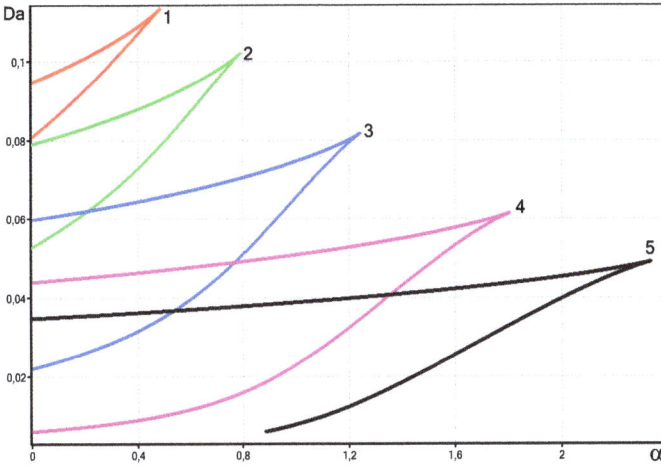

Fig. 4.32: Curves of multiplicity $L_\Delta(Da, \alpha)$ at $s = 2.6$, $\gamma = 39$, 1) $\beta = 0.35$; 2) 0.4; 3) 0.45; 4) 0.5; 5) 0.55; 6) 0.65; 7) 0.75

Explicit expressions for the curves of neutrality can be written in the following planes of the parameters:

$$L_\sigma(Da, \beta): \quad \beta(y) = \frac{D}{2}\left(\frac{1}{K} + sA\right) \pm \sqrt{\left(\frac{D}{2}\left(\frac{1}{K} + sA\right)\right)^2 - \alpha sA\left(\frac{2}{K} + sA\right)},$$

$$Da = Da(y) \,;$$

$$L_\sigma(Da, \alpha): \quad \alpha(y) = \frac{\beta(C(s(\gamma A - y^2) - y^2) - sAy^2)}{s^2 A(C(\gamma A - y^2) - A)} \,,$$

$$Da = Da(y) \,;$$

$$L_\sigma(Da, \gamma): \quad \gamma(y) = \frac{y^2(s+1)}{sA} + \frac{y^2}{C} + \frac{\alpha y^2}{E} \,,$$

$$Da = Da(y) \,, \tag{4.72}$$

where $A = y - 1$, $C = \beta - sA$, $D = 1 + \alpha$, $E = \beta - \alpha sA$, $K = (\gamma sA - y^2(s+1))/(y^2 sA)$.

To plot L_Δ and L_σ in other parametric planes we should apply the standard computational procedure of the method of continuation on a parameter [5, 35, 48] or the proposed a graphical method. An example of the curves L_σ using the formulas (4.72) is given in Fig. 4.33.

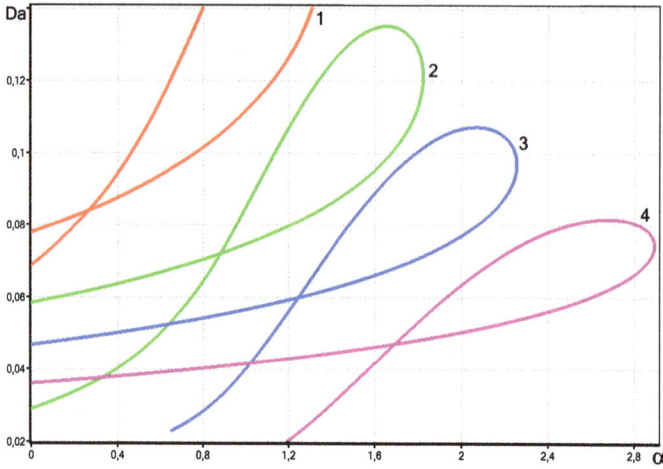

Fig. 4.33: Curves of neutrality $L_\sigma(\alpha, Da)$ at $\gamma = 39$, $s = 2.6$, 1) $\beta = 0.5$; 2) 0.6; 3) 0.7; 4) 0.8

4.3.4 Reaction with arbitrary kinetics

The Aris–Amundson mathematical model for arbitrary kinetic functions is written as follows:

$$\frac{dx}{d\tau} = f(y)g(x) - x = f_1(x, y) \,,$$

$$\frac{dy}{d\tau} = \beta f(y)g(x) - s(y - 1) = f_2(x, y) \,, \qquad (4.73)$$

where $g(x)$ is an arbitrary kinetic function.

Steady states. The system of stationarity has the form

$$f(y)g(x) - x = 0 \,,$$
$$\beta f(y)g(x) - s(y - 1) = 0 \,.$$

After excluding x the stationarity equation for y takes the form

$$\beta f(y)g\left(\frac{s}{\beta}(y - 1)\right) = s(y - 1) \,.$$

Parametric dependencies. For the selected parameters of model (4.73) we have

$$Da(y) = \frac{s(y - 1)}{\beta \exp\left(\gamma(1 - 1/y)\right)g(x)} \,,$$

$$y(y) = \frac{y}{y - 1} \ln \frac{s(y - 1)}{Da\beta g(x)} \,,$$

where $x = (y - 1)s/\beta$. One of the variants of the calculations is given in Fig. 4.34.

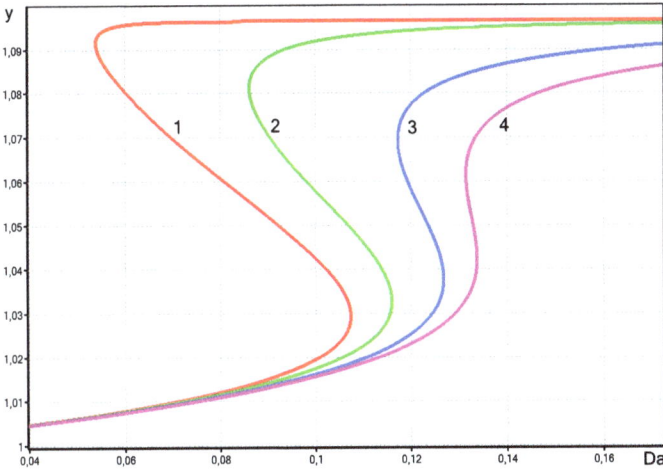

Fig. 4.34: Dependencies of temperature on parameter Da at $\beta = 0.25$, $s = 2.6$, $\gamma = 39$, $\alpha = 0.1$,
1) $n = 0.1$; 2) 0.3; 3) 0.5; 4) 0.6

The coefficients of the characteristic equation have the form

$$\sigma = a_{11} + a_{22} = f(y)g'(x) + \frac{y}{y^2}\beta f(y)g(x) - s - 1 ,$$

$$\Delta = a_{11}a_{22} - a_{12}21 = s(1 - f(y)g'(x)) - \frac{y}{y^2}\beta f(y)g(x) .$$

Curves of multiplicity L_Δ can be obtained from the system

$$\beta f(y)g(x) - s(y - 1) = 0 ,$$

$$s(1 - f(y)g'(x)) - \frac{y}{y^2}\beta f(y)g(x) = 0 .$$

The curves L_Δ in the plane of the parameters (Da, y) have the form

$$Da(y, y(y)) = \frac{s(y - 1)}{\beta \exp{(y(1 - 1/y))}g(x)} ,$$

$$y(y) = \frac{y}{y - 1} - \frac{y^2 sg'(x)}{\beta g(x)} .$$

An example of plotting the curves of multiplicity for the kinetic function $g(x) = (1 - x)^n(1 - \alpha x)$ is given in Fig. 4.35.

Curves of neutrality L_σ can be given by the system

$$\beta f(y)g(x) - s(y - 1) = 0 ,$$

$$f(y)g'(x) + \frac{y}{y^2}\beta f(y)g(x) - s - 1 = 0 .$$

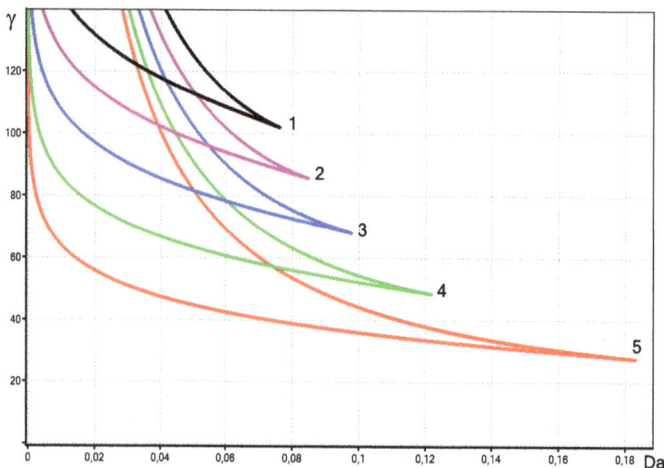

Fig. 4.35: Curves of multiplicity $L_\Delta(Da, \gamma)$ at $\beta = 0.25$, $s = 2.6$, $\alpha = 0.2$, 1) $n = 4$; 2) 3; 3) 2; 4) 1; 5) 0.2

Hence, for the curve L_σ in the plane of parameters (Da, γ) we have the following expression:

$$Da(y, \gamma(y)) = \frac{s(y-1)}{\beta \exp\left(\gamma(1 - 1/y)\right)g(x)},$$

$$\gamma(y) = \frac{y^2(s+1)}{s(y-1)} - \frac{y^2 g'(x)}{\beta g(x)}.$$

Examples of L_σ are given in Fig. 4.36. For each real kind of the kinetic function, its own parameters can be considered and the corresponding parametric dependencies plotted by the above-described algorithms.

Thus, consistent application of the procedure of parametric analysis to the Aris–Amundson model allowed us to obtain the curves of local bifurcations of steady states in various combinations of planes of the dimensionless parameters and to investigate the influence on them of other options. This gives you the opportunity to study the influence of other parameters on the dimensions of the regions of multiplicity of steady states and auto-oscillations. The obtained results are useful from methodical and practical points of view, because they adequately describe the features of one of the basic models of chemical reactors. In particular, there is the opportunity to plot the bifurcation curves in the plane of the dimensional parameters which correspond to the specific geometry and thermophysical characteristics of a real exothermic processes in CSTR.

Here, besides the first order reaction, we considered the cases of oxidation reactions and reactions of the n-th order, as well as reactions with arbitrary kinetic dependencies. In the planes of the dimensionless parameters the bifurcation curves of

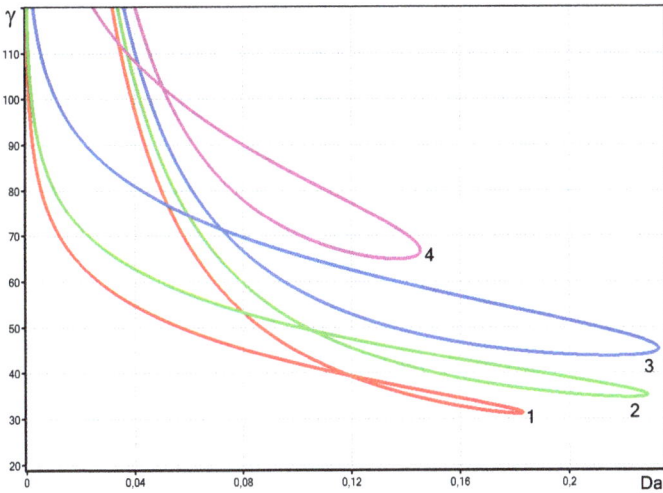

Fig. 4.36: Curves of neutrality $L_\sigma(Da, \gamma)$ at $\beta = 0.25$, $s = 2.6$, $n = 0.5$, 1) $\alpha = 0$; 2) 0.5; 3) 1; 4) 1.7

the change in the number and type of stability of the steady states were plotted by analytical formulas or numerical methods.

Comparative analysis of the parametric portraits for different kinetic functions shows that the kind of kinetics significantly affects the conditions of the existence of multiplicity of steady states and auto-oscillations in a chemical reactor.

The technique of parametric analysis, demonstrated here for the mathematical model of CSTR, can naturally be extended to other nonlinear dynamic models. Our experience shows that as soon as the conditions of stationarity can be reduced to one or two equations, the general procedure of the method of continuation on a parameter is greatly simplified. So, it is possible to obtain not only numerical, but also analytical results.

4.3.5 Andronov–Hopf bifurcations

An important challenge in chemical engineering systems design is the analysis of possible dynamic behavior of chemical reactors [58]. This behavior can be quite diversified [59]. In particular, the dynamics of a physicochemical process can be characterized by undamped oscillations. The conditions for the emergence of oscillating operating modes of chemical reactors and the oscillation amplitude and period are substantially dependent on the process parameters. These conditions can be determined by the so-called Andronov–Hopf bifurcations [55], which take place when, with varying the parameters, the real parts of the (complex-conjugate) eigenvalues of the matrix of the linearized set of equations in the neighborhood of a steady state become zero. Study of such bifurcations of steady states enables one to analyze the evolution of limit cycles with variations in the system parameters. Knowing features of this evolution is not only of theoretical but also of practical significance. Technologically, oscillating modes are as a rule undesirable, and one should be capable of predicting them.

In this section, we obtained the parametric portraits of local bifurcations of steady states for the classical model of the Aris–Amundson and determined the conditions for emergence of the Andronov–Hopf bifurcations. The critical conditions were found for both a first-order reaction and a reaction with arbitrary kinetics. A general physicochemical significance of the obtained diagrams consists of the fact that they enable one to reveal the conditions for the emergence of critical phenomena when varying the parameters over a wide range and to predict emergency operating modes of chemical reactors.

Steady states of system (4.49) are determined by the conditions $f_1(x, y) = f_2(x, y) = 0$ and are found from the stationarity equation

$$\beta \frac{f(y)}{1 + f(y)} = s(y - 1).$$

The stability of the steady states is determined by the roots of the characteristic equation

$$\lambda^2 - \sigma\lambda + \Delta = 0,$$

where $\sigma = a_{11} + a_{22}$, $\Delta = a_{11}a_{22} - a_{12}a_{21}$; and a_{ij} are the elements of the Jacobian matrix for the right sides of system (4.49) at the steady states:

$$a_{11} = -f(y) - 1, \quad a_{12} = \frac{y}{y^2}f(y)(1 - x),$$

$$a_{21} = -\beta f(y), \quad a_{22} = \beta y f(y)(1 - x)/y^2 - s.$$

The changes in the number of steady states and in the type of their stability are vividly illustrated by the curve L_Δ of multiplicity of the steady states (at $\sigma = 0$). The specificity of model (4.49) enables the plotting of these bifurcation curves in an explicit form. For example, in the plane of parameters Da, β, these curves are given by the equations:

$$L_\Delta(Da, \beta): \quad Da(y) = \frac{y(y - 1) - y^2}{e(y)y^2};$$

$$\beta(y, Da(y)) = \frac{s(y - 1)(1 + f(y))}{f(y)},$$

$$L_\sigma(Da, \beta): \quad Da(y) = \left(\frac{sy(y - 1)}{y^2} - s - 1\right)\frac{1}{e(y)};$$

$$\beta(y, Da(y)) = \frac{s(y - 1)(1 + f(y))}{f(y)}. \tag{4.74}$$

Figure 4.37 offers an example of a plot of the parametric portrait by (4.74). The Andronov–Hopf bifurcation curve L_{AH} (the circle-traced curve in Fig. 4.37) is determined by the conditions $\operatorname{Re}\lambda = 0$ and $\sigma^2 - 4\Delta < 0$, when $\lambda = \pm i\operatorname{Im}\lambda$ and $\operatorname{Im}\lambda = \sqrt{\Delta - (\sigma^2/4)}$.

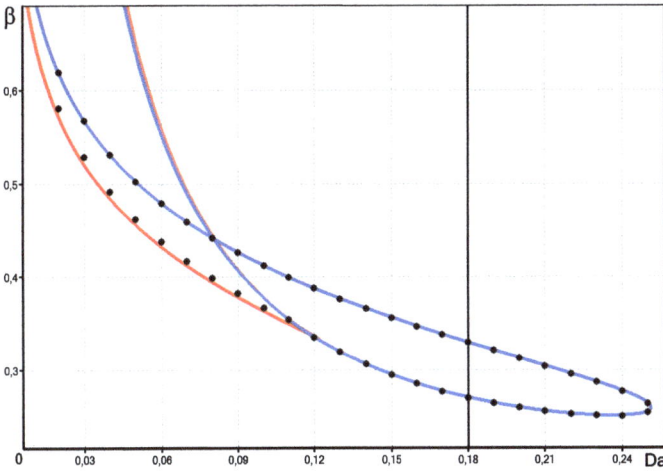

Fig. 4.37: Parametric portrait for model (4.49) on the plane (Da, β) at $\gamma = 40$, $s = 3$

Once a representative point moving along the section $Da = 0.18$ intersects this curve and enters its bounding parameter region, there emerges a stable limit cycle which has a small size and therefore corresponds to oscillations with a small amplitude. As the representative point moves farther this limit cycle initially grows, but then it decreases after exceeding a certain value of the parameter β. Once the representative point crosses the other boundary of the parameter region, the steady state is unique and stable, and there are no limit cycles. Figure 4.38 illustrates the described evolution of the limit cycles.

Near the curve L_{AH} outside the region bounded by the curve L_σ, the steady state is unique and stable. However, the motion of a representative point toward this stable state in the phase portrait is executed with a very small damping factor (Fig. 4.38a). Near the curve L_{AH} inside the region bounded by the curve L_σ, the steady state is unique and unstable and is within a small limit cycle (Fig. 4.38b). The motion of a representative point toward the limit cycle is also characterized by a small damping factor. Far from the curve L_{AH}, the limit cycle grows, and the solutions approach it rapidly (Fig. 4.38d).

An analysis of the functions $y(t)$ and $x(t)$, one of which is plotted in Fig. 4.39, showed that the oscillation period is estimated with acceptable accuracy at $\tau = 2\pi / \operatorname{Im} \lambda$, where λ is calculated at the corresponding steady state. For example, at $\beta = 0.28$ (a large limit cycle), the oscillation period is estimated at $\tau = 5.25$, and direct computations give $\tau^* \approx 4.2$. At $\beta = 0.327$ (a small limit cycle), the respective values of the period are $\tau = 2.08$ and $\tau^* \approx 2.1$.

The above technique of parametric analysis, including construction of the bifurcation curves L_Δ, L_σ, and L_{AH}, can be extended to the general case of system (4.73) with an arbitrary kinetic function. For example, the curves of multiplicity and neutrality of

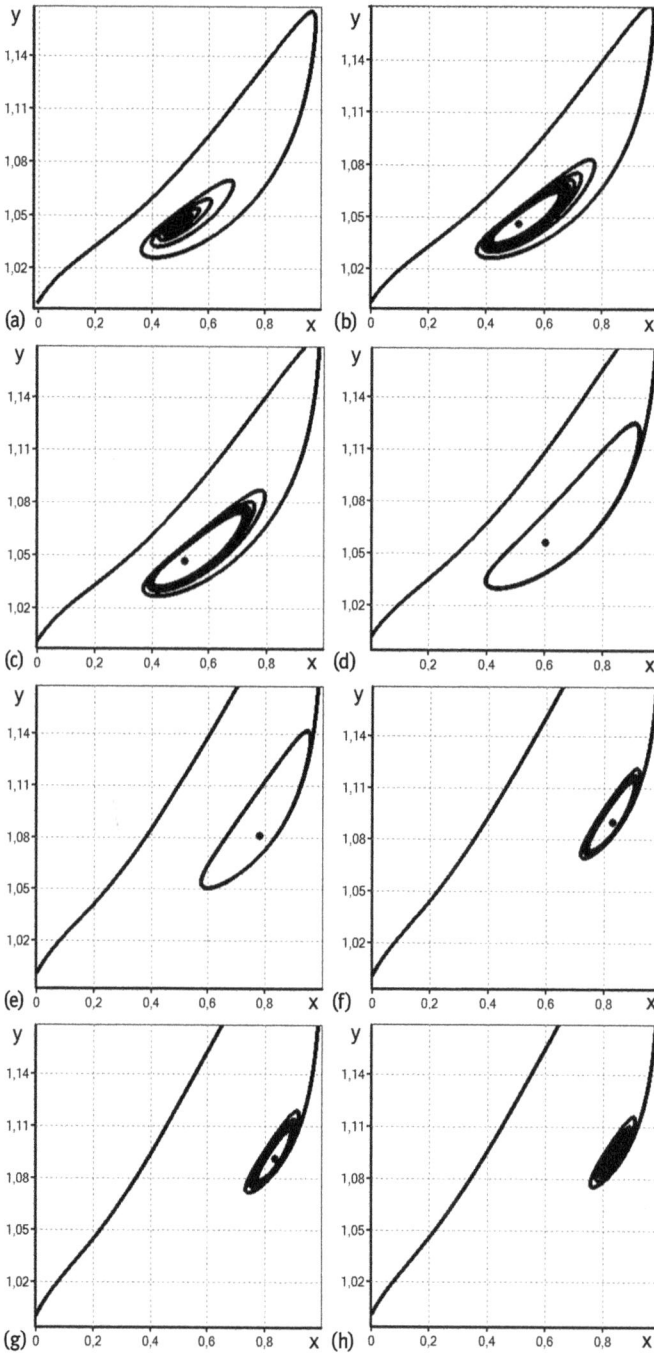

Fig. 4.38: Evolution of the phase portraits of system (4.49) at $\gamma = 40$, $s = 3$, $Da = 0.18$ (a) $\beta = 0.268$; (b) 0.2701; (c) 0.2707; (d) 0.28; (e) 0.31; (f) 0.325; (g) 0.327; (h) 0.33

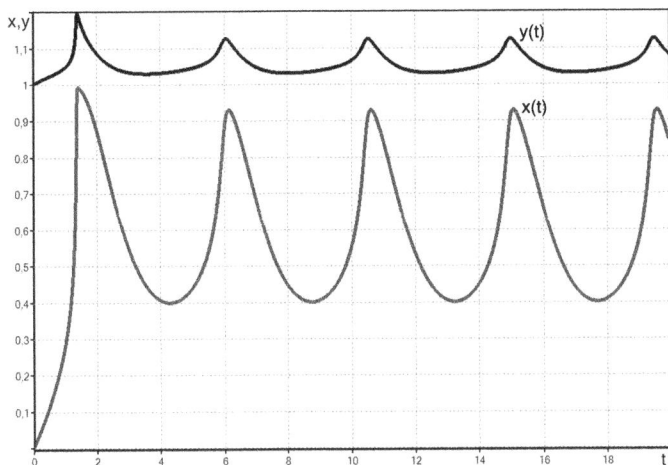

Fig. 4.39: Time dependencies of temperature and concentration at $Da = 0.18$, $\beta = 0.28$, $s = 3$, $\gamma = 40$

steady states are written as

$$L_{\Delta}(Da, \gamma): \quad Da(y, \gamma(y)) = \frac{s(y-1)}{\beta e(y) g(x)} \ ;$$

$$\gamma(y) = \frac{y}{y-1} - \frac{y^2 sg'(x)}{\beta g(x)} \ ,$$

$$L_{\sigma}(Da, \gamma): \quad Da(y, \gamma(y)) = \frac{s(y-1)}{\beta e(y) g(x)} \ ,$$

$$\gamma(y) = \frac{y^2(s+1)}{s(y-1)} - \frac{y^2 g'(x)}{\beta g(x)} \ .$$

Thus, knowing the conditions for the emergence of the Andronov–Hopf bifurcation enables one to estimate the period and the amplitude of possible oscillations. The curve L_{AH} in the plane of the corresponding parameters separates the region of technologically safe operating modes. An analysis of the mutual position of the curves L_{Δ} and L_{σ} allows one to obtain a classification of the main bifurcations of steady states and to partition the parameter region into subregions with uniform phase portraits of dynamic system (4.49). Such a partition performed, *inter alia*, by the Andronov–Hopf bifurcation curve gives quite a complete description of possible dynamic operating modes of a reactor. In modeling a specific reactor, the condition for the existence of critical modes can be evaluated by calculating dimensionless parameters.

4.3.6 Safe and unsafe boundaries of regions of critical phenomena

An important challenge in ensuring the safe operation of chemical reactors in which exothermic reactions are performed is the study of the features of their dynamics with variation of their parameters [53, 54]. One of the emergency scenarios can be such a self-organization of a high energy process that a chain of nonlinear critical phenomena leads to an accident [60]. Variation of the parameters of dynamic systems over wide ranges can give rise to both soft and hard bifurcations, which correspond to safe and unsafe boundaries of regions of critical phenomena [53]. Although parameter ranges where the dynamics is complex are often quite narrow, variation of the external conditions over wide ranges causes a high probability of loss of reactor stability and, as a consequence, accident. The nonlinearity, the unsteadiness, and the high energy of exothermic processes substantially increase the accident risk and require detailed analysis of critical conditions for chemical reactions already at the kinetic level.

In this section, for a model of a well-stirred continuous reactor in which a single exothermic reaction takes place, we determined, for the first time, safe and unsafe boundaries of regions of critical phenomena. In addition, we revealed such initial conditions that the stabilization toward a low-temperature steady state can occur with a considerable dynamic overshoot, which can be considered as emergency. The results obtained are of fundamental importance in evaluating possible emergency modes of exothermic processes in chemical reactors.

The set of nonlinear equations describing the dynamic system to be considered has the form

$$\dot{x} = f(y)(1 - x) - x \,,$$
$$\dot{y} = \beta f(y)(1 - x) - s(y - 1) \,, \tag{4.75}$$

where x and y are the dimensionless concentration and temperature, respectively; $f(y) = Da \exp(y(1 - 1/y))$ and Da, β, y, s are dimensionless parameters. The parametric analysis of system (4.75) enables the construction of its parametric portrait, which is a partition of a plane of some dimensionless parameters by bifurcation curves (the curve L_Δ of multiplicity of steady states and the curve L_σ of neutrality of steady states) into regions with different numbers of steady states and different types of their stability [54]. Figure 4.40 gives an example of such a portrait in the plane (Da, β).

At various values of one of the parameters, e.g., β (and, then, at a fixed Da value), system (4.49) can have from one to three steady states, with their stability changing at the boundary L_σ. We have already studied the soft Andronov–Hopf bifurcation, when crossing the boundary L_σ gives rise to a small limit cycle [53, 54]. Here, we analyze the hard Andronov–Hopf bifurcation, which occurs at the so-called unsafe boundary of the region of critical phenomena (Fig. 4.41). For example, with an increase in the parameter β from 0.3771 to 0.3776, a large limit cycle emerges at once, which can encircle as many as three steady states. In this case, the dynamics of the system results in a considerable high-temperature overshoot. Of interest is the case of the emergence of two limit cycles, one of which is stable and encompasses three steady states, and the

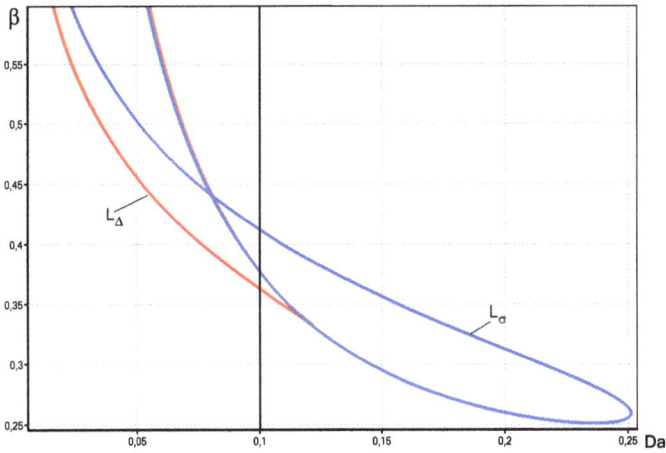

Fig. 4.40: Parametric portrait of dynamic model (4.75) at $s = 3$ and $\gamma = 40$

other is unstable and surrounds a single stable steady state (Fig. 4.41d). With a further increase in β, two steady states merge and vanish, and the limit cycle encircles a single unstable steady state (Fig. 4.41c). Crossing the upper boundary L_σ causes a soft bifurcation (the transition from Fig. 4.41b to Fig. 4.41a), when the limit cycle gradually contracts into a point and collapses.

A practical significance of the plotted curves L_Δ and L_σ and the soft and hard Andronov–Hopf bifurcations emerging on crossing these curves consists of the possibility of determining safe and unsafe critical conditions for each specific exothermic reaction. Crossing the safe boundary gives rise to oscillations with a small amplitude, which, with a further variation of the parameter, initially increases, then decreases, and eventually (on crossing the curve L_σ the second time) vanishes. Such a scenario of gradual development of oscillating modes allows one to take necessary preventive measures against emergency. In a hard bifurcation, auto-oscillations emerge at a jump. Such a mode is technologically unsafe. At the same time, an analysis of the phase portraits of the system (see, e.g., Fig. 4.41) enables one to see a sort of forerunner of hard bifurcation: this is a considerable dynamic overshoot, whose time characteristics are close to the amplitude and the period of the limit cycle to emerge. Thus, a detailed study of the parametric and phase portraits of a dynamic system can give complete information on possible emergencies.

Critical phenomena of sorts can also be induced by varying the initial conditions $x(0)$ and $y(0)$. Figure 4.41f shows the phase portrait of system (4.49) that illustrates the possibility of considerable dynamic overshoots during stabilization of the solutions toward the singular, low-temperature steady state. There is a region of safe initial points: the stabilization from them toward the steady state proceeds quite gradually, without high reactor overheats. However, outside this region, the transition into the steady state occurs with high overheats: albeit short-time, they can be undesirable

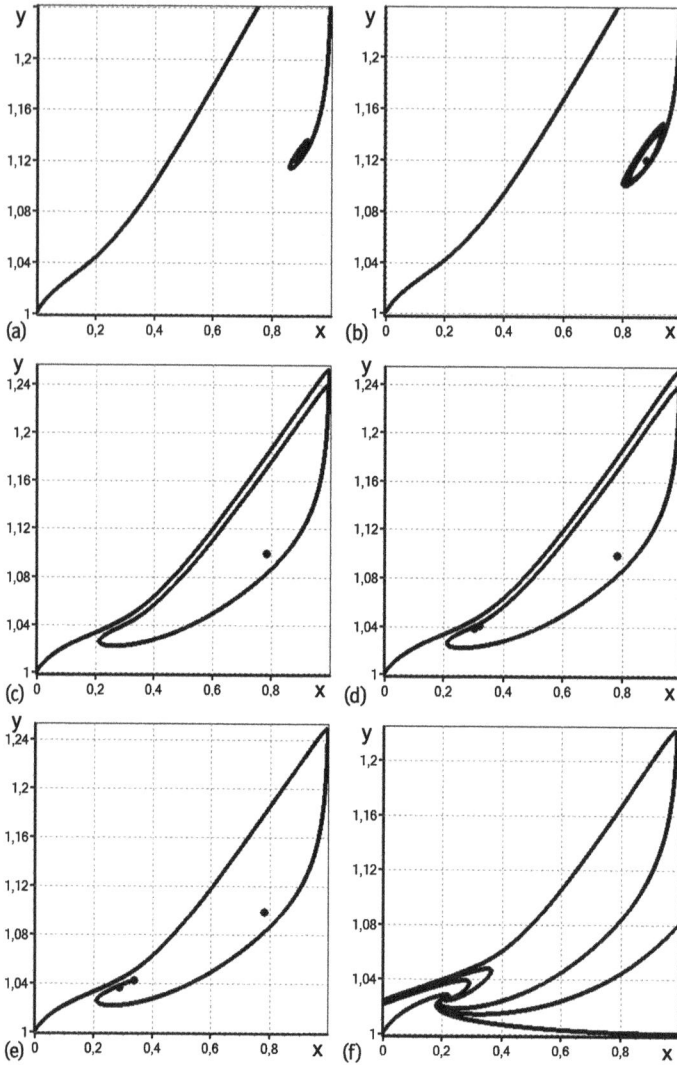

Fig. 4.41: Phase portraits of system (4.49) at $\gamma = 40$, $s = 3$, $Da = 0.1$; and β: (a) 0.42, (b) 0.41, (c) 0.3778, (d) 0.3776, (e) 0.3771, and (f) 0.36

from a safety standpoint. At specific values of the parameters, an analysis of the phase portrait in Fig. 4.41f suggested that the range of safe initial dimensionless temperatures can be estimated at $y(0) \in [1, 1.02]$ (in this case, $x(0) \in [0, 1]$). At $y(0) > 1.02$, the stabilization toward the low-temperature steady state is accompanied by a substantial temperature increase. Naturally, the critical value of the initial data $y(0)$ depends on the values of the dimensionless parameters.

Thus, a necessary step of modeling in assessing the emergency risk in performing a nonlinear exothermic process is the analysis of the dynamics of the process while varying its parameters over wide ranges. Such a possibility is provided by the modern theory of dynamic systems and parametric analysis. As our investigations showed, an important role in finding the critical conditions is played by the form of the kinetic functions. Modern computational experiment facilities enable one to study dynamic models that are more complicated than model (4.49), e.g., models describing a single reaction with arbitrary kinetics or complex reactions. In addition, in modeling specific processes and reactors, one can construct unsafe boundaries in the planes of not only dimensionless parameters but also dimensional parameters, which characterize geometrical, chemical, and thermophysical features of a system. In this case, the parametric analysis can be applied directly to a specific model with dimensional parameters, or the critical conditions can be found indirectly by studying bifurcations in the model with dimensionless parameters. Under the physicochemical assumptions made, a full guarantee of the impossibility of critical phenomena can be given only by a detailed investigation of the parametric and phase portraits of the corresponding dynamic models [56].

4.4 Volter–Salnikov model

Sections 4.2 and 4.3 were devoted to the basic models of the theory of combustion and the theoretical foundations of chemical technology (Zel'dovich–Semenov model and Aris–Amundson model). In this section we will consider one more class of dimensionless models, which was introduced in the works of B. V. Volter and I. E. Salnikov about modeling the dynamics of polymerization. Note that these three classes of models correspond to three time scales of the studied processes. Combustion processes are characterized by times of order 10^{-3} s, chemical-technological processes – 10^0 s, and polymerization – 10^3 s. This significant difference leads to the three systems of nondimensionalization, where each of the models most adequately corresponds to the physical chemistry and dynamics of the process.

Following Volter–Salnikov [33], we will perform some other nondimensionalization of the CSTR (4.1) and carry out its parametric analysis.

Mathematical model. When studying the reaction of the first order $A \to P$ occurring in the CSTR, the authors [33] proposed the following dimensionless variables and parameters:

$$\mu = \frac{hS + C_p\rho q}{k^0 V C_p\rho}, \quad y_0 = \frac{R(hST_x + C_p\rho q T_0)}{E(hS + C_p\rho q)}, \quad \lambda = \frac{q}{k^0 V},$$

$$x_0 = \frac{(-\Delta H)R}{E C_p\rho}X^0, \quad x = \frac{X}{X^0}, \quad y = \frac{RT}{E}, \quad \tau = k^0 t. \tag{4.76}$$

Using the formula (4.76), we write down the model of the CSTR (4.1) in a dimensionless form:

$$\frac{dx}{d\tau} = -x \exp(-1/y) + \lambda(x_0 - x) = f_1(x, y),$$

$$\frac{dy}{d\tau} = x \exp(-1/y) + \mu(y_0 - y) = f_2(x, y). \tag{4.77}$$

According to the scheme of parametric analysis described above, we define the steady states and investigate their stability.

Steady states of system (4.77) are determined as solutions of the equations:

$$f_1(x, y) = 0,$$
$$f_2(x, y) = 0.$$

After excluding x the system of stationarity can be written in the following form:

$$\left(x_0 + \frac{\mu}{\lambda}(y_0 - y)\right) e(y) + \mu(y_0 - y) = 0 \tag{4.78}$$

or, as usual, in terms of a balance of heat production and heat removal

$$F(y) = P(y),$$

where $e(y) = \exp(-1/y)$. The functions of heat production and heat removal are

$$F(y) = \left(x_0 + \frac{\mu}{\lambda}(y_0 - y)\right) e(y),$$
$$P(y) = \mu(y - y_0).$$

Conditions of stability of steady states. Let us write the elements of the Jacobian matrix for (4.77):

$$a_{11} = \frac{\partial f_1}{\partial x} = -e(y) - \lambda,$$

$$a_{12} = \frac{\partial f_1}{\partial y} = -\frac{x}{y^2} e(y),$$

$$a_{21} = \frac{\partial f_2}{\partial x} = e(y),$$

$$a_{22} = \frac{\partial f_2}{\partial y} = \frac{x}{y^2} e(y) - \mu.$$

The coefficients of the characteristic equation are defined as

$$\sigma = a_{11} + a_{22} = \left(x_0 + \frac{\mu}{\lambda}(y_0 - y)\right) \frac{e(y)}{y^2} - e(y) - \lambda - \mu,$$

$$\Delta = a_{11}a_{22} - a_{12}a_{21} = \mu(e(y) + \lambda) - \frac{\lambda e(y)}{y^2}\left(x_0 + \frac{\mu}{\lambda}(y_0 - y)\right).$$

Parametric dependencies. We write down the parametric dependence from equation (4.78) in explicit form:

$$x_0(y) = (y - y_0)\left(\frac{\mu}{\lambda} + \frac{\mu}{e(y)}\right),$$

$$\mu(y) = \frac{\lambda x_0 e(y)}{(y - y_0)(e(y) + \lambda)},$$

$$\lambda(y) = \frac{\mu e(y)(y - y_0)}{x_0 e(y) - \mu(y - y_0)},$$

$$y_0(y) = \frac{\mu y(e(y) + \lambda) - \lambda x_0 e(y)}{\mu(e(y) + \lambda)}. \tag{4.79}$$

To plot the dependencies of the stationary concentrations on parameters let us write the expression as a system of two equations:

$$(x, x_0): \quad \begin{aligned} x_0 &= x_0(y), \\ x(y, x_0(y)) &= x_0(y) + \frac{\mu}{\lambda}(y_0 - y), \end{aligned}$$

$$(x, y_0): \quad \begin{aligned} y_0 &= y_0(y), \\ x(y, y_0(y)) &= x_0 + \frac{\mu}{\lambda}(y_0(y) - y)), \end{aligned}$$

$$(x, \mu): \quad \begin{aligned} \mu &= \mu(y), \\ x(y, \mu(y)) &= x_0 + \frac{\mu(y)}{\lambda}(y_0 - y)), \end{aligned}$$

$$(x, \lambda): \quad \begin{aligned} \lambda &= \lambda(y), \\ x(y, \lambda(y)) &= x_0 + \frac{\mu}{\lambda(y)}(y_0 - y)), \end{aligned}$$

An example of a plot of parametric dependencies (4.79) analyzing the impact of other parameters on them is shown in Fig. 4.42.

The use of explicit expressions of the dependencies of steady states on the parameters of type (4.79) allows us to plot the dependencies of the eigenvalues of the Jacobian matrix in the steady state on the parameters.

Curves of multiplicity. To plot the curves of multiplicity, for example, in the parameter plane (μ, λ), it is necessary to solve the following system:

$$L_\Delta: \quad \begin{aligned} G(y, \mu, \lambda) &= 0, \\ \Delta(y, \mu, \lambda) &= 0, \end{aligned} \tag{4.80}$$

which can be written in the following form:

$$L_\Delta: \quad \begin{aligned} \left(x_0 + \frac{\mu}{\lambda}(y_0 - y)\right)e(y) + \mu(y_0 - y) &= 0, \\ \mu(e(y) + \lambda) - \frac{\lambda e(y)}{y^2}\left(x_0 + \frac{\mu}{\lambda}(y_0 - y)\right) &= 0. \end{aligned}$$

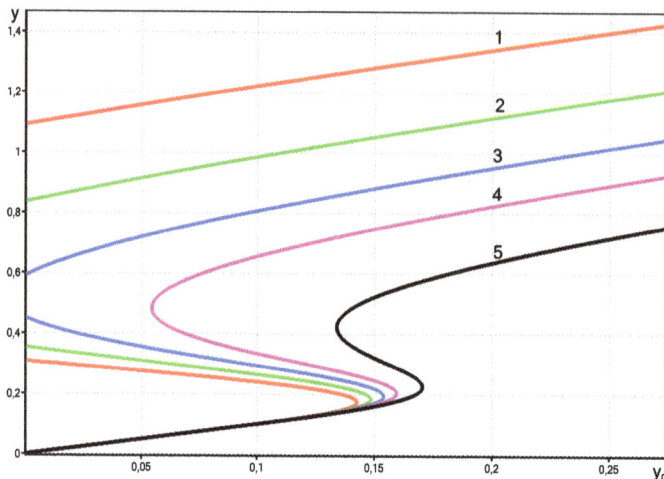

Fig. 4.42: Dependencies of stationary temperature y on parameter y_0 at $\lambda = 0.15$, $x_0 = 2.5$, 1) $\mu = 0.25$; 2) 0.3; 3) 0.35; 4) 0.4; 5) 0.5

Substituting the explicit expression $\lambda(y, \mu)$ of (4.79) into the second equation of system (4.80), we get the expression $\mu(y)$:

$$L_\Delta(\mu, \lambda): \qquad \mu(y) = \frac{x_0 y^2 e(y)}{C^2}, \tag{4.81}$$

$$\lambda(y) = \lambda(y, \mu(y)) .$$

Similarly we write L_Δ for other planes of parameters:

$$L_\Delta(y_0, \lambda): \qquad y_0(y) = y\left(1 - \sqrt{x_0 e(y)/\mu}\right) ,$$

$$\lambda(y) = \lambda(y, y_0(y)) .$$

$$L_\Delta(x_0, \lambda): \qquad x_0(y) = \frac{\mu C^2}{y^2 e(y)} ,$$

$$\lambda(y) = \lambda(y, x_0(y)) .$$

$$L_\Delta(y_0, \mu): \qquad y_0(y) = \frac{y(\lambda(1 - y) - y e(y))}{\lambda} ,$$

$$\mu(y) = \mu(y, y_0(y)) .$$

$$L_\Delta(x_0, y_0): \qquad y_0(y) = \frac{y(\lambda(1 - y) - y e(y))}{\lambda} , \tag{4.82}$$

$$x_0(y) = x_0(y, y_0(y)) ,$$

where $C = y - y_0$.

Examples of the curves of multiplicity (4.81)–(4.82) in different planes of the dimensionless parameters are given in Fig. 4.43–4.44.

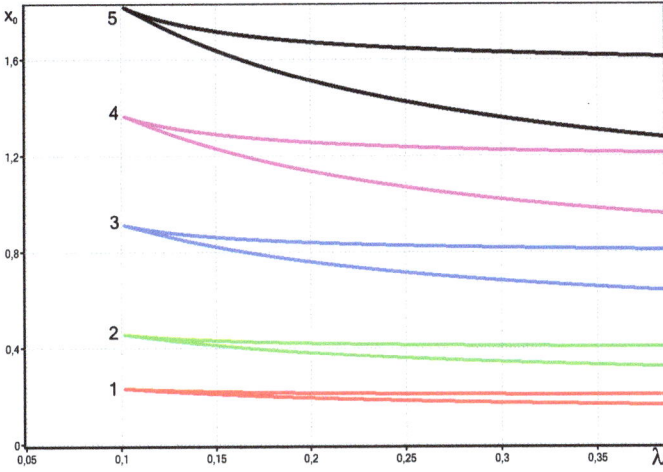

Fig. 4.43: Curves of multiplicity $L_\Delta(\lambda, x_0)$ at $y_0 = 0.18$, 1) $\mu = 0.05$; 2) 0.1; 3) 0.2; 4) 0.3; 5) 0.4

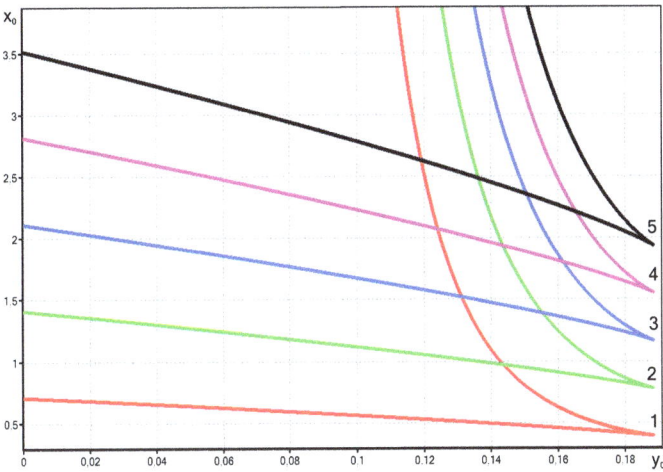

Fig. 4.44: Curves of multiplicity $L_\Delta(y_0, x_0)$ at $\lambda = 0.15$, 1) $\mu = 0.1$; 2) 0.2; 3) 0.3; 4) 0.4; 5) 0.5

Curves of neutrality. Let us write the expression of the neutrality curve in the parameter plane (μ, λ):

$$L_\sigma: \quad \begin{aligned} G(y, \mu, \lambda) &= 0 \,, \\ \sigma(y, \mu, \lambda) &= 0 \,. \end{aligned}$$

The expressions used to plot the curve L_σ for different planes of the parameters have the following form:

$$L_\sigma(x_0, \lambda): \quad \begin{aligned} x_0(y) &= \frac{\mu^2 C(y - y_0 - y^2)}{e(y)(\mu C - y^2(\mu + e(y)))} \,, \\ \lambda(y) &= \lambda(y, x_0(y)) \,. \end{aligned}$$

$$L_\sigma(y_0, x_0): \quad \begin{aligned} y_0(y) &= y\left(1 - \frac{y(e(y) + \lambda + \mu)}{\mu}\right) \,, \\ x_0(y) &= x_0(y, y_0(y)) \,. \end{aligned}$$

$$L_\sigma(x_0, \mu): \quad \begin{aligned} x_0(y) &= \frac{y^2 C(e(y) + \lambda)^2}{\lambda e(y)(y - y_0 + y^2)} \,, \\ \mu(y) &= \mu(y, x_0(y)) \,. \end{aligned}$$

$$L_\sigma(y_0, \mu): \quad \begin{aligned} y_0(y) &= \frac{y(\lambda x_0 r(y)(1 - y) - y^2(e(y) + \lambda)^2)}{\lambda x_0 e(y) - y^2(e(y) + \lambda)^2} \,, \\ \mu(y) &= \mu(y, y_0(y)) \,. \end{aligned}$$

$$L_\sigma(\lambda, \mu): \quad \begin{aligned} \lambda(y) &= \frac{e(y)(CD - x_0 y^2)}{2y^2 C} \pm \sqrt{\left(\frac{e(y)(CD - x_0 y^2)}{2y^2 C}\right)^2 - e^2(y)} \,, \\ \mu(y) &= \mu(y, \lambda(y)) \,, \end{aligned} \tag{4.83}$$

where $C = y - y_0$ and $D = (x_0 - 2y^2)$.

Examples of plots of the curves of neutrality (4.83) while varying the third parameter are given in Fig. 4.45–4.46.

Parametric portraits. One of the variants of the parametric portraits for the Volter–Salnikov model in accordance with formulas (4.81)–(4.82), (4.83) is shown in Fig. 4.47. The explicit form of the corresponding dependencies gives the possibility to simply study the impact of third parameters on the appearance of curves L_Δ, L_σ.

Phase portraits. In accordance with the parametric portrait of system (4.77) we can identify, as above, several types of phase portraits. The most characteristic phase portraits for the one stable and one unstable steady states are presented in Fig. 4.48 and Fig. 4.49. In the second case there are auto-oscillations in the system.

As seen, the phase portraits corresponding to the singular stable (Fig. 4.48) and the singular unstable steady state (Fig. 4.49), are significantly different. In the first case the dynamic system from any initial data stabilizes to the stable steady state. In this case, however, this stabilization is not always monotone. It can occur with significant dynamic casts (the boomerang effect). In the second case, from any initial data the solutions of the model lead to the limit cycle, which characterizes undamped oscillations. This limit cycle is stable. The calculation of the time dependencies $x(t)$, $y(t)$

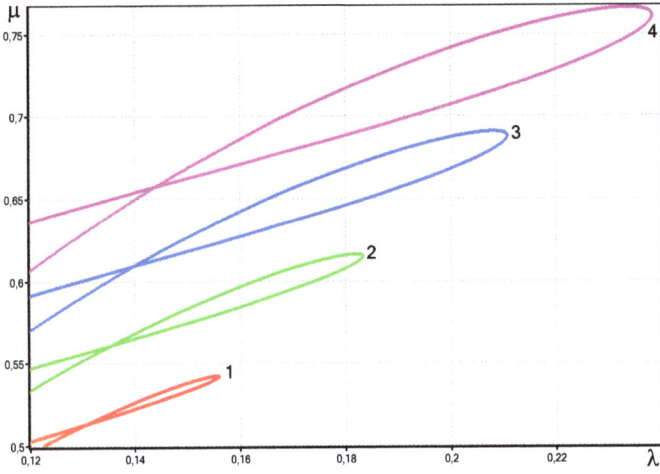

Fig. 4.45: Curves of neutrality $L_\sigma(\lambda, \mu)$ at $y_0 = 0.18$, 1) $x_0 = 2.2$; 2) 2.4; 3) 2.6; 4) 2.8

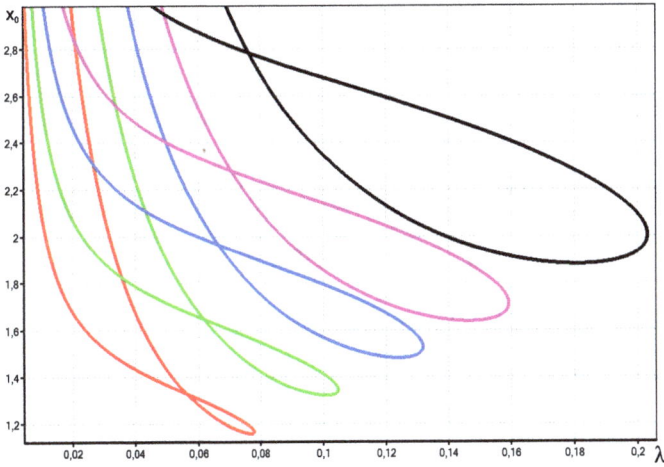

Fig. 4.46: Curves of neutrality $L_\sigma(\lambda, x_0)$ at $y_0 = 0.16$, 1) $\mu = 0.2$; 2) 0.25; 3) 0.3; 4) 0.35; 5) 0.43

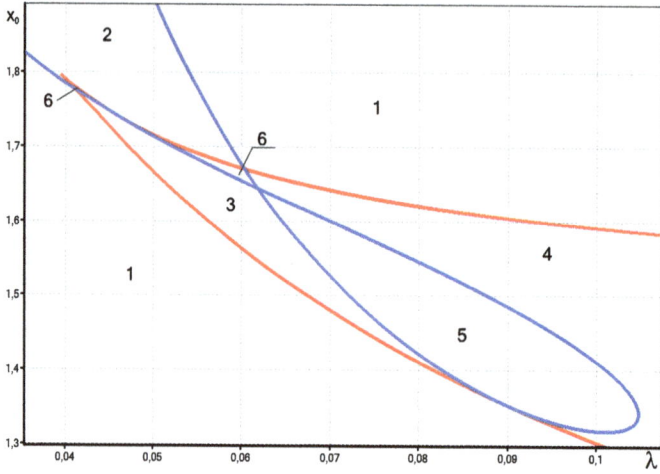

Fig. 4.47: Parametric portrait on the plane (λ, x_0) at $y_0 = 0.16$, $\mu = 0.25$

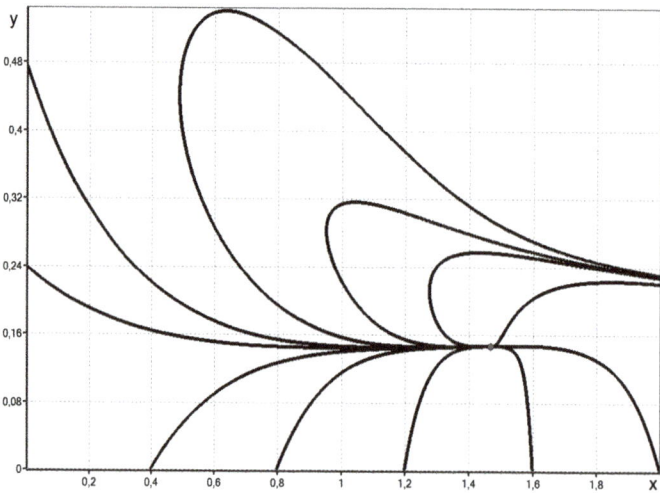

Fig. 4.48: Phase portrait for the single stable steady state at $y_0 = 0.14$, $\mu = 0.25$, $\lambda = 0.05$, $x_0 = 1.5$

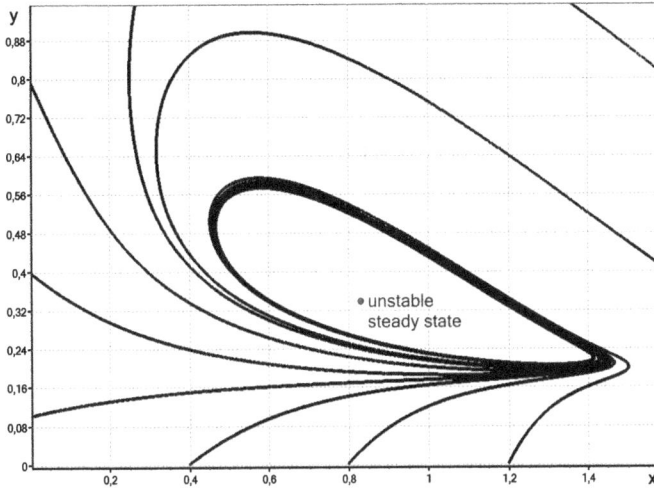

Fig. 4.49: Phase portrait for the single unstable steady state at $y_0 = 0.165, \mu = 0.25, \lambda = 0.05$, $x_0 = 1.7$

allows us to obtain estimates of the oscillation period. Their amplitude can be estimated based on the corresponding phase portraits.

4.5 Models of a continuous stirred tank reactor and a tube reactor

A parametric analysis of models with dimensionless parameters was performed above. The obtained results are of methodical and theoretical interest. However, in modeling specific processes and reactors, it is necessary to analyze models with parameters possessing a definite physicochemical meaning by numerical and qualitative methods. Therefore, a parametric analysis should also be performed for specific mathematical models that correspond to the geometrical, thermophysical characteristics of a reactor considered.

In this section, we illustrate the parametric analysis procedure using a dimensional model of a continuous stirred tank reactor (CSTR), where only one first-order exothermic reaction occurs. The relationship between the results of the parametric analysis of dimensionless and dimensional models is established. Moreover, we compare the dynamic properties of a continuous stirred tank (spherical) reactor and continuous tube (cylindrical) reactor characterized by similar physical and technological parameters in which the same chemical reaction occurs [46].

4.5.1 Parametric analysis of a dimensional model

The mathematical model of a spherical continuous stirred tank reactor (Frank-Kamenetskii reactor) has the form

$$V\frac{dX}{dt} = -Vk(T)X + q(X^0 - X) = f(X, T) ,$$

$$C_p\rho V\frac{dT}{dt} = (-\Delta H)Vk(T)X + qC_p\rho(T_0 - T) + hS(T_w - T) = g(\ldots X, T) , \quad (4.84)$$

where the notation is similar to that in Section 4.1.

A parametric analysis of the dimensional model (4.84) is performed according to the procedure applied to dimensionless models (4.2), (4.49), (4.77).

Steady states. Equating the right sides of system (4.84), we obtain the stationary conditions

$$- Vk(T)X + q(X^0 - X) = 0 ,$$

$$(-\Delta H)Vk(T)X + qC_p\rho(T_0 - T) + hS(T_w - T) = 0 , \quad (4.85)$$

which imply that

$$X = \frac{qC_p\rho(T_0 - T) + hS(T_w - T)}{q(-\Delta H)} + X^0 . \quad (4.86)$$

Inserting (4.86) into the second equation in (4.85), we obtain the equation for determining the stationary values of temperatures:

$$(-\Delta H)Vk(T)X^0 + \frac{Vk(T)}{q}\left(qC_p\rho(T_0 - T) + +hS(T_w - T)\right)$$

$$+ qC_p\rho(T_0 - T) + hS(T_w - T) = 0 . \quad (4.87)$$

Tab. 4.1: Basic set of parameters of model (4.84)

Physical parameters	Values
Inlet temperature, T_0	583 K
Inlet concentration of the substance, X^0	$1.4 \cdot 10^{-6}$ mol cm^{-3}
Density of the mixture, ρ	$3 \cdot 10^{-5}$ mol cm^{-3}
Specific heat of the mixture, C_p	2166.2 J mol^{-1} K^{-1}
Volume of reactor, V	588.75 cm^3
Area of the reactor surface, S	471 cm^2
Coefficient of heat transfer, h	$8 \cdot 10^{-3}$ J cm^{-2} s^{-1} K^{-1}
Thermal effect of the reaction, $(-\Delta H)$	$4,87 \cdot 10^6$ J mol^{-1}
Volume flow rate, q	180 cm^3 s^{-1}
Temperature of the reactor walls, T_w	583 K
Activation energy, E	$1.89 \cdot 10^5$ J mol^{-1}
Universal gas constant, R	8.31 J mol^{-1} K^{-1})
Pre-exponent, k^0	$2.3 \cdot 10^{15}$ s^{-1}

As the basic parameters of model (4.84), we use quantities corresponding to experimental data on the oxidation of hydrocarbons in a continuous stirred tank reactor [61–63]. In the parametric analysis, some parameters given in Tab. 4.1 were slightly varied. The varied parameters include the inlet temperature (T_0) and concentration (X^0) and the heat transfer coefficient (h).

Parametric dependencies. Following the general procedure of parametric analysis [46], from (4.87) we obtain the dependencies inverse to the desired dependencies of the stationary temperature and concentration on the parameters of the system:

$$X^0(T) = \frac{C_p\rho(T - T_0)(q + Vk(T)) + hS(T - T_w)(1 + Vk(T)/q)}{(-\Delta H)Vk(T)} ,$$

$$T_0(T) = T - \frac{(-\Delta H)Vk(T)X^0 - hS(T - T_w)(1 + Vk(T)/q)}{C_p\rho(q + Vk(T))} ,$$

$$V(T) = \frac{qC_p\rho(T - T_0) + hS(T - T_w)}{k(T)\left((-\Delta H)X^0 - C_p\rho(T - T_0) - (hS/q)(T - T_w)\right)} ,$$

$$S(T) = \frac{(-\Delta H)Vk(T)X^0 - C_p\rho(T - T_0)(q + Vk(T))}{h(T - T_w)(1 + Vk(T)/q)} ,$$

$$\rho(T) = \frac{(-\Delta H)Vk(T)X^0 - hS(T - T_w)(1 + Vk(T)/q)}{C_p(T - T_0)(q + Vk(T))} ,$$

$$h(T) = \frac{(-\Delta H)Vk(T)X^0 - C_p\rho(T - T_0)(q + Vk(T))}{S(T - T_w)(1 + Vk(T)/q)} . \tag{4.88}$$

Similarly to (4.88), one can easily obtain dependencies on other parameters. Figures 4.50 and 4.51 show some parametric dependencies in accordance with the formulas (4.88). These curves are characterized by a region of multivalued solutions, which corresponds to multiple steady states. With a second parameter varied (curves 3–5 in Fig. 4.51), the region of multiple steady states changes significantly or even disappears (curve 1). These parametric dependencies allow one to estimate the conditions under which hysteresis and critical phenomena exist for variation in various parameters of the reactor considered. Parametric dependencies of the stationary concentrations X are constructed with the use of relation (4.86) where T is replaced by the corresponding dependencies $T(X^0)$, $T(T_0)$, etc. Here and further the values of the parameters are taken from the Tab. 4.1.

For multiple steady states, the parametric dependence of the stationary temperature (for example, curve 2 in Fig. 4.50) has three branches of low, high, and intermediate temperatures. Transition from one branch to another is performed by jumps. Upon attainment of a certain critical state, the temperature of the reactor is changed abruptly, i.e., its "ignition" or "extinction" occurs.

Stability of steady states is determined by the roots of the characteristic equation

$$\lambda^2 - \sigma\lambda + \Delta = 0 ,$$

where

$$\sigma = a_{11} + a_{22}, \quad \Delta = a_{11}a_{22} - a_{12}a_{21} ,$$

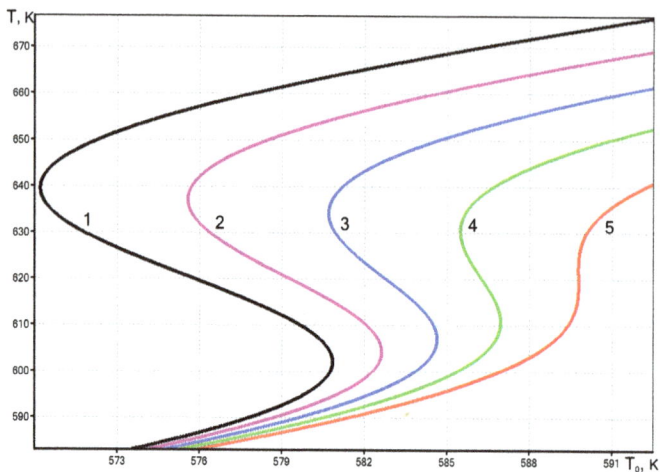

Fig. 4.50: Dependencies of stationary temperature on inlet temperature T_0 at 1) $X^0 = 1.6 \cdot 10^{-6}$; 2) $1.5 \cdot 10^{-6}$; 3) $1.4 \cdot 10^{-6}$; 4) $1.3 \cdot 10^{-6}$; 5) $1.2 \cdot 10^{-6}$

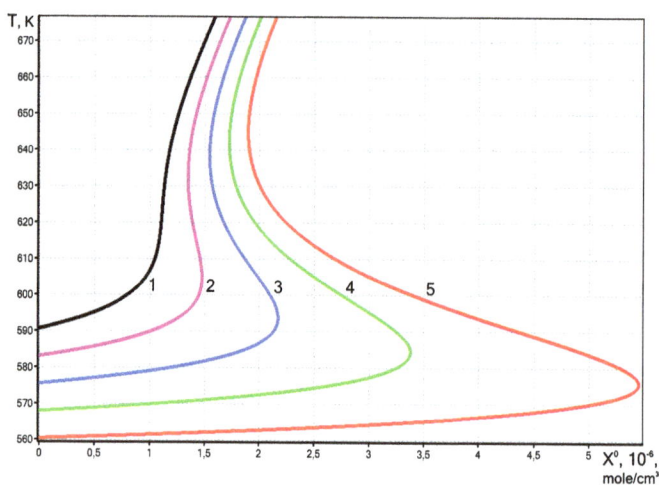

Fig. 4.51: Dependencies of stationary temperature on inlet concentration X^0 at 1) $T_0 = 593$; 2) 583; 3) 573; 4) 563; 5) 553

a_{ij} are the elements of the Jacobian matrix of the right sides of system (4.84), calculated in the steady state:

$$a_{11} = \partial f/\partial X, \quad a_{12} = \partial f/\partial T, \quad a_{21} = \partial g/\partial X, \quad a_{22} = \partial g/\partial T.$$

The steady state is unstable if $\Delta < 0$. If $\Delta > 0$, the steady state is stable for $\sigma < 0$ and unstable for $\sigma > 0$. Consequently, for $\Delta = 0$ or $\sigma = 0$, the type of stability of the steady states and their number change, i.e., a bifurcation occurs. Therefore, the conditions that Δ and σ vanish yield important information on the number and type of steady states. Combined with the stationary condition (4.87), these equalities allow one to construct bifurcation curves for various combinations of parameters: curves of multiplicity ($\Delta = 0$) and neutrality ($\sigma = 0$) of steady states.

Bifurcation curves. For model (4.84), as in the case of dimensionless models [28], we write equations for constructing local bifurcation curves L_Δ ($\Delta = 0$) and L_σ ($\sigma = 0$) in various parameter planes. Combining, for example, the equation $\Delta(X^0, V) = 0$ with (4.87), one can obtain the dependence $X^0(T, V)$, which together with (4.88) defines a multiplicity curve in the parameter plane (X^0, V).

Multiplicity Curves:

$$L_\Delta(X^0, V): \quad X^0 = X^0(T, V),$$

$$V(T) = \frac{E}{RT^2}\left(\frac{q^2(T - T_0) + qA(T - T_w)}{k(T)(q + A)}\right) - \frac{q}{k(T)},$$

$$L_\Delta(X^0, T_0): \quad X^0 = X^0(T, T_0),$$

$$T_0(T) = T + \frac{A}{q}(T - T_w) - \frac{RT^2(q + Vk(T))(q + A)}{Eq^2},$$

$$L_\Delta(T_0, \rho): \quad T_0 = T_0(T, \rho),$$

$$\rho(T) = \frac{E}{RT^2}\frac{qk(T)(-\Delta H)X^0}{C_p(q + Vk(T))(k(T) + q/V)} - \frac{hS}{qC_p}, \tag{4.89}$$

where $A = hS/(C_p\rho)$.

Neutrality Curves:

$$L_\sigma(X^0, T_0): \quad X^0 = X^0(T, T_0),$$

$$T_0(T) = T - \frac{A}{q}\left(\frac{1}{B} - (T - T_w)\right) - \frac{1}{qB}(2q + Vk(T)),$$

$$L_\sigma(X^0, V): \quad X^0 = X^0(T, V),$$

$$V(T) = qB(T - T_0) - A(1 - B(T - T_w) - 2q)/k(T),$$

$$L_\sigma(T_0, \rho): \quad T_0 = T_0(T, \rho),$$

$$\rho(T) = \frac{Bq(-\Delta H)k(T)X^0}{C_p(q + Vk(T))(k(T) + 2q/V)} - \frac{hS}{C_p(2q + Vk(T))}, \tag{4.90}$$

where $B = E/(RT^2)$.

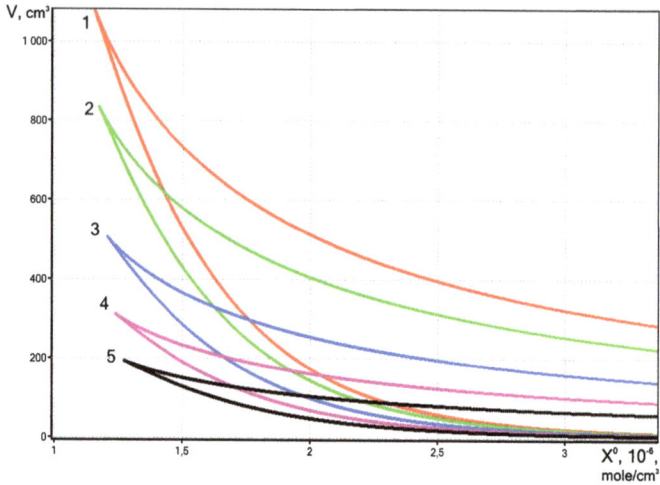

Fig. 4.52: Multiplicity curves $L_\Delta(X^0, V)$ at 1) $T_0 = 578$; 2) 583; 3) 593; 4) 603; 5) 613

Figures 4.52–4.55 show some bifurcation curves obtained by formulas (4.89) and (4.90). Several multiplicity curves L_Δ in the (X^0, V) plane obtained by varying a third parameter (T_0) are shown in Fig. 4.52. There are three steady states inside the region bounded by the curve L_Δ, and there is one steady state outside this region. One can see from Fig. 4.52 that the region of multiple steady states has the largest size at relatively low inlet temperatures and at all other factors being equal. The size of the region bounded by the neutrality curve L_σ (see Fig. 4.55) also depend strongly on the parameter T_0. In passing through the boundary, the steady state changes its type of stability.

Aligning the curves L_Δ and L_σ on the same parameter plane, one obtains a so-called parametric portrait of the system. The mutual position of the curves L_Δ and L_σ (see Fig. 4.56–4.57) determines the division of this region into subregions with different number of steady states (one or three) and different type of stability (stable or unstable steady state). In Fig. 4.57 one can see the region with a unique and unstable steady state (inside the loop of the neutrality curve L_σ), which is responsible for self-oscillatory solutions of system (4.84). Outside the regions bounded by the curves L_Δ and L_σ, the steady state is unique and stable.

Each subregion of the parametric portrait of system (4.84) (see, for example, Fig. 4.56) has its type of phase portrait which reflects the specific features of the dynamics of the system and the relative position of its phase trajectories for varied initial data. Figure 4.58 shows a phase portrait corresponding to the case of a unique and unstable steady state, which ensures the existence of a limiting cycle for system (4.84). If the initial data lie outside this cycle, all phase trajectories are "wound" on it as $t \to \infty$.

If it is necessary to construct bifurcation curves of the type (4.89) and (4.90) in other parameter planes, one can either use the geometrical procedure proposed

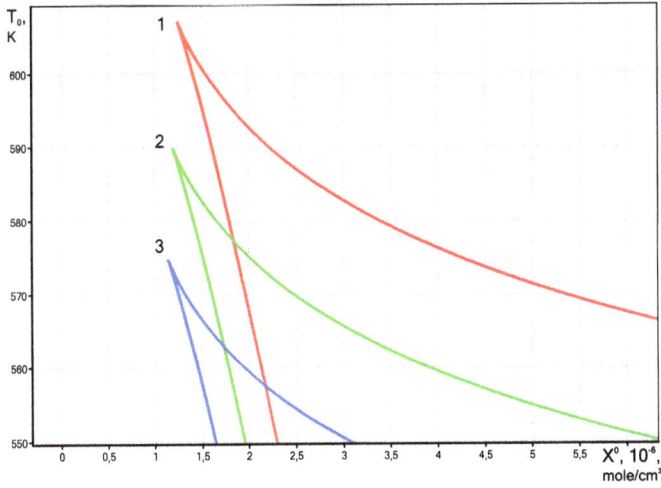

Fig. 4.53: Multiplicity curves $L_\Delta(X^0, T_0)$ at 1) $k^0 = 10^{15}$; 2) $2.3 \cdot 10^{15}$; 3) $5 \cdot 10^{15}$

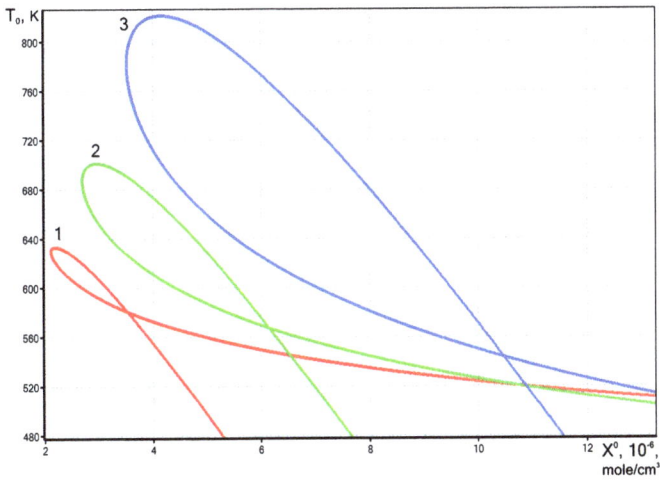

Fig. 4.54: Neutrality curves $L_\sigma(X^0, T_0)$ at 1) $h = 0.05$; 2) 0.09; 3) 0.15

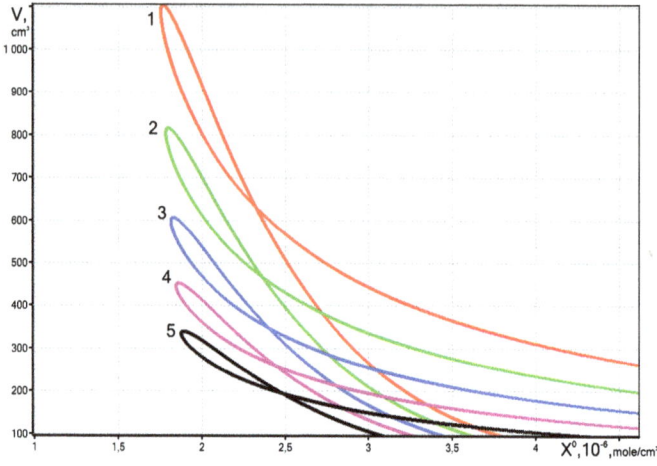

Fig. 4.55: Neutrality curves $L_\sigma(X^0, V)$ at $h = 0.03$, 1) $T_0 = 583$; 2) 593; 3) 603; 4) 613; 5) 623

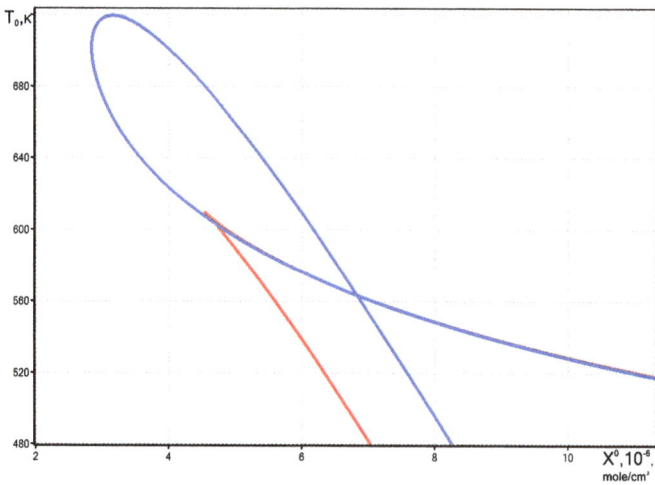

Fig. 4.56: Parametric portrait in the (X^0, T_0) plane at $h = 0.1$

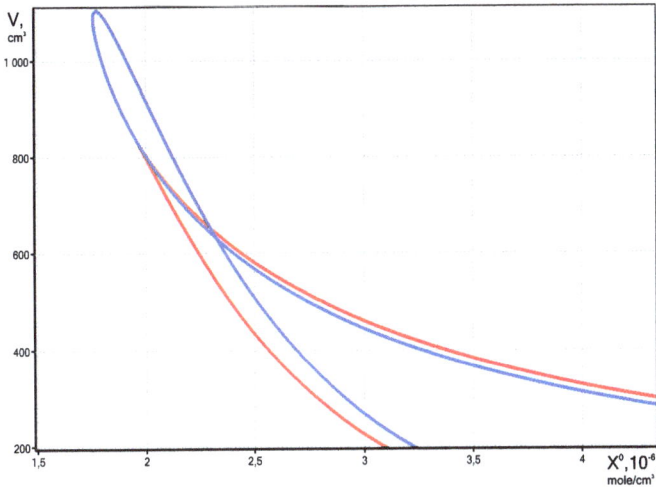

Fig. 4.57: Parametric portrait of model (4.84) in the (X^0, V) plane at $h = 0.03$

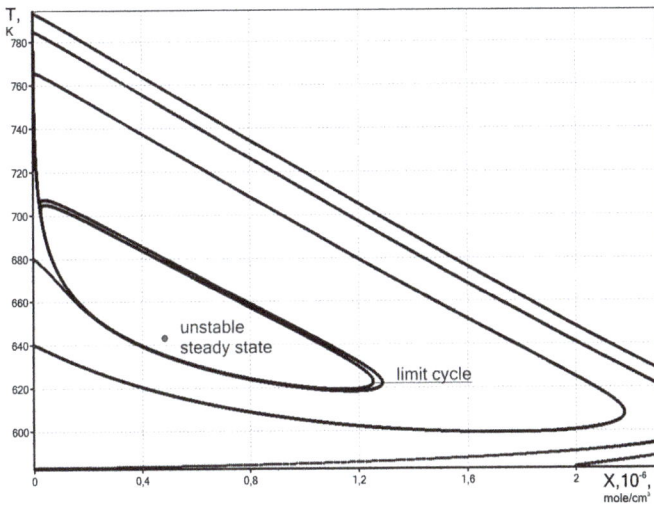

Fig. 4.58: Phase portrait of system (4.84) for the unique and unstable steady state at $h = 0.1$, $V = 1,500$, $X^0 = 4.5 \cdot 10^{-6}$. [inside the limit cycle, there is one unstable steady state (thick curve)]

above, which is based on explicit parametric relations of the type (4.88), or the numerical parameter-continuation method, which we modified for the case of one equation with a parameter $G(x, \alpha) = 0$ [44].

4.5.2 Relation between dimensionless and dimensional models

With the parametric portrait of a model with dimensionless parameters, one can readily construct the parametric portrait in the plane of dimensional parameters. For illustration, we consider the Aris–Amundson model widely used in the theory of chemical reactors [45]:

$$\frac{dx}{dt} = f(y)(1 - x) - x ,$$

$$\frac{dy}{dt} = \beta f(y)(1 - x) - s(y - 1) . \tag{4.91}$$

Here x and y are the dimensionless concentration and temperature, respectively, and $f(y) = Da \exp(y(1 - 1/y))$. Introducing the dimensionless parameters

$$y = \frac{E}{RT_0}, \quad s = 1 + \frac{hS}{q\rho C_p} ,$$

$$Da = \frac{V}{q} k(T_0), \quad \beta = \frac{(-\Delta H)X^0}{\rho C_p T_0}$$

one can obtain, for example,

$$V = \frac{q}{k(T_0)} Da , \quad X^0 = \frac{\rho C_p T_0}{(-\Delta H)} \beta . \tag{4.92}$$

To construct the parametric portrait in the (V, X^0) plane with the use of the parametric portrait in the dimensional (Da, β) plane and formulas (4.92), it is necessary to specify the dimensionless parameters y and s for which the last portrait is obtained. Substituting the calculated values of Da and β into the relation between dimensionless and dimensional parameters (4.92), we obtain the bifurcation values of the dimensional parameters V and X^0. For these values, we construct the parametric portrait in the (V, X^0) plane.

Figure 4.59 shows the parametric portrait for model (4.84) constructed in the (V, X^0) plane by analyzing the corresponding dimensionless model (4.91) in the (Da, β) plane. In contrast to Fig. 4.57, the parametric portrait in Fig. 4.59 is constructed for slightly different values of the parameters. In this case, the region of self-oscillatory regimes (inside the loop of the curve L_σ) is large; the curves L_Δ and L_σ are constructed with the use of the parametric portrait of system (4.91) given in the Section 4.3.1. The parametric portrait can be similarly constructed in other planes of dimensional parameters; in some cases, however, technical difficulties arise due to the fact that the same dimensional parameter can enter several dimensionless parameters. In this case, it is necessary to perform a direct parametric analysis of the dimensional model (4.84).

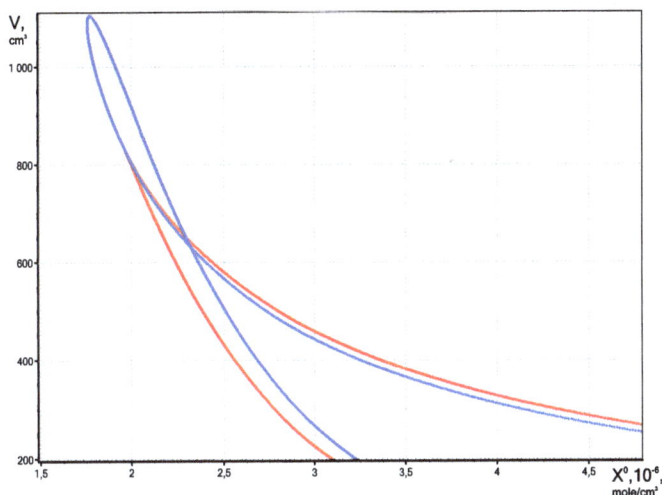

Fig. 4.59: Parametric portrait of model (4.84) in the (X^0, V) plane at $h = 0.03$

4.5.3 Determination of ignition boundaries

To ensure the technological safety of chemical processes, it is necessary to determine, among other factors, the limits of existence of low- and high-temperature regimes. An analysis of time dependences and corresponding phase portraits shows that a steady state can be attained with a considerable dynamic "throw." Even though a low-temperature steady state exists, there is a wide range of the initial data for which the transition of the system to the steady states is accompanied by intense heating during the reaction. Therefore, we consider dimensional and dimensionless models to find a region of technologically safe regimes characterized by monotonic transition to steady states in one of the parameter planes.

We consider the Zel'dovich–Semenov model for a first-order reaction (4.2). The corresponding parametric portrait is shown in Fig. 4.6. An analysis of the phase portrait of the dynamic system with a low-temperature steady state shows that there exists a region of the initial data for which the transition to the steady state is monotonic. Outside this region, the transition of the system from the initial data to the steady state is accompanied by considerable heating. It is quite possible that abrupt heating of the system can lead to an explosion or fire. Therefore, the line that separates regimes with low- and high-temperature steady state may be called the boundary of technological or fire safety.

For illustration, we also calculated the dimensional model of a CSTR (4.84) in which an exothermic reaction occurs. Figure 4.57 shows the corresponding parametric portrait in the (V, X^0) plane. The technological safety boundary in the chosen parameter plane corresponds to relatively low values of the inlet concentration X^0 and volume of the reactor V, i.e., it lies below and to the left of the curves L_Δ and L_σ. Since the bifurcation curves can be constructed by explicit formulas of the type (4.89)–(4.90), one can study the effect of various characteristics of the reactor on the position of this curve.

Thus, an analysis of the parametric and phase portraits of the corresponding mathematical model allows one to determine regimes of technological safety of the processes.

4.5.4 Continuous tube reactor

In addition to the model of a CSTR considered above, the model of a continuous tube reactor describing processes in tube reactors is widely used in mathematical modeling of combustion processes. It is, therefore, of interest to compare the dynamic character-istics of a continuous stirred reactor and continuous tube reactor in which the same re-action occurs, provided the thermal and geometrical characteristics of the reactors are similar. As an example, we consider the mathematical models for a first-order reaction and find the relation between the dynamic characteristics of a continuous stirred re-actor and the concentration and temperature profiles in a stationary continuous tube reactor.

In the stationary case, the mathematical model of a continuous tube reactor in which an exothermic reaction occurs has the form

$$u \frac{dX}{d\ell} = -k(T)X ,$$

$$C_p \rho u \frac{dT}{d\ell} = (-\Delta H)k(T)X + \frac{4h}{d}(T_w - T) , \qquad (4.93)$$

where $k(T) = k^0 \exp(-E/(RT))$ is the reaction rate constant; k^0 is the pre-exponent; E is an activation energy; R is the universal gas constant; X is the current concentration; T_w is the temperature of the reactor walls; T is the current temperature; ℓ is the current length of the reactor; h is the coefficient of heat transfer through the reactor wall; d is diameter of the tube reactor; ρ, C_p, u are the density of the mixture, the specific heat of the reactive mixture and the velocity of injection of the reactive mixture; $(-\Delta H)$ is the thermal effect of the reaction.

In model (4.93), the current length of the reactor ℓ varies within the limits

$$0 \leq \ell \leq L ,$$

where L is the total length of the tube reactor. The inlet conditions have the form

$$X(0) = X^0 , \qquad T(0) = T_0 ,$$

where X^0 is the inlet concentration of the reactive mixture and T_0 is the inlet temper-ature. The volume flow rate is related to the velocity of the mixture by the formula

$$u = 4q/(\pi d^2) .$$

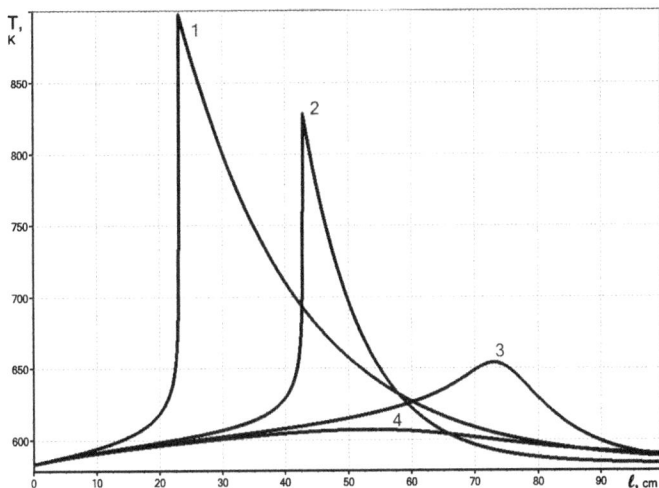

Fig. 4.60: Temperature distribution along the reactor at $u = 9.171975$, $T_0 = 583$, $X^0 = 4.5 \cdot 10^{-6}$,
1) $h = 0.04$; 2) 0.08; 3) 0.087; 4) 0.09

The values of thermal and physical parameters and the ranges of their variation are taken to be the same as for the model of a continuous stirred tank reactor (see Table 4.1).

The diameter and length of the tube reactor are chosen from the condition that its volume is equal to the volume of the continuous stirred tank reactor. Figures 4.60 and 4.61 compare the auto-oscillatory regime in the continuous stirred tank reactor with the corresponding temperature and concentration profiles in the tube reactor. The auto-oscillations in the continuous stirred reactor are characterized by a periodic increase and decrease in temperature, and the corresponding temperature profile in the tube reactor has a pronounced hot spot. The calculations show that the position of the hot spot can be determined approximately by the formula $\ell_* = ut_*$, where t_* is the period of auto-oscillations in the stirred reactor. Thus, the knowledge of the dynamic characteristics of the stirred reactor allows one to estimate the position of the hot spot in the tube reactor. However, our calculations show that the quantity ℓ_* may be used as a first approximation to determine the position of the hot spot.

It is noteworthy that the temperature of the hot spot is highly sensitive to variation in the parameters. There exists a peculiar bifurcation value of parameters for which a smooth temperature profile becomes a profile with a sharp rise in temperature, which can be interpreted as a thermal explosion in the tube. These profiles are shown in Fig. 4.60 and 4.61, where the coefficient of heat transfer through the reactor wall is the varied parameter. A similar effect is produced by varying the inlet concentration of the reacting substance.

Mathematical models of specific physicochemical processes contain dimensional variables and parameters that characterize a given geometry of the reactor and partic-

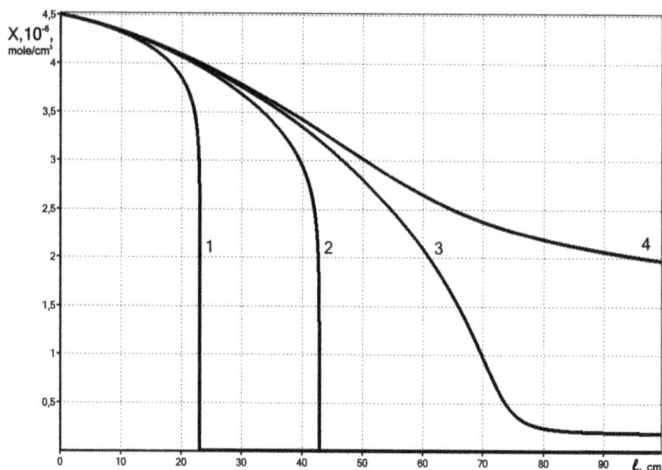

Fig. 4.61: Concentration distribution along the reactor at $u = 9.171975$, $T_0 = 583$, $X^0 = 4.5 \cdot 10^{-6}$, 1) $h = 0.04$; 2) 0.08; 3) 0.087; 4) 0.09

ular physical and kinetic properties of the reaction. In this case, a parametric analysis of the dimensional model can be performed directly for various combinations of dimensional parameters or indirectly by using the results of the parametric analysis of dimensionless models. For example, at least the following three dimensionless models correspond to the basic model of combustion theory (4.84), which describes one exothermic reaction in a continuous stirred tank reactor: the Zel'dovich–Semenov model [28], the Aris–Amundson model [45], and the Volter–Salnikov model [33].

The results of bifurcation analysis of these models can be used to estimate critical conditions in the initial dimensional model (4.1). Moreover, the knowledge of the dynamic processes occurring in a continuous stirred tank reactor allows one to estimate the temperature profiles and critical conditions in a tube reactor. The position of the hot spot and the magnitude of heating in the tube reactor correlate well with the amplitude and period of oscillation of the temperature in the corresponding spherical continuous stirred reactor. Consequently, the laboratory data on the dynamics of combustion processes, which are usually obtained in the Frank-Kamenetskii well-stirred reactor, can be used for approximate estimation of the characteristics of the same process in the tube reactor.

4.6 Combustion model of hydrocarbon mixture

It is well known that critical phenomena such as multiplicity of steady states, hysteresis, oscillations, etc. are possible in the low-temperature oxidation of hydrocarbons [63–80]. For example, in the cool flames of mixtures of iso-octane and n-heptane, oscillation modes were discovered [63] whose characteristics depend on the quality of the mixture. In particular, the amplitude and period of oscillations of the reagent concentrations change with a change in the composition of a hydrocarbon mixture. This fact allows one to offer an express method for the determination of the octane number of real fuels [63].

Here a parametric analysis of the dimensionless mathematical model of combustion of a mixture of the two hydrocarbons in a CSTR will be performed. The analyzed model is a system of three nonlinear ordinary differential equations, which requires the development of special software.

As an example of calculations illustrating the general scheme of parametric analysis, we consider two reactions of the cool flame oxidation of a mixture of n-heptane and iso-octane. The use of bifurcation curves found for the dimensionless model allowed us to build regions of multiplicity of steady states and auto-oscillations for various combinations of real parameters (for example, the initial temperature of the mixture and the concentration of oxygen). The variation of the mixture quality allows us to set its clear correlation with the characteristics of oscillations, which corresponds well with observed experimental data.

Dimensional model

A considerable number of works are devoted to the study of dynamic processes of the combustion of mixtures of hydrocarbons. For example, in cool flames a number of critical phenomena were discovered experimentally. In the works [63, 80] a mathematical model qualitatively describing the process of the cool flame combustion of a mixture of two hydrocarbons (n-heptane and iso-octane) occurring in the oscillation mode is proposed. In [63] the conditions for the existence of auto-oscillations was found from the requirement that the only stationary point of the corresponding mathematical model was unstable. Parametric analysis of the proposed model, developed in [35–54] for plane systems, requires some modification.

Suppose the process occurs according to the scheme

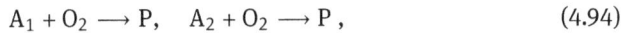

$$A_1 + O_2 \longrightarrow P, \quad A_2 + O_2 \longrightarrow P, \tag{4.94}$$

where A_1 and A_2 are iso-octane and n-heptane, respectively; O_2 is oxygen, P are combustion products. Here and further, it is assumed that oxygen is in excess, i.e., O_2 concentration is constant.

A mathematical model of CSTR for (4.94) has the form

$$V\frac{dX_1}{dt} = -Vk_1(T)X_1 + q(X_1^0 - X_1),$$

$$V\frac{dX_2}{dt} = -Vk_2(T)X_2 + q(X_2^0 - X_2),$$

$$\rho C_p V\frac{dT}{dt} = (-\Delta H_1)Vk_1(T)X_1 + (-\Delta H_2)Vk_2(T)X_2 + q\rho C_p(T_0 - T) + hS(T_x - T),$$
$$\tag{4.95}$$

where $k_i(T) = k_i^0 \exp(-E_i/RT)$, $i = 1, 2$ are the reaction rate constants; k_i^0 is the pre-exponential factor; E_i is the activation energy; R is the universal gas constant; X_i and X_i^0 are the current and inlet concentration of iso-octane and n-heptane; T_w is the temperature of the reactor walls; T and T_0 are the current and inlet temperature of reaction mixture; V is the reactor volume; q, ρ and C_p are the volume flow rate, density, specific heat of the reaction mixture, respectively; S and h are the area of the reactor surface and the coefficient of heat transfer through the reactor wall; $(-\Delta H_i)$ is the thermal effect of the reaction, and t is the astronomical time.

Dimensionless model

For the two irreversible first-order reactions the scheme is

$$X_1 \to P, \quad X_2 \to P,$$

where X_1 and X_2 are the reagents (hydrocarbons), P are the products of oxidation (oxygen concentration is assumed to be constant). A dimensionless model of CSTR can be written as

$$\frac{dx_1}{d\tau} = f_1(y)(1 - x_1) - x_1,$$

$$\frac{dx_2}{d\tau} = f_2(y)(1 - x_2) - x_2,$$

$$\frac{dy}{d\tau} = \beta_1 f_1(y)(1 - x_1) + \beta_2 f_2(y)(1 - x_2) + (1 - y) - s(y - \bar{y}), \qquad (4.96)$$

where x_1 and x_2 are the concentrations of X_1 and X_2; y is the temperature; τ is the dimensionless time; Da_i, γ_i, β_i, s, $\bar{y}$ are the dimensionless parameters, which correspond to the procedure of nondimensionalization of Frank-Kamenetsky [30]:

$$f_i(y) = Da_i \exp(\gamma_i(1 - 1/y)), \quad i = 1, 2.$$

Assume the initial conditions with the characteristics at the inlet to the reactor coincide. Then the initial data for model (4.95) can be written as:

$$x_1(0) = x_2(0) = 0,$$

$$y(0) = (1 + s\bar{y})/(1 + s).$$

In particular, when $\bar{y} = 1$ (which corresponds to the equality of temperatures at the inlet and the walls of the reactor) we have $y(0) = 1$.

Mathematical model (4.96) is a system of three differential equations with parameters. In accordance with the procedure of parametric analysis it is necessary to study the dependencies of steady states on the parameters: to plot the bifurcation curves, which allow in different planes of two selected parameters to highlight regions with various numbers and stability of steady states; to analyze changes in these regions with variation of the third parameter; to plot phase portraits of the original dynamic system; to study temporal dependencies. Compared to dynamic systems on a plane (two phase variables) the parametric analysis of three-dimensional systems is more difficult and the procedure requires significant modifications [5].

Steady states. Equating to zero the right sides of equations of system (4.96), we obtain the following expressions for the stationary values of concentrations and temperature:

$$x_1 = \frac{f_1(y)}{1 + f_1(y)}, \quad x_2 = \frac{f_2(y)}{1 + f_2(y)},$$

$$\frac{\beta_1 f_1(y)}{1 + f_1(y)} + \frac{\beta_2 f_2(y)}{1 + f_2(y)} + 1 - y - s(y - \bar{y}) = 0. \tag{4.97}$$

The stationary values of the dimensionless temperature are found as solutions of the heat balance equation (4.97), which is an equality of functions of heat production and heat removal.

Parametric dependencies. The dependencies of steady states on the parameters can be plotted in accordance with the procedure of parametric analysis developed above. It is enough to write the stationarity equation (4.97) regarding the parameters in the form $p = p(y)$, i.e., to plot the dependencies which are reverse to the desired $y = y(p)$. From (4.97) we have explicitly:

$$\beta_1(y) = \left(s(y - \bar{y}) + y - 1 - \frac{\beta_2 f_2(y)}{1 + f_2(y)} \right) \cdot (1 + f_1(y)) / f_1(y),$$

$$Da_1 = \frac{s(y - \bar{y}) + y - 1 - \beta_2 f_2(y)/(1 + f_2(y))}{e_1(y) \left[\beta_1 + \beta_2 f_2(y)/(1 + f_2(y)) + 1 - y - s(y - \bar{y}) \right]},$$

$$y_1 = \frac{y}{y - 1} \ln \frac{s(y - \bar{y}) + y - 1 - \beta_2 f_2(y)/(1 + f_2(y))}{Da_1 \left(\beta_1 + \beta_2 f_2(y)/(1 + f_2(y)) + 1 - y - s(y - \bar{y}) \right)},$$

$$s = \left(\frac{\beta_1 f_1(y)}{1 + f_1(y)} + \frac{\beta_2 f_2(y)}{1 + f_2(y)} + 1 - y \right) / (y - \bar{y}),$$

$$\bar{y} = \frac{\beta_1 f_1(y)}{1 + f_1(y)} + \frac{\beta_2 f_2(y)}{1 + f_2(y)} + \frac{1 - y}{s} - y, \tag{4.98}$$

where $e_1(y) = \exp(y_1/(1 - 1/y))$. By analogy, we can write down expressions for parameters Da_2, β_2 and y_2. Examples of plotted parametric dependencies are given in Fig. 4.62–4.63, varying the second parameter. The dependencies of the stationary temperature y on $\bar{y}$ according to (4.98) while varying the parameter y_1 are presented in Fig. 4.62. Note that there are up to five steady states, which arise from a superposition of two exponents in the temperature dependence. Fig. 4.62 shows the dependence $y(y_1)$ while varying of the parameter s.

Stability of steady states. As it is known [33], the type of stability of steady states of a dynamical system is determined by the roots of the characteristic equation

$$\lambda^3 + \sigma\lambda^2 + \delta\lambda + \Delta = 0, \tag{4.99}$$

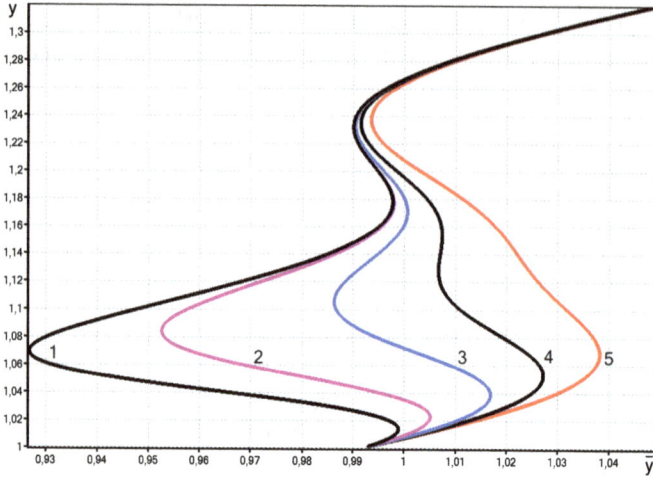

Fig. 4.62: Parametric dependencies $y(\bar{y})$ at $\beta_1 = 0.39$, $\beta_2 = 0.5$, $s = 2$, $\gamma_2 = 40$, $Da_1 = 0.04$, $Da_2 = 0.001$, 1) $\gamma_1 = 80$; 2) 65; 3) 50; 4) 43; 5) 38

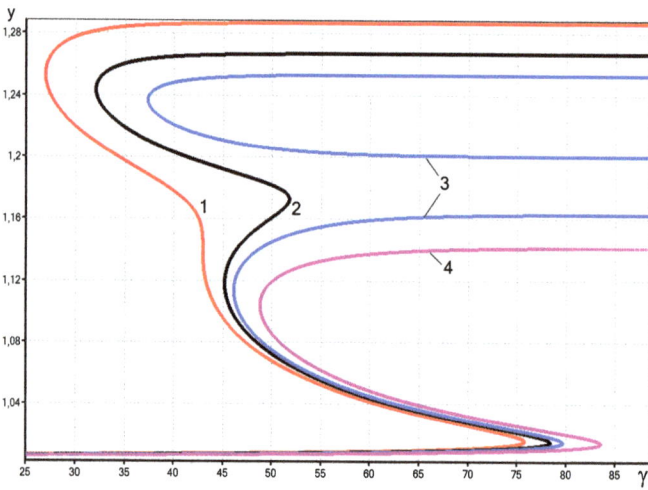

Fig. 4.63: Parametric dependencies $y(\gamma_1)$ at $\beta_1 = 0.39$, $\beta_2 = 0.5$, $\gamma_2 = 40$, $Da_1 = 0.04$, $Da_2 = 0.001$, $\bar{y} = 1$, 1) $s = 1.9$; 2) 2; 3) 2.05; 4) 2.2

where

$$\sigma = -(a_{11} + a_{22} + a_{33}),$$

$$\delta = \begin{vmatrix} a_{11} & a_{12} \\ a_{21} & a_{22} \end{vmatrix} + \begin{vmatrix} a_{11} & a_{13} \\ a_{31} & a_{33} \end{vmatrix} + \begin{vmatrix} a_{22} & a_{23} \\ a_{32} & a_{33} \end{vmatrix},$$

$$\Delta = - \begin{vmatrix} a_{11} & a_{12} & a_{13} \\ a_{22} & a_{22} & a_{23} \\ a_{31} & a_{32} & a_{33} \end{vmatrix},$$

an a_{ij} are the elements of the Jacobian matrix of the right sides of a dynamic system. In our case, a_{ij} are of the form:

$$a_{11} = -f_1(y) - 1,$$
$$a_{12} = 0,$$
$$a_{13} = \gamma_1 f_1(y)(1 - x_1)/y^2,$$
$$a_{21} = 0,$$
$$a_{22} = -f_2(y) - 1,$$
$$a_{23} = \gamma_2 f_2(y)(1 - x_2)/y^2,$$
$$a_{31} = -\beta_1 f_1(y),$$
$$a_{32} = -\beta_2 f_2(y),$$
$$a_{33} = \gamma_1 \beta_1 f_1(y)(1 - x_1)/y^2 + \gamma_2 \beta_2 f_2(y)(1 - x_2)/y^2 - 1 - s.$$

The type of stability of steady states depends on the sign of the real parts of the roots of (4.99). According to the Routh–Hurwitz criterion [33] the number of roots of (4.99) with positive real part (instability) equals the number of changes of sign in the sequence

$$1, \quad \sigma, \quad \xi = \sigma\delta - \Delta, \quad \Delta.$$

Therefore, the number and stability of steady states varies with the change of the sign of values Δ, σ and ξ. For example, if there is a single and unstable steady state, there are oscillations because of the specifics of model (4.96).

Multiplicity curves. A change of the sign of Δ means a change in the number of steady states. The equality $\Delta = 0$ corresponds to the multiple roots of stationarity equation (4.97). In the plane of two parameters, for example (β_1, s), we find regions differing in the number of steady states. To plot the boundaries of these regions we should solve the system of equations:

$$F(y, s, \beta_1) = 0,$$
$$\Delta(y, s, \beta_1) = 0. \tag{4.100}$$

From the stationarity equation (4.97) we have $\beta_1 = \beta_1(y, s)$ in form (4.98). After substituting into the second equation of system (4.100) and selecting parameter s, we can get $s = s(y)$, so the multiplicity curve of steady states L_Δ in the parameter plane (β_1, s) can be written in explicit form:

$$L_\Delta(\beta_1, s): \quad s = \frac{A - (\gamma_1/y^2)(1 + f_2(y))(y - 1 - B) - (\gamma_2/y^2)(1 + f_1(y))B}{(\gamma_1/y^2)(y - \bar{y})(1 + f_2(y)) - A},$$

$$\beta_1 = \beta_1(y, s),$$

where $A = (1 + f_1(y))(1 + f_2(y))$ and $B = (\beta_2 f_2(y))/(1 + f_2(y))$. The variable y is changed as a parameter.

Multiplicity curves in other parameter planes can similarly be written down in a parametric form:

$$L_\Delta(\beta_1, Da_1): \quad Da_1 = \frac{(s + 1)(1 + f_2(y)) - (\gamma_2/y^2)B}{e_1(y)\left((\gamma_1/y^2)B - (s + 1)(1 + f_2(y))\right)} - $$
$$- \frac{(\gamma_1/y^2)(1 + f_2(y))(s(y - \bar{y}) + y - 1 - B)}{e_1(y)\left((\gamma_1/y^2)B - (s + 1)(1 + f_2(y))\right)},$$

$$\beta_1 = \beta_1(y, Da),$$

$$L_\Delta(\beta_1, \beta_2): \quad \beta_1 = \frac{A(s + 1) - (\gamma_2/y^2)(1 + f_1(y))(s(y - \bar{y}) + y - 1)}{f_1(y)(\gamma_1 D - \gamma_2)/y^2},$$

$$\beta_2 = \beta_2(y, \beta_1),$$

$$L_\Delta(\beta_1, \bar{y}): \quad \bar{y} = y + \frac{1}{s}(y - 1 - B) + (\gamma_2/y^2)s(1 + f_1(y))B - $$
$$- \frac{y^2}{\gamma_1 s}(s + 1)(1 + f_1(y)),$$

$$\beta_1 = \beta_1(y, \bar{y}),$$

$$L_\Delta(s, \bar{y}): \quad \bar{y} = y - \frac{\beta_1 f_1(y)(1 + f_2(y)) + \beta_2 f_2(y)(1 + f_1(y)) + A(1 - y)}{(\gamma_1/y^2)C(1 + f_2(y)) + (\gamma_2/y^2)B(1 + f_1(y)) - A},$$

$$s = s(y, \bar{y}),$$

where

$$C = (\beta_1 f_1(y))/(1 + f_1(y)), \qquad D = (1 + f_2(y))/(1 + f_1(y)).$$

It is also possible to obtain the multiplicity curves in the parameter planes $L_\Delta(\beta_2, s)$, $L_\Delta(\beta_2, Da_2)$ and $L_\Delta(\beta_2, \bar{y})$ in explicit form. For other combinations of parameters it is difficult to obtain explicit expressions of the curves L_Δ. But we can plot them graphically: a series of parametric dependencies of type (4.98) are plotted with the variation of any parameter, and in parallel the value of Δ is calculated. When changing the sign of Δ the corresponding values of parameters are plotted into the selected plane. An example of a plotted multiplicity curve is given in Fig. 4.64 for the plane of two parameters.

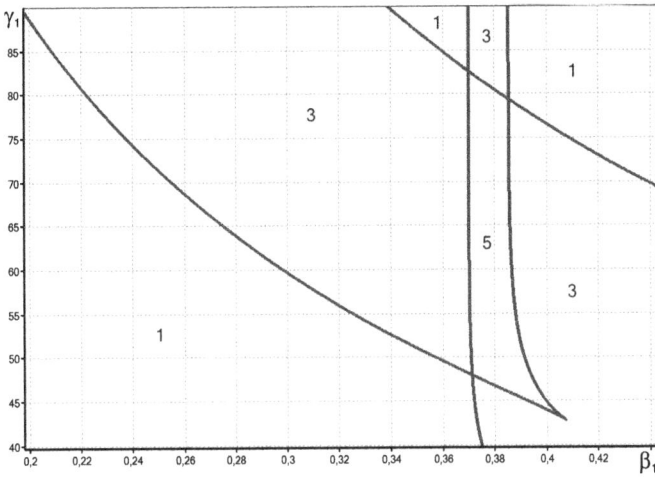

Fig. 4.64: Multiplicity curve (L_Δ) in the parameter plane (β_1, γ_1) at $\beta_2 = 0.5, \gamma_2 = 40, Da_1 = 0.04$, $Da_2 = 0.001, s = 2, \bar{y} = 1$

Numbers in Fig. 4.64 show the regions with one, three, and five steady states. Note that the region with five steady states is quite narrow. Three steady states are the most typical situation of a multiplicity of steady states. While varying the third parameter the region with five steady states can appear or disappear.

Neutrality curves. The change of type of stability of steady states occurs where the values of σ, ξ and Δ cross zero. So along with L_Δ it is important to build neutrality curves L_σ and L_ξ. Using the specifics of system (4.96), the expressions of neutrality curves L_σ can be written like L_Δ in explicit form:

$$L_\Delta(\beta_1, s): \quad s = \frac{A - (\gamma_1/y^2)(1 + f_2(y))(y - 1 - B) - (\gamma_2/y^2)(1 + f_1(y))B}{(\gamma_1/y^2)(y - \bar{y})(1 + f_2(y)) - A},$$

$$\beta_1 = \beta_1(y, s),$$

$$L_\sigma(\beta_1, Da_1): \quad Da_1 = \frac{\left(\frac{\gamma_1(s(y - \bar{y}) + y - 1}{y^2} - \frac{\beta_2 f_2(y)(\gamma_1 - \gamma_2)}{(1 + f_2(y))} - f_2(y) - s - 3\right)}{e_1(y)},$$

$$\beta_1 = \beta_1(y, Da), \qquad\qquad (4.101)$$

$$L_\sigma(\beta_1, \beta_2): \quad \beta_1 = \frac{A(s + 1) - (\gamma_2/y^2)(1 + f_1(y))(s(y - \bar{y}) + y - 1)}{f_1(y)(\gamma_1 C - \gamma_2)/y^2},$$

$$\beta_2 = \beta_2(y, \beta_1),$$

$$L_\sigma(\beta_1, \bar{y}): \quad \bar{y} = y + \frac{1}{s}(y - 1 - B) + \frac{\gamma_2}{\gamma_1 s}(1 + f_1(y))B - \frac{y^2}{\gamma_1 s}(s + 1)(1 + f_1(y)),$$

$$\beta_1 = \beta_1(y, \bar{y}),$$

The neutrality curves $L_\sigma(\beta_2, s)$, $L_\sigma(\beta_2, Da_2)$ and $L_\sigma(\beta_2, \bar{y})$ can be written in a form similar to (4.101). It is difficult to obtain the expressions of curves L_σ for other com-

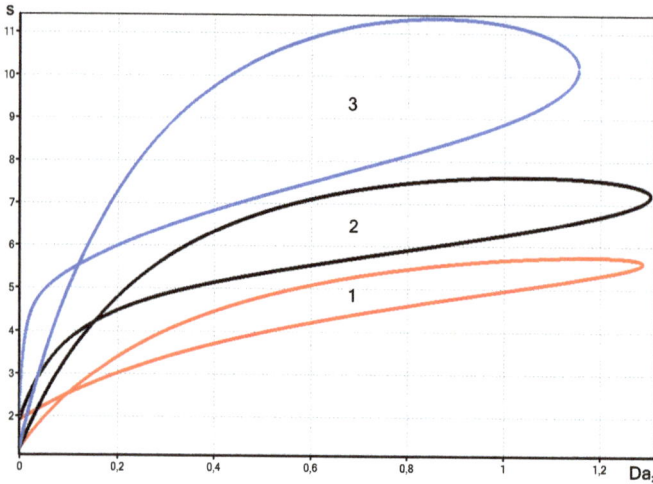

Fig. 4.65: Neutrality curves (L_σ) in the parameter plane (Da_2, s) at $\beta_1 = 0.39$, $\beta_2 = 0.5$, $\gamma_1 = 50$, $Da_1 = 0.04$, $\bar{y} = 1$: 1) $\gamma_2 = 5$; 2) 20; 3) 40

binations of parameters and bifurcation curves L_ξ in an explicit form. So they can be plotted as above, graphically. One of the calculation variants of L_σ is presented in Fig. 4.65. Inside the loop of the neutrality curve the steady state may be unique and unstable.

Parametric portraits. The plotting of curves of local bifurcations of L_Δ, L_σ and L_ξ in one of the selected planes of dimensionless parameters allows us to obtain the parametric portrait of model (4.96). The mutual position of these curves divides the parameter plane into regions which are characterized by a different number and type of stability of steady states. The calculations show that auto-oscillations in system (4.96) appear with a greater probability for parameter values at which there exists a single and unstable steady state. There are other conditions for the existence of oscillations, for example, the presence of three unstable steady states. However, such regions of parameters are quite narrow.

The explicit form of the bifurcation curves L_Δ and L_σ in various planes of dimensionless parameters allows one to quite simply plot these curves in the planes of real dimensional parameters [47]. For example, having the curve $L_\Delta(Da_i, \beta_i)$, it is easy to plot a similar bifurcation curve $L_\Delta(T_0, p_0)$, where T_0 is the initial temperature in the reactor and p_0 is the concentration of oxygen in the fuel-air mixture of n-heptane and iso-octane, because

$$Da_i = Vk_i(T_0)p_0/q , \quad \beta_i = (\Delta H_i)x_i^0/(\rho C_p T_0) ,$$

where V is the reactor volume, k_i is the step rate constant, ΔH_i is the thermal effect of reactions, ρ is the density of the mixture, C_p is the specific heat of the mixture.

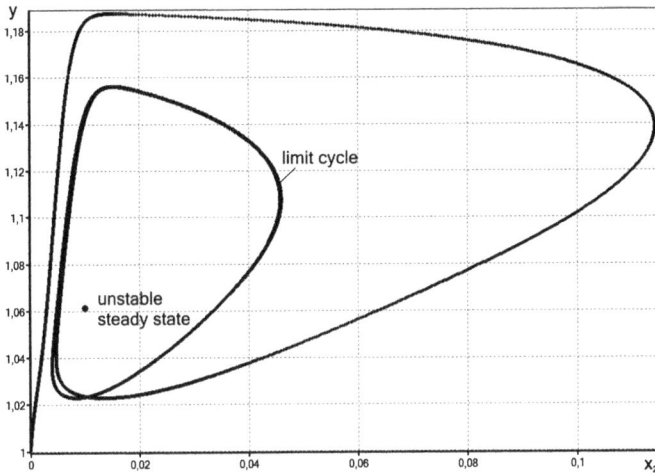

Fig. 4.66: Projection of the phase portrait on the plane (x_2, y) for a single and unstable steady state at $\beta_1 = 0.25$, $\beta_2 = 0.5$, $\gamma_1 = 50$, $\gamma_2 = 40$, $Da_1 = 0.14$, $Da_2 = 0.001$, $s = 2$, $\bar{y} = 1$

Calculations show that *ceteris paribus* a multiplicity of steady states is possible at low temperatures T_0 and a significant excess of oxygen.

Phase portraits. A visual representation of the dynamics of model (4.96) serve as its phase portraits. One of the projections of the three-dimensional phase space on the phase plane (x_2, y) is presented in Fig. 4.66. In the case of a single and unstable steady state with any initial data the phase trajectories are wound on the limit cycle, which corresponds to the existence of undamped oscillations of the concentrations and temperature.

The specificity of the dynamics exist for three and five of steady states. In these cases, all the phase space splits into the regions of attraction of the stable steady states. Such "stratification" of concentrations and temperatures leads to a need for special regulation of initial conditions for the purpose of implementing one or another steady state.

Time dependencies. Characteristics of relaxation processes in system (4.96) can be estimated while plotting the time dependencies $x_i(\tau)$ and $y(\tau)$. In particular, oscillatory solutions are characterized by slow and fast movements (see Fig. 4.67). The short-time flashes of temperature alternate with relatively slow changes of the concentrations of the reactants.

Thus, for the thermokinetic model of oxidation of a mixture of two hydrocarbons (4.96) the parametric analysis of steady states was performed. The dependences of the stationary concentrations of reagents and temperature of the mixture on the various dimensionless parameters are plotted. The conditions of the existence of one, three and five steady states are found. On the basis of the specificity of the model the equations of the bifurcation curves of the change in the number and type of stability of

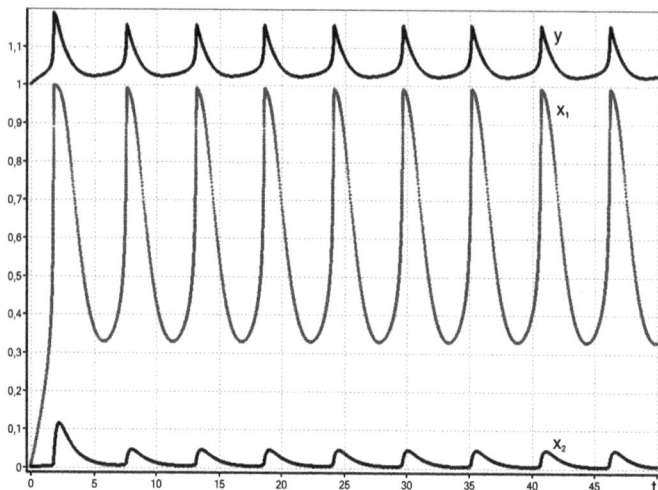

Fig. 4.67: Time dependencies $x_1(\tau)$, $x_2(\tau)$ and $y(\tau)$. The parameter values similar to Fig. 4.66

steady states are written out explicitly. The efficient procedure of numerical plots of the parametric portrait of the local bifurcations in various parameter planes is proposed. It allows one to find regions where a multiplicity of steady states and auto-oscillations exist. The dimensionless form of the dynamic model of (4.96) and its corresponding parametric portraits allow us to easily convert the conditions of existence of critical phenomena for real conditions of implementation of the experiment in the reactor of ideal mixing. In particular, it was found that the quality of the hydrocarbon mixture (the ratio of the initial concentrations) significantly affects the period and amplitude of possible oscillations.

4.7 Thermocatalytic triggers and oscillators

As was shown in a series of works of V. V. Azatyan and coworkers [82–87], the kinetics of chain reactions can substantially influence nonisothermal modes of these processes. Competition, reaction chain branching, and termination can play a predominant role in the formation of concentration limits of flame propagation. This all can be used to control detonation of mixtures by small chemically active admixtures.

On the other hand, a wide class of homogeneous-heterogeneous reactions is characterized by noticeable thermal effects. For this reason, studies of the dynamics of chain and heterogeneous catalytic processes should generally be performed taking into account the special features of interactions between their kinetics and temperature dependencies. Temperature and kinetic nonlinearities in combination can substantially complicate the physicochemical picture of process. For example, if the kinetic subsystem has several steady states, taking temperature changes into account leads to auto-oscillations, and if the kinetic subsystem admits auto-oscillations the system as a whole can exhibit complex aperiodic modes. Such nonlinear and nonstationary phenomena require the use of thermokinetic equations, which are systems of ordinary differential equations with parameters.

A study of solutions to these equations allows us to find the conditions of the appearance of critical phenomena, such as multiplicity of steady states, loss of stability by these states, the appearance of oscillation modes, etc. A special direction (a parametric analysis of the corresponding mathematical models) appears in the context of the general mathematical modeling approach.

In this section, we consider a series of the simplest thermokinetic models, in which the kinetic subsystem corresponds to traditional catalytic mechanisms and autocatalytic mechanisms with multiplicity of steady states and auto-oscillations in the kinetic region [46]. Studies of the basic thermokinetic models were performed using the method of continuation along a parameter [45]. The parametric analysis performed allowed us to identify parameter regions with the existence of critical phenomena (multiplicity of steady states and auto-oscillations).

In the ideal mixing approximation, the thermokinetic model of a heterogeneous catalytic reaction can be written as

$$\frac{dT}{dt} = \sum_{j=1}^{m} h_j w_j(T, \mathbf{x}) + \alpha(T_0 - T), \tag{4.102}$$

$$\frac{dx_i}{dt} = \sum_{j=1}^{m} \gamma_{ji} w_j(T, \mathbf{x}), \quad i = 1, \ldots, n, \tag{4.103}$$

where T is the temperature, $\mathbf{x} = (x_1, \ldots, x_n)$ is the vector of concentrations of intermediate substances adsorbed on the surface of the catalyst, γ_{ji} denotes the stoichiometric coefficients of the reactions of the intermediates, w_j are the rates of these reactions, h_j is the heat effect of the j-th reaction per unit surface area, α is the coefficient characterizing the intensity of gas–solid heat exchange, and t is the time.

Model (4.102), (4.103) is simplified by ignoring reactions in the gas phase and the assumption that the concentrations of the substances observed are constant.

For a complex catalytic reaction, we have

$$\sum_{i=1}^{n} \alpha_{ji} X_i \rightarrow \sum_{i=1}^{n} \beta_{ji} X_i, \quad j = 1, \ldots, m, \tag{4.104}$$

where X_i denotes intermediate substances, α_{ji} and β_{ji} are the stoichiometric coefficients. The rates of separate stages of mechanism (4.104), for example, for the law of mass action, are written in the form

$$w_j(T, \mathbf{x}) = k_j(T) \prod_{i=1}^{n} x_i^{\alpha_{ji}},$$

$$k_j(T) = k_j^0 \exp\left(-\frac{E_j}{RT}\right),$$

where k_j, k_j^0 and E_j are the rate constant, pre-exponential factor, and activation energy of the j-th reaction, respectively, and R is the universal gas constant.

In terms of dimensionless parameters and variables, the thermokinetic model (4.102), (4.103) takes the form

$$\frac{dT}{dt} = Q(T, \mathbf{x}) + \alpha(T_0 - T) , \tag{4.105}$$

$$\frac{d\mathbf{x}}{dt} = \mathbf{f}(T, \mathbf{x}) , \tag{4.106}$$

where Q is the function characterizing heat release in the reaction on the surface of the catalyst and $\mathbf{f}$ is the vector function corresponding to the kinetics of transformations of intermediate substances. The separation of the heat balance equation (4.105) and kinetic subsystem (4.106) corresponding to a complex chemical reaction on the surface of the catalyst is the special feature of dynamic system (4.105), (4.106).

It is well known [3, 35–54, 88–95] that the kinetic subsystem itself (at $T \equiv$ const) can exhibit substantially nonlinear (multiplicity of steady states) and nonstationary (auto-oscillations) behavior. For this reason, an important problem in an analysis of thermokinetic models of catalytic reactions is the study of the special features of the interaction of the thermal and kinetic subsystems. Even the classic Semenov heat removal–heat release diagram has special features,

$$Q(T, \mathbf{x}) = \alpha(T - T_0) ,$$

$$\mathbf{f}(T, \mathbf{x}) = 0 , \tag{4.107}$$

where the stationary heat balance equation is added to equations for the stationarity of intermediate substances. The elimination of unknowns $\mathbf{x}$ from (4.107) in the form $\mathbf{x} = \mathbf{x}(T)$ gives

$$Q(T, \mathbf{x}(T)) = \alpha(T - T_0) ,$$

where the $\mathbf{x}(T)$ function in the case of multiple steady states can be characterized by a hysteresis. Moreover, the kinetic subsystem itself can be an oscillator. The interaction of the thermal and kinetic model constituents is characterized by complex dynamic behavior. We studied the special features of such interactions for series of the so-called basic thermokinetic models, in which the heat balance equation is added to kinetic subsystems corresponding to standard mechanisms of catalytic transformations (the Eley–Rideal and Langmuir–Hinshelwood mechanisms, systems with auto-catalysis, etc.). The use of the techniques of parametric analysis of dynamic systems allows us to perform a quite thorough analysis of the special features of interactions between the thermal and kinetic subsystems.

In this section, a parametric analysis is performed in the simplest form for a series of so-called basic thermokinetic models of catalytic reactions. As a rule the models studied contain a minimum number of variables and variable parameters. Generally, the procedure of complete parametric analysis is a quite laborious mathematical and computational problem. We implement only the initial parametric analysis stages. The results, however, give quite abundant information about the possible dynamic behaviors of the thermokinetic models under consideration. In particular, they make it clear that the interaction of temperature nonlinearity with kinetic features of chemical reactions provides a variety of dynamic characteristics of physicochemical processes. Moreover, parametric analysis allows us to obtain quantitative characteristics of the chemical dynamics, which is important not only theoretically but especially from the point of view of applications. For example, in chemical technology of catalytic processes, it is important to know dangerous and safe boundaries of varied parameters which can determine emergency and technologically undesirable modes [53, 54].

4.7.1 The Eley–Rideal monomolecular mechanism

The scheme of conversion is as follows

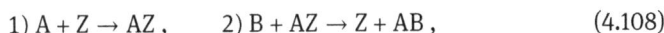

$$\text{1) A} + \text{Z} \rightarrow \text{AZ}, \qquad \text{2) B} + \text{AZ} \rightarrow \text{Z} + \text{AB}, \qquad (4.108)$$

Here, A, B and AB are the gas-phase substances, Z is the catalyst, and AZ is the intermediate substance on the catalyst surface. The thermokinetic model (4.102), (4.103) that corresponds to scheme (4.108) has the form

$$\frac{dT}{dt} = h_1 w_1 + h_{-1} w_{-1} + \alpha(T_0 - T),$$

$$\frac{dx}{dt} = w_1 - w_{-1} - w_2, \qquad (4.109)$$

where $w_1 = k_1 p_A z$, $w_{-1} = k_{-1} x$, $w_2 = k_2 p_B x$. For simplicity, we assume that only the second step depends on temperature. The introduction of dimensionless parameters and variables transforms (4.109) into the dimensionless model

$$\frac{dx}{d\tau} = k_1 p_A z - k_{-1} x - f(y) x,$$

$$\frac{dy}{d\tau} = \beta f(y) x + s(1 - y), \qquad (4.110)$$

where $z = 1 - x$; x and y are the dimensionless concentration and temperature; Da, β, γ, s, k_1, k_{-1} and p_A are dimensionless parameters; and $f(y) = Da \exp(\gamma(1 - 1/y))$ is the temperature dependence of the reaction rate.

Steady states of system (4.110) are solutions to the equations

$$k_1 p_A z - k_{-1} x - f(y) x = 0,$$

$$\beta f(y) x + s(1 - y) = 0.$$

After the exclusion of variable y, we obtain one equation for steady states,

$$\beta f(y(x))x + s(1 - y(x)) = 0 , \tag{4.111}$$

where $y(x) = 1 + (\beta/s)(k_1 p_A(1 - x) - k_{-1}x)$.

Parametric dependencies. For (4.111), the dependences, which are inverse desired dependences of steady states on parameters, can explicitly be written as

$$Da(x) = \frac{s(y(x) - 1)}{\beta x \exp(y(1 - 1/y(x)))} ,$$

$$y(x) = \frac{y(x)}{y(x) - 1} \ln \frac{s(y(x) - 1)}{\beta Dax} .$$

To plot dependences of the stationary concentration and temperature on the other parameters, it is necessary to use the numerical procedure of continuation on a parameter [35–52]. Given these dependences, it is comparatively easy to plot the main bifurcation curves, the multiplicity curve L_Δ, and the neutrality curve L_σ. For example, for the parameters Da and y, after calculation of the elements of the Jacobian matrix for (4.110),

$$a_{11} = -k_1 p_A - k_{-1} - f(y) , \quad a_{12} = -\frac{y}{y^2} f(y)x ,$$

$$a_{21} = \beta f(y) , \quad a_{22} = \frac{\beta y}{y^2} f(y)x - s ,$$

we obtain the following equations for the multiplicity and neutrality curves on the (Da, y) plane of parameters:

$$L_\Delta(Da, y): \quad y(x) = \frac{y^2(x)}{y(x) - 1} \left(1 + \frac{s(y(x) - 1)}{\beta x(k_1 p_A + k_{-1})} \right) ,$$

$$Da = Da(x, y(x)) .$$

$$L_\sigma(Da, y): \quad y(x) = \frac{y^2(x)}{y(x) - 1} \left(1 + \frac{(y(x) - 1)}{\beta x} + \frac{(k_1 p_A + k_{-1})}{s} \right) ,$$

$$Da = Da(x, y(x)) . \tag{4.112}$$

An example of the plotted bifurcation curves (4.112) is given in Fig. 4.68. The mutual arrangement of the bifurcation curves allows us to identify six regions in the parameter plane with a different number and stability of steady states. For example, in the intervals of the varying parameters $0.01 \le Da \le 1$ and $25 \le y \le 70$, there exists a quite extended region with a single and unstable steady state. It guarantees the presence of auto-oscillations in the dynamic system (4.110). Auto-oscillations can also exist with three steady states, but the corresponding region of parameters is quite narrow. The developed procedure and program-mathematical software for the parametric analysis of thermokinetic models of type (4.110) can be used to study the influence of other parameters on the conditions for the existence of critical effects. The bifurcation curves for the considered model were plotted at different parameter combinations, and the influence of a third parameter on the parametric portrait in the plane of two parameters was studied.

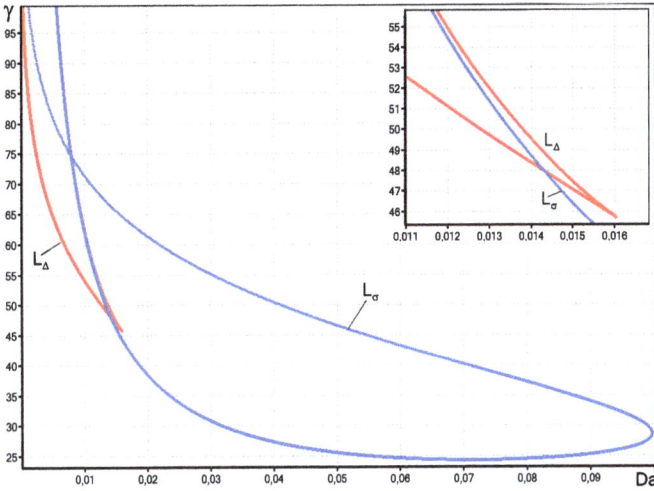

Fig. 4.68: Multiplicity (L_Δ) and neutrality (L_σ) curves of steady states at $\beta = 0.8$, $s = 1$, $k_1 = 0.12$, $k_{-1} = 0.01$, $p_A = 1$

4.7.2 The Eley–Rideal bimolecular mechanism

Along with (4.108), one of the basic catalytic schemes of transformations is the following two-step nonlinear mechanism of the CO oxidation

$$1)\ O_2 + 2Z \rightleftharpoons 2ZO,$$
$$2)\ CO + ZO \rightarrow Z + CO_2. \tag{4.113}$$

The dimensionless thermokinetic model similar to (4.110) that corresponds to this mechanism has the form

$$\frac{dx}{d\tau} = 2k_1 p_A z^2 - 2k_{-1}x^2 - f(y)x,$$

$$\frac{dy}{d\tau} = \beta f(y)x + s(1-y), \tag{4.114}$$

where the main denotations are the same as in (4.110). The equation for steady states corresponding to (4.114) is similar to (4.111), where

$$y(x) = 1 + \frac{2\beta}{s}(k_1 p_A(1-x)^2 - k_{-1}x^2). \tag{4.115}$$

The parametric dependencies with the function $y(x)$ are calculated according to (4.115). The Jacobian matrix elements for system (4.114) are

$$a_{11} = -4k_1 p_A(1-x) - 4k_{-1}x - f(y), \quad a_{12} = -\frac{y}{y^2}f(y)x,$$

$$a_{21} = \beta f(y), \quad a_{22} = \frac{\beta y}{y^2}f(y)x - s.$$

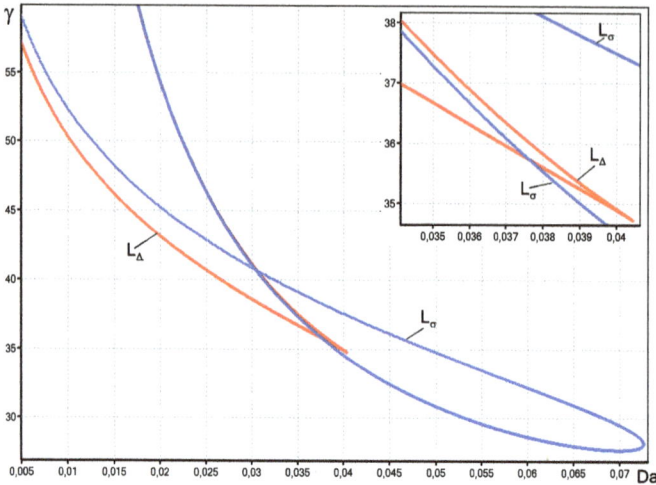

Fig. 4.69: Multiplicity (L_Δ) and neutrality (L_σ) curves of steady states at $\beta = 0.6$, $s = 1$, $k_1 = 0.12$, $k_{-1} = 0.01$, $p_A = 1$

The curves of multiplicity and neutrality of steady states in the (Da, y) plane of parameters have the form

$$L_\Delta(Da, y): \quad y(x) = \frac{y^2(x)}{y(x) - 1}\left(1 + \frac{s(y(x) - 1)}{4\beta x(k_1 p_A(1 - x) + k_{-1}x)}\right),$$
$$Da = Da(x, y(x)).$$

$$L_\sigma(Da, y): \quad y(x) = \frac{y^2(x)}{y(x) - 1}\left(1 + \frac{(y(x) - 1)}{\beta x} + \frac{4}{s}(k_1 p_A(1 - x) + k_{-1}x)\right),$$
$$Da = Da(x, y(x)).$$

An example of a plotted parametric portrait for the Eley–Rideal bimolecular mechanism (4.113) is shown in Fig. 4.69. As previously, we give only one parametric portrait variant in the (Da, y) plane of parameters.

The developed procedure and software allow us quite simply to plot bifurcation curves in arbitrary parameter planes and study the influence of all the other parameters on their arrangement. This all makes it possible to determine the critical conditions of the ongoing reaction and study the influence of various thermophysical and kinetic parameters on them. A comparative analysis of the parametric portraits for bimolecular and monomolecular mechanisms (Fig. 4.69) shows that the regions of the existence of various modes (multiplicity of steady states and auto-oscillations) depend on the kinetic characteristics of the mechanism of the reaction. Parametric analysis allows us to give the quantitative characteristics of this influence.

4.7.3 The linear catalytic cycle

Consider a three-step catalytic cycle

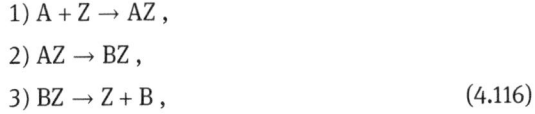

$$
\begin{aligned}
&1)\ A + Z \rightarrow AZ\,, \\
&2)\ AZ \rightarrow BZ\,, \\
&3)\ BZ \rightarrow Z + B\,,
\end{aligned}
\qquad (4.116)
$$

where Z is the catalyst, and AZ and BZ are intermediate substances. The assumption of constant gas phase and temperature dependence of the rate of the third step is similar to the previously considered thermokinetic model for (4.116) in the dimensionless form:

$$
\begin{aligned}
\frac{dx_1}{d\tau} &= k_1 p_A z - k_2 x_1\,, \\
\frac{dx_2}{d\tau} &= k_2 x_1 - f(y)x_2\,, \\
\frac{dy}{d\tau} &= \beta f(y)x_2 + s(1 - y)\,,
\end{aligned}
\qquad (4.117)
$$

where $z = 1 - x_1 - x_2$; and x_1 and x_2 are the degrees of coating of the catalytic surface by the intermediate substances AZ and BZ. At constant temperature the kinetic subsystem (4.117) is linear, so it has a unique solution $x_1 = x_1(y)$, $x_2 = x_2(y)$. After substituting them into the third equation, we obtain one equation for the dimensionless temperature

$$
\beta f(y)x_2(y) + s(1 - y) = 0\,.
\qquad (4.118)
$$

The use of (4.118) allows us relatively simply to plot the appropriate parametric dependencies. However, the specificity of model (4.117) is that the dimension of its phase space is equal to 3. This significantly complicates the overall procedure of parametric analysis of dynamic models of the type (4.117). We partially implement it below for autocatalytic schemes of reactions.

4.7.4 The Langmuir–Hinshelwood Mechanism

Consider the nonlinear three-step scheme of transformations:

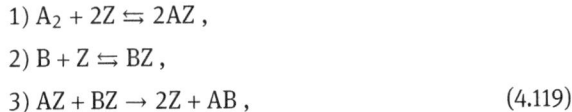

$$
\begin{aligned}
&1)\ A_2 + 2Z \leftrightarrows 2AZ\,, \\
&2)\ B + Z \leftrightarrows BZ\,, \\
&3)\ AZ + BZ \rightarrow 2Z + AB\,,
\end{aligned}
\qquad (4.119)
$$

where Z is the catalyst, and AZ and BZ are intermediate substances. The corresponding dimensionless thermokinetic model has the form

$$\frac{dx_1}{d\tau} = 2k_1 p_A z^2 - 2k_{-1} x_1^2 - f(y) x_1 x_2 \, ,$$

$$\frac{dx_2}{d\tau} = k_2 p_B z - k_{-2} x_2 - f(y) x_1 x_2 \, ,$$

$$\frac{dy}{d\tau} = \beta f(y) x_1 x_2 + s(1 - y) \, , \tag{4.120}$$

where the kinetic subsystem is nonlinear in comparison with (4.117). It is well known that, under isothermal conditions, system (4.120) is a catalytic trigger. It contains three steady states, where two of them are stable and one is unstable. The dependence of the stationary rate of the third reaction can have a hysteresis on the parameter y; that is, in the steady state, the heat release function in the stationary heat balance equation (4.120)

$$\beta f(y) x_1(y) x_2(y) + s(1 - y) = 0$$

can have a hysteresis character. In the general case, it is difficult to obtain the corresponding explicit equations. However, in this case a visual geometric interpretation can be given when the adsorption stages in (4.119) are irreversible. Indeed, for $k_{-1} = k_{-2} = 0$, from (4.120) we have

$$2k_1 p_A z^2 - f(y) x_1 x_2 = 0 \, ,$$

$$k_2 p_B z - f(y) x_1 x_2 = 0 \, . \tag{4.121}$$

The system of the two equations (4.121) has four solutions at y = const, two boundary solutions: 1) $x_1 = 1$, $x_2 = 0$ and 2) $x_1 = 0$, $x_2 = 1$, for which the reaction rate is zero, and two internal steady states at certain parameter values, for which

$$z = \frac{k_2 p_B}{2k_1 p_A} \, . \tag{4.122}$$

The stationary values x_1 and x_2 are determined by solving the quadratic equation and can be found in explicit form. We do not use it and therefore do not give it here. According to (4.122), the Semenov diagram (4.121) in the steady state has the form

$$\frac{(k_2 p_B)^2}{2k_1 p_A} = s(y - 1) \, . \tag{4.123}$$

In addition to (4.123), the zero branch of the stationary reaction rate should be taken into account. The stationary temperature for it is $y = 1$. Calculations show that, at small k_{-1} and k_{-2}, the temperature dependence of the stationary reaction rate is changed insignificantly, except at the bifurcation points of jumps from one branch of stable steady states to another.

The thermokinetic model (4.117) is a three-dimensional dynamic system. It combines a trigger and an oscillator, which is a special feature of the model. The kinetic subsystem has trigger properties, and the "temperature + concentration" subsystem can have auto-oscillations. The dynamics of system (4.120) can generally be quite complex, and its detailed study requires special investigation.

4.7.5 Autocatalytic schemes of transformations

Here we study several thermokinetic models corresponding to autocatalytic reaction schemes. They are in a certain sense the simplest nonlinear schemes with essentially nonlinear properties (multiplicity of steady states and auto-oscillations). In isothermal formulation, the corresponding kinetic models are considered in Chapter 3.

Autocatalytic trigger. Let us consider the autocatalytic system:

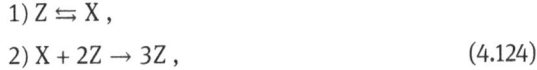

$$1)\ Z \leftrightarrows X,$$
$$2)\ X + 2Z \to 3Z, \tag{4.124}$$

where Z is the active site on the catalyst surface and X is the intermediate substance. The thermokinetic model corresponds to scheme (4.124) and can be written as follows:

$$\frac{dT}{dt} = \sum_i h_i w_i(x, T) + \alpha(T_0 - T),$$

$$\frac{dx}{dt} = w_1(x, T) - w_{-1}(x, T) - w_2(x, T), \tag{4.125}$$

where

$$w_1(x, T) = k_1(T)(1 - x),$$
$$w_{-1}(x, T) = k_{-1}(T)x,$$
$$w_2(x, T) = k_2(T)x(1 - x)^2.$$

Let us introduce dimensionless variables and parameters using the Frank–Kamenetskii scheme for this purpose and assume for simplicity that only k_{-1} depends on temperature: ($E_1 = E_2 = 0, h_1 = h_2 = 0, E_{-1} \neq 0, h_{-1} \neq 0$). The system (4.125) can be rewritten as:

$$\frac{dy}{d\tau} = \beta f(y)x + s(1 - y) = f_1(x, y),$$

$$\frac{dx}{d\tau} = k_1 z - f(y)x - k_2 x z^2 = f_2(x, y), \tag{4.126}$$

where $z = 1 - x$ and $f(y) = Da \exp(y(1 - 1/y))$ as previously. x and y are the dimensionless concentration and temperature, respectively; β, s, Da, y, k_1 and k_2 are dimensionless parameters; and τ is the dimensionless time. As mentioned, our goal is to find the region of auto-oscillations, study the influence of various parameters on the characteristics of the system, determine the dynamics of the process, and divide the parametric space into regions corresponding to qualitatively different dynamic behavior types.

Steady states of system (4.126) are the solutions to the equations:

$$\beta f(y)x + s(1 - y) = 0,$$
$$k_1 z - f(y)x - k_2 x z^2 = 0.$$

By eliminating y and substituting it into the first equation, we obtain the nonlinear equation:

$$\beta Da \exp(y(1 - 1/y(x)))x + s(1 - y(x)) = 0, \tag{4.127}$$

where $y(x) = 1 + (k_1 - k_2 x(1 - x))(1 - x)\beta/s$.

The stability type of a steady state is determined by the roots of the characteristic equation:

$$\lambda^2 - \sigma\lambda + \Delta = 0,$$

where σ and Δ are found by using the Jacobian matrix elements:

$$a_{11} = \frac{\partial f_1}{\partial y} = \frac{y}{y^2}\beta x f(y) - s,$$

$$a_{12} = \frac{\partial f_1}{\partial x} = \beta f(y),$$

$$a_{21} = \frac{\partial f_2}{\partial y} = -\frac{y}{y^2}x f(y),$$

$$a_{22} = \frac{\partial f_2}{\partial x} = -k_1 - f(y) - k_2(1 - x)(1 - 3x),$$

namely

$$\sigma = a_{11} + a_{22},$$

$$\Delta = a_{11}a_{22} - a_{12}a_{21}. \tag{4.128}$$

Parametric dependencies. The stationarity equation (4.127) can be used to obtain parametric dependencies in explicit form, for example:

$$Da(x) = \frac{s(y(x) - 1)}{\beta x \exp(y(1 - 1/y(x)))}. \tag{4.129}$$

Calculations show that the region of multiplicity of steady states decreases as parameter s increases. Similarly, we can obtain the $y(x)$ dependence

$$y(x) = \frac{y(x)}{y(x) - 1} \ln \frac{s(y(x) - 1)}{\beta x Da}. \tag{4.130}$$

In this plane of parameters, the region of multiplicity decreases as β increases (Fig. 4.70). Figure 4.70 illustrates only one example of a plot of parametric dependencies of steady states. The $x(Da)$ plot shows that, over the interval $0.02 \leq Da \leq 0.03$, five steady states can exist on the third curve. When varying the second parameter, the intervals of the existence of three and five steady states change. The presence of five steady states in this model corresponds to the interaction of kinetic nonlinearity (under isothermal conditions, this autocatalytic process is a trigger) with temperature nonlinearity.

Multiplicity curves L_Δ ($\Delta = 0$) are the boundaries separating the region of parameters into regions with different number of steady states. For example, let us consider

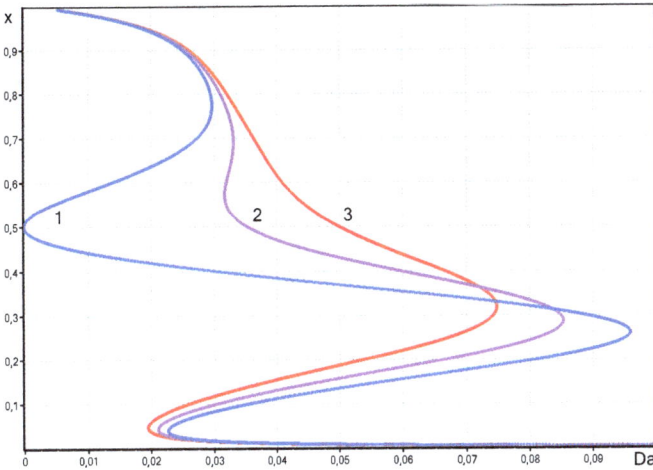

Fig. 4.70: Parametric dependencies $x(Da)$ at $\beta = 0.375$, $s = 2$, $\gamma = 75$, $k_1 = 0.6$, 1) $k_2 = 2$; 2) 2.2; 3) 2.4

the (Da, y) plane of parameters and write an equation for the multiplicity curve L_Δ in this plane. For this purpose, we must solve the system

$$f_1(x, y, Da, \gamma) = 0 ,$$
$$f_2(x, y, Da, \gamma) = 0 ,$$
$$\Delta(x, y, Da, \gamma) = 0 , \tag{4.131}$$

where Δ is determined by using the Jacobian matrix elements (4.128).

Substituting the explicit equation (4.129) for $Da(x, y)$ into the third equation of (4.131), we obtain the boundaries of the region of multiplicity of steady states in the (Da, y) plane of parameters:

$$L_\Delta(Da, \gamma): \quad y(x) = \frac{y(x)^2}{y(x) - 1}\left(1 + \frac{s(y(x) - 1)}{\beta x(k_1 + k_2(1 - x)(1 - 3x))}\right) ,$$
$$Da = Da(x, y(x)) .$$

An example of the plotted multiplicity curves by using explicit equations in the (Da, y) plane of parameters is given in Fig. 4.71. Calculations show that in the parameter plane (Da, y) the region of multiplicity increases with decreasing values of the parameter s.

We were unable to obtain explicit equations for L_Δ in other parameter planes. For this reason, we used the procedure of graphical plotting L_Δ. This procedure involves the plotting of a series of parametric dependences with respect to one of the parameters while the second parameter is varied. Note that we were able to plot the corresponding dependencies for all the parameters. The region of multiplicity of steady states lies between the turning points of the parametric curves (their maximum and

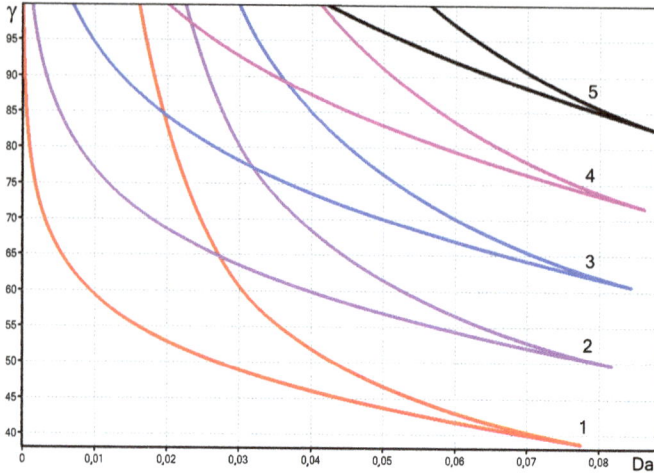

Fig. 4.71: Multiplicity curves $L_\Delta(Da, \gamma)$ at $\beta = 0.375$, $k_1 = 0.6$, $k_2 = 1$:
1) $s = 1.5$; 2) 2; 3) 2.5; 4) 3; 5) 3.5

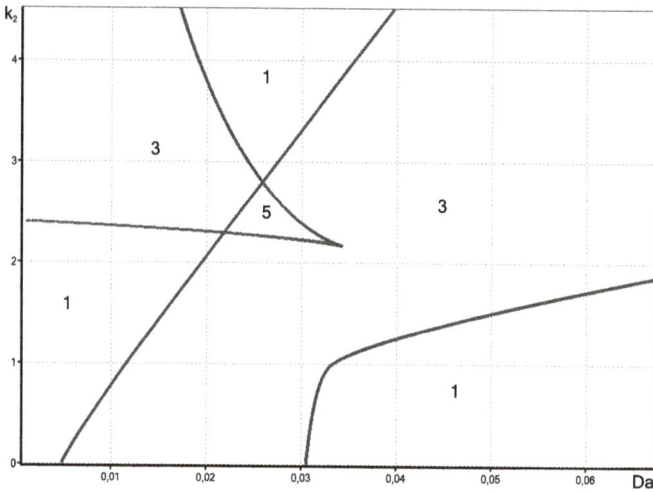

Fig. 4.72: Multiplicity curves $L_\Delta(Da, k_2)$ for model (4.126) with five steady states; numbers 1–5 correspond to the number of steady states in the corresponding parameter regions at $\beta = 0.375$, $s = 2$, $\gamma = 75$, $k_1 = 0.6$

minimum values). When varying the second parameter, the boundaries of this region limit the region of multiplicity of steady states on the corresponding parameter plane. The result of applying the graphical procedure is given in Fig. 4.72.

The numbers in the figure show the number of steady states in this region of parameters. The region with five steady states is quite narrow. Three steady states are observed over a wide region of parameters, and this region consists of two parts, which

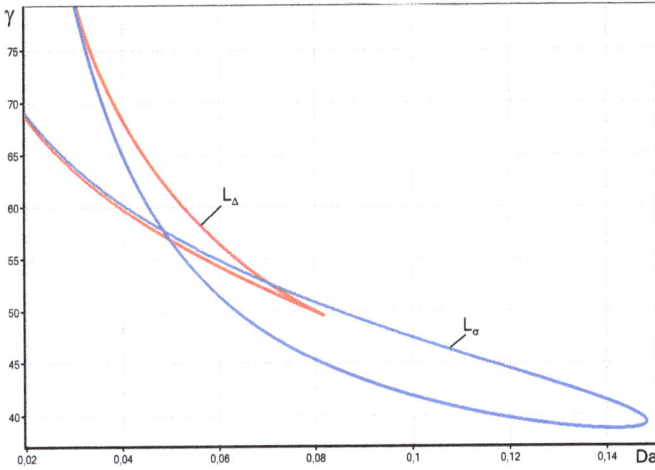

Fig. 4.73: Parametric portrait in the plane (Da, γ) at $s = 2$, $\beta = 0.375$, $k_1 = 0.6$, $k_2 = 1$

correspond to the two types of nonlinearities present in the model under consideration, temperature and kinetic nonlinearities.

Neutrality curves L_σ ($\sigma = 0$) determine the type of stability of steady states. To
plot L_σ, for example, in the plane (Da, y), we need to solve the following system of
equations:

$$f_1(x, y, Da, \gamma) = 0 \,,$$
$$f_2(x, y, Da, \gamma) = 0 \,,$$
$$\sigma(x, y, Da, \gamma) = 0 \,. \tag{4.132}$$

Substituting the explicit expression of the parametric dependence $Da(x)$ into the third
equation, we obtain the equation of the neutrality curve in the explicit form:

$$L_\sigma(Da, \gamma): \quad \gamma(x) = \frac{y(x)^2}{y(x) - 1}\left(1 + \frac{y(x) - 1}{\beta x} + \frac{k_1 + k_2(1 - x)(1 - 3x)}{s}\right),$$
$$Da = Da(x, y(x)) \,.$$

It was difficult to obtain the explicit equations for plotting L_σ in other parameter planes. For this reason, we had to use a graphic procedure for plotting neutrality
curves too. We constructed a series of parametric dependences with respect to one of
the parameters, while varying the second parameter. Values corresponding to the sign
change of σ, i.e., $\sigma = 0$, were plotted on the parameter plane.

The mutual arrangement of multiplicity (L_Δ) and neutrality (L_σ) curves determines the parametric portrait of the system. It describes the various parameter regions
with different number and type of stability of steady states. In the system under consideration, there can be six such regions (see Fig. 4.73). Here there are the regions with
a single stable steady state, with a single unstable steady state, three steady states, two

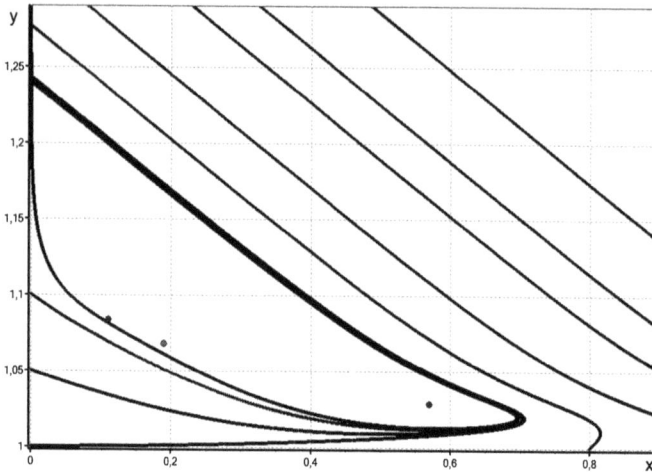

Fig. 4.74: Phase portrait in (x, y) at $Da = 0.058$, $s = 2$, $\beta = 0.375$, $\gamma = 55$, $k_1 = 0.6$, $k_2 = 1$

of which can be stable, three unstable steady states, and so on. As above, the region bounded by the multiplicity curve consists of two subregions, and the loop of the neutrality curve is situated between them.

Phase portrait. In accordance with the parametric portrait (Fig. 4.73) one of the possible phase portraits is plotted for the case of three unstable steady states (see Fig. 4.74) which are situated inside a stable limit cycle.

Time dependencies. Numerical integration of the original dynamic system (4.126) allows us to obtain time dependencies $x(t)$ and $y(t)$ at different initial conditions and parameters. The results of the study of time dependencies show that the increase of the parameter s leads to growth as the amplitude and period of oscillations. When increasing the parameter Da, the amplitude and period of oscillations are reduced.

On the basis of the technique of parametric analysis for system (4.126) the parameter regions can be plotted, where there are a multiplicity of steady states and auto-oscillations. When varying s in the interval (3.25–24.29) there are auto-oscillations with sufficiently large amplitude on the concentration of x in the system. The feature of thermokinetic model (4.126)) is that its kinetic subsystem at a constant temperature $y = $ const allows only a multiplicity of steady states. Under a joint change of concentration and temperature the combination of kinetic and thermal nonlinearities leads to the appearance of auto-oscillations.

Autocatalytic trigger (a modification of the model). Consider a modification of model (4.125), assuming a temperature dependence only for k_2 ($E_1 = E_{-1} = 0$,

$h_1 = h_{-1} = 0$, $E_2 \neq 0$, $h_2 \neq 0$). The corresponding dimensionless model has the form:

$$\frac{dy}{d\tau} = \beta f(y)xz^2 + s(1 - y) = f_1(x, y),$$

$$\frac{dx}{d\tau} = k_1 z - k_{-1}x - f(y)xz^2 = f_2(x, y),$$ (4.133)

where $f(y) = Da \exp(y(1 - 1/y))$.

In the system of two differential equations (4.133), x and y are the variables, and β, γ, Da, s, k_1, k_{-1} are parameters. As above, the goal of our study is to find the region of parameters of model (4.133), where there are oscillations, to investigate the effect of different parameters on the characteristics of these oscillations, and to define the structure of the partition of the parameter space β, γ, Da, s, k_1, k_{-1} on the regions corresponding to qualitatively different types of dynamic behavior.

The stationarity equation for system (4.133) has the form:

$$\beta f(y(x))xz^2 + s(1 - y(x)) = 0,$$

where $z = 1 - x$, $y(x) = 1 + (k_1 z - k_{-1}x)\beta/s$.

Stability conditions of steady states. We repeat one more time that the type of stability of steady states is determined by the roots of the characteristic equation $\lambda^2 - \sigma\lambda + \Delta = 0$, where σ and Δ are defined by the elements of the Jacobian matrix of the right sides of system (4.133). We write down the elements of this matrix:

$$a_{11} = \frac{\partial f_1}{\partial y} = \frac{\gamma}{y^2}\beta xz^2 f(y) - s,$$

$$a_{12} = \frac{\partial f_1}{\partial x} = \beta f(y)z(z - 2x),$$

$$a_{21} = \frac{\partial f_2}{\partial y} = -\frac{\gamma}{y^2}xz^2 f(y),$$

$$a_{22} = \frac{\partial f_2}{\partial x} = -(k_1 + k_{-1}) - f(y)z(z - 2x),$$

where $f(y) = Da \exp(y(1 - 1/y))$.

The studied steady state is unstable if at least one of the inequalities is violated:

$$\sigma < 0, \qquad \Delta > 0.$$

If $\Delta < 0$, the stationary point is a saddle (unstable point). If $\Delta > 0$ and $\sigma > 0$, the steady state will be an unstable node or unstable focus. If the steady state is single and unstable at some given set of the system parameters then there are oscillations in the system.

Similarly to (4.126) for system (4.133) some dependencies of the steady states on parameters can be written down explicitly, for example:

$$Da(x) = \frac{s(y(x) - 1)}{\beta xz^2 \exp(y(1 - 1/y(x)))},$$

$$y(x) = \frac{y(x)}{y(x) - 1} \ln \frac{s(y(x) - 1)}{\beta Daxz^2}.$$ (4.134)

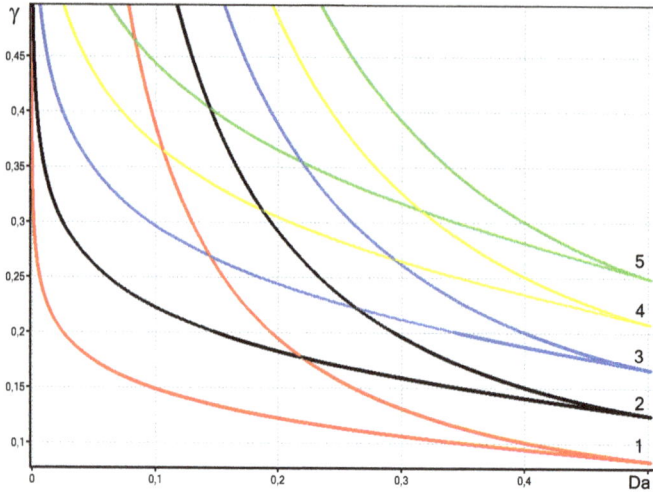

Fig. 4.75: Multiplicity curves $L_\Delta(Da, \beta)$ at $\gamma = 65$, $k_1 = 0.6$, $k_{-1} = 1$,
1) $s = 1$; 2) 1.5; 3) 2; 4) 2.5; 5) 3

As shown by calculations of the parametric dependence of $x(Da)$, the region of the multiplicity of steady states is reduced at increasing values of the parameter s.

Multiplicity curves. There are six parameters, β, γ, Da, s, k_1, k_{-1}, in system (4.133). Select two of them, for example, Da and y, considering the others fixed. We will plot a curve of multiplicity of steady states in the selected parameter plane.

Using the explicit expression $Da(x)$ of (4.134), we write the equation of the boundary region of the multiplicity of steady states L_Δ in the parameter plane (Da, y) in the form:

$$L_\Delta(Da, y): \quad y(x) = \frac{y^2(x)}{y(x) - 1} + \frac{sy^2(x)(z - 2x)}{\beta xz(k_1 + k_{-1})},$$
$$Da = Da(x, y(x)).$$

It was difficult to obtain explicit expressions for the other combinations of parameters. Therefore, to plot the multiplicity curves we used the graphical procedure which was described above. For curves of multiplicity (Da, β) the region of existence of three steady states is decreasing with increasing s (see Fig. 4.75).

Neutrality curves. The curve L_σ corresponds to the bifurcation values of the parameters, i.e., it allows us to analyze the change of stability of steady state. The curve L_σ, for example, in the parameter plane (Da, y), is defined as follows:

$$L_\sigma(Da, y): \quad y(x) = \frac{y^2(x)}{y(x) - 1}\left(1 + \frac{k_1 + k_{-1}}{s} + \frac{(y(x) - 1)(z - 2x)}{\beta xz}\right),$$
$$Da = Da(x, y(x)).$$

Curves of neutrality in other planes of parameters were plotted graphically. The curves of neutrality can be plotted for all combinations of parameters in this way.

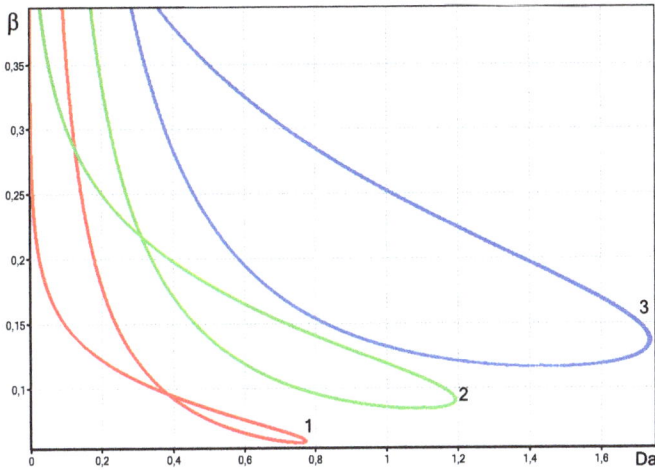

Fig. 4.76: Neutrality curves $L_{\Delta}(Da, \beta)$ at $\gamma = 65$, $k_1 = 0.6$, $k_{-1} = 1$:
1) $s = 1$; 2) 2; 3) 3.5

An example of curves of neutrality which were plotted graphically is presented in Fig. 4.76.

Parametric portrait. As before, the mutual position of curves of multiplicity (L_{Δ}) and neutrality (L_{σ}) gives the parametric portrait of system (4.133). It defines the various regions of parameters, which are different by number and type of stability of steady states.

The calculations allow us to set the interval of change of parameters at which oscillations exist. For example, if $\beta \in (0.374; 0.411)$ the system has limit cycles. Comparison of thermokinetic oscillator (4.133) with the same model (4.126) shows that the character of the temperature dependencies of the rate constants of the reactions and the corresponding thermal effects has a significant impact on the conditions of occurrence of critical phenomena in the kinetics of heterogeneous catalytic reactions. One of the features of the proposed models of thermokinetic oscillators is that of the kinetic subsystem being the trigger. Therefore, a multiplicity of steady states can be observed at any sufficiently large intensity of the heat transfer of gas with the surface.

4.7.6 Autocatalytic oscillator

Complete the reaction scheme (4.124) with a buffer step:

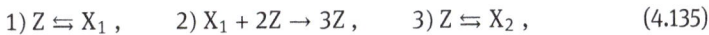

$$1) \ Z \leftrightarrows X_1 , \qquad 2) \ X_1 + 2Z \rightarrow 3Z , \qquad 3) \ Z \leftrightarrows X_2 , \qquad (4.135)$$

where X_1, X_2 are intermediate substances on the catalyst surface Z. As is known, the kinetic model corresponding to the autocatalytic scheme of transformations (4.135) is the simplest autocatalytic oscillator which provides the auto-oscillatory modes. As above, we complete the kinetic model, corresponding to (4.135), by the heat balance equation and consider the obtained system in a dimensionless form:

$$\frac{dy}{d\tau} = \beta f(y) x_1 z^2 + s(1 - y),$$

$$\frac{dx_1}{d\tau} = k_1 z - k_{-1} x_1 - f(y) x_1 z^2,$$

$$\frac{dx_2}{d\tau} = k_3 z - k_{-3} x_2, \tag{4.136}$$

where $z = 1 - x_1 - x_2$; x_1, x_2, y are dimensionless concentrations and temperature; and Da, β, γ, s, k_1, k_{-1}, k_3, k_{-3} are dimensionless parameters of the system.

For system (4.136) steady states are determined from the equation (in x_1):

$$\beta f(y(x_1)) x_1 z^2 + s(1 - y(x_1)) = 0, \tag{4.137}$$

where

$$x_2 = \frac{k_3(1 - x_1)}{k_3 + k_{-3}}; \qquad \alpha = \frac{k_3}{k_3 + k_{-3}}; \qquad z = (1 - x_1)(1 - \alpha);$$

$$y(x_1) = 1 + \frac{\beta}{s}(k_1 z - k_1 x_1); \qquad f(y(x_1)) = Da \exp(\gamma(1 - 1/y)).$$

Stability conditions of steady states. There are many criteria allowing us to estimate the sign of the eigenvalues of the Jacobian matrix without computing the roots themselves. One of these criteria is the criterion of Routh–Hurwitz. For example, for a system of three equations (4.136) the criterion is formed in the following way. The characteristic equation of the third order has the form:

$$\lambda^3 - \sigma \lambda^2 + \delta \lambda - \Delta = 0,$$

where σ, Δ, δ are expressed through the elements of the Jacobian matrix.

According to the criterion of Routh–Hurwitz a steady state is stable if the inequalities $\sigma > 0$, $\theta = \sigma \delta - \Delta > 0$, $\Delta > 0$ are true. Find the elements of the Jacobian matrix for system (4.136):

$$a_{11} = -(k_1 + k_{-1}) - f(y)(z - 2x_1), \qquad a_{21} = -k_3,$$

$$a_{12} = -k_1 + 2f(y)x_1 z, \qquad a_{22} = -(k_3 + k_{-3}),$$

$$a_{13} = -\frac{\gamma}{y^2} f(y) x_1 z^2, \qquad a_{23} = 0,$$

$$a_{31} = \beta f(y) z(z - 2x_1), \qquad a_{32} = -2\beta f(y) x_1 z,$$

$$a_{33} = \frac{\gamma}{y^2} \beta f(y) x_1 z^2 - s.$$

Similar to the above discussed models it is possible to obtain explicit expressions for the parametric dependencies from the conditions of stationarity (4.137):

$$Da(x_1) = \frac{s}{\beta} \frac{y(x_1) - 1}{x_1 z^2 \exp(\gamma(1 - 1/y(x_1)))},$$

$$y(x_1) = \frac{y(x_1)}{y(x_1) - 1} \ln \frac{s(y(x_1) - 1)}{\beta Da x_1 z^2}.$$

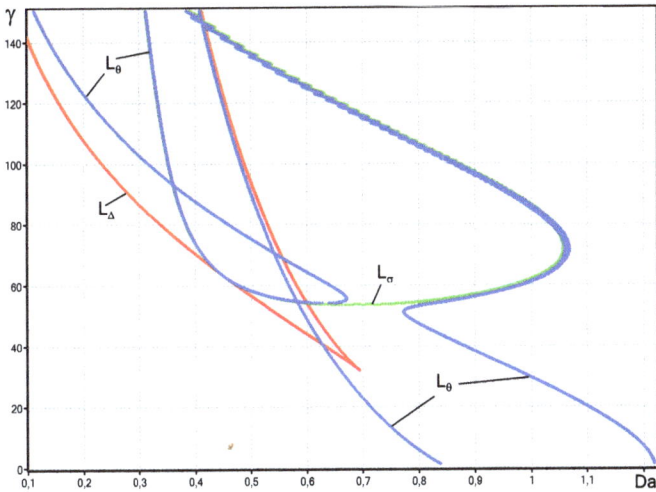

Fig. 4.77: Parametric portrait in the (Da, γ) plane at $\beta = 0.6, s = 1, k_1 = 0.12, k_{-1} = 0.01$, $k_3 = 0.0032, k_{-3} = 0.002$

Knowledge of the explicit form of parametric dependence allows us to plot the bifurcation curves according to the procedure of parametric analysis of model (4.136). The bifurcation curves define a multiplicity of steady states, their stability, and conditions for the existence of auto-oscillations.

Parametric portrait. A parametric portrait of model (4.136), in contrast to the earlier models, will consist of three curves $\sigma = 0$, $\Delta = 0$, $\theta = 0$. For example, in the parameter plane (Da, γ), these curves divide the parametric plane into a large number of regions with different number and type of stability of steady states (see Fig. 4.77).

Phase portraits. In accordance with the parametric portrait of the system several types of phase portraits can be distinguished. There are regions of parameters with a single and unstable steady state which provide the existence of auto-oscillations in (4.136). The regions with three steady states are distinguished, where one of them is stable and two are unstable or two of them are stable and one is unstable. If there are three steady states, one of them is a low temperature steady state, the second is a high temperature steady state and the third is an intermediate one.

Thus, the combination of kinetic and thermal nonlinearities can lead to a significant complication of the dynamics of catalytic systems. So, if there is a multiplicity of steady states in the kinetic subsystem, the temperature change can cause auto-oscillations, which are characterized by complicated dynamics at three degrees of freedom. The discussed mathematical models are the simplest in a sense. They are minimal both in nonlinearity and in dimension of the phase space. In thermokinetic oscillators the degree of nonlinearity of the kinetic component does not exceed three, and the temperature dependence traditionally has exponential form $\exp(\gamma(1 - 1/y(x)))$. Our research shows the proposed model can be taken as a base model for the description and interpretation of the complex dynamics of thermoki-netic processes on the catalyst surface.

The technique of primary parametric analysis, demonstrated on the basic models of the thermokinetic processes occurring in flow reactors of ideal mixing, can natu-rally be extended to other nonlinear dynamic models of physical chemistry. Our expe-rience shows that the development of mathematical software of parametric analysis of nonlinear models allows us to obtain important numerical and analytical results.

4.8 Parallel scheme

In Sections 4.8 and 4.9 we study mathematical models of CSTR, in which two parallel or two consistent reactions occur. These models are the traditional object of paramet-ric analysis in the theory of chemical reactors [9–14, 95]. Here we present in a quite condensed form the results of their parametric analysis, containing only the analysis of bifurcations of steady states.

For two parallel reactions occurring in a CSTR

$$1)\ A \to B\,, \qquad 2)\ A \to C\,,$$

a dimensionless mathematical model is written as:

$$\frac{dx}{d\tau} = (f_1(y) + f_2(y))(1 - x) - x\,,$$

$$\frac{dy}{d\tau} = (\beta_1 f_1(y) + \beta_2 f_2(y))(1 - x) - s(y - 1)\,, \tag{4.138}$$

where x and y are the dimensionless concentration and temperature;

$$f_1(y) = Da_1 \exp(\gamma_1(1 - 1/y))\,, \quad f_2(y) = Da_2 \exp(\gamma_2(1 - 1/y))\,;$$

β_1, β_2, Da_1 and Da_2, γ_1, γ_2 and s are the dimensionless parameters. The stationarity equation can be written as:

$$\frac{\beta_1 f_1(y) + \beta_2 f_2(y)}{(1 + f_1(y) + f_2(y))} - s(y - 1) = 0\,. \tag{4.139}$$

The elements of the Jacobian matrix for (4.138) are

$$a_{11} = -(f_1(y) + f_2(y)) - 1\,,$$

$$a_{12} = \frac{\gamma_1 f_1(y) + \gamma_2 f_2(y)}{y^2(1 + f_1(y) + f_2(y))}\,,$$

$$a_{21} = -(\beta_1 f_1(y) + \beta_2 f_2(y))\,,$$

$$a_{22} = \frac{\beta_1 \gamma_1 f_1(y) + \beta_2 \gamma_2 f_2(y)}{y^2(1 + f_1(y) + f_2(y))} - s\,.$$

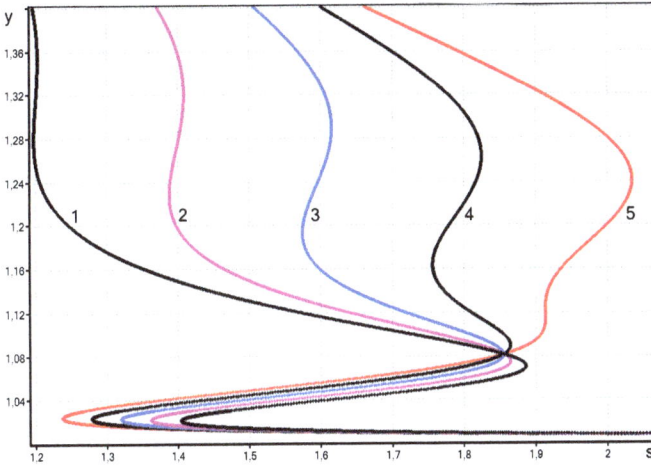

Fig. 4.78: Parametric dependencies $y(s)$ at $\gamma_2 = 75$, $Da_1 = 0.065$, $Da_2 = 0.001$, $\beta_1 = 0.15$, $\beta_2 = 0.7$, 1) $\gamma_1 = 59$; 2) 57; 3) 55; 4) 53; 5) 591

Parametric dependencies of steady states can be written of the stationarity equation (4.139) as follows:

$$s(y) = \frac{\beta_1 f_1(y) + \beta_2 f_2(y)}{(y-1)(1 + f_1(y) + f_2(y))},$$

$$\beta_1(y) = \frac{s(y-1)(1 + f_1(y) + f_2(y)) - \beta_2 f_2(y)}{f_1(y)},$$

$$\beta_2(y) = \frac{s(y-1)(1 + f_1(y) + f_2(y)) - \beta_1 f_1(y)}{f_2(y)},$$

$$Da_1(y) = \frac{s(y-1)(1 + f_2(y)) - \beta_2 f_2(y)}{e_1(y)(\beta_1 - s(y-1))},$$

$$Da_2(y) = \frac{s(y-1)(1 + f_1(y)) - \beta_1 f_1(y)}{e_2(y)(\beta_2 - s(y-1))},$$

$$\gamma_1(y) = \frac{y}{y-1} \ln \frac{s(y-1)(1 + f_2(y)) - \beta_2 f_2(y)}{Da_1(\beta_1 - s(y-1))},$$

$$\gamma_2(y) = \frac{y}{y-1} \ln \frac{s(y-1)(1 + f_1(y)) - \beta_1 f_1(y)}{Da_2(\beta_2 - s(y-1))}, \tag{4.140}$$

where $e_1(y) = \exp(\gamma_1(1 - 1/y))$ and $e_2(y) = \exp(\gamma_2(1 - 1/y))$. Some examples of plots of parametric dependencies are given in Figs. 4.78–4.80. Note that model (4.138) can have up to five steady states, because there are two exhibitors f_1 and f_2 in it.

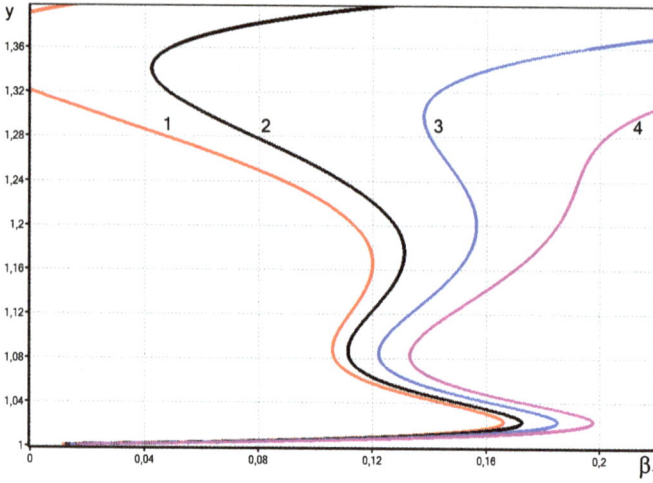

Fig. 4.79: Parametric dependencies $y(\beta_1)$ at $\gamma_1 = 55$, $\gamma_2 = 75$, $Da_1 = 0.065$, $Da_2 = 0.001$, $\beta_2 = 0.7$, 1) $s = 1.45$; 2) 1.5; 3) 1.6; 4) 1.7

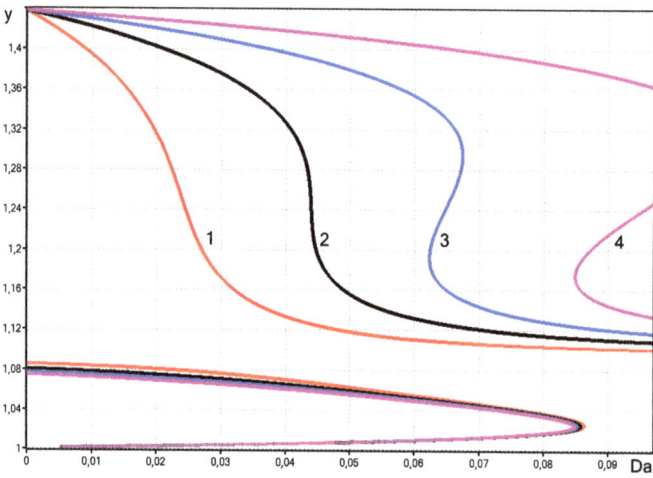

Fig. 4.80: Parametric dependencies $y(Da_1)$ at $\gamma_1 = 55$, $Da_2 = 0.001$, $s = 1.6$, $\beta_1 = 0.15$, $\beta_2 = 0.7$, 1) $\gamma_2 = 70$; 2) 73; 3) 75; 4) 77

Multiplicity curves.

$L_\Delta(\beta_1, \beta_2)$: $\quad \beta_2(y) = \dfrac{s(y-1)(\gamma_1 A - (\gamma_1 f_1(y) + \gamma_2 f_2(y)))}{f_2(y)(\gamma_1 - \gamma_2)}$,

$\qquad\qquad\qquad \beta_1 = \beta_1(y, \beta_2(y))$,

$L_\Delta(s, \beta_1)$: $\quad \beta_1(y) = \dfrac{\beta_2 f_2(y)[A(\gamma_2(y-1) - y^2) - (y-1)(\gamma_1 f_1(y) + \gamma_2 f_2(y))]}{f_1(y)[A(y^2 - \gamma_1(y-1)) + (y-1)(\gamma_1 f_1(y) + \gamma_2 f_2(y))]}$,

$\qquad\qquad\qquad s = s(y, \beta_1(y))$;

$L_\Delta(s, \beta_2)$: $\quad \beta_2(y) = \dfrac{\beta_1 f_1(y)[A(y^2 - \gamma_1(y-1)) + (y-1)(\gamma_1 f_1(y) + \gamma_2 f_2(y))]}{f_2(y)[A(\gamma_2(y-1) - y^2) - (y-1)(\gamma_1 f_1(y) + \gamma_2 f_2(y))]}$,

$\qquad\qquad\qquad s = s(y, \beta_2(y))$;

$$(4.141)$$

where $A = 1 + f_1(y) + f_2(y)$.

Neutrality curves.

$L_\sigma(\beta_1, \beta_2)$: $\quad \beta_2(y) = \dfrac{[y^2(1 + f_1(y) + f_2(y)) + s(y^2 - \gamma_1(y-1))](1 + f_1(y) + f_2(y))}{f_2(y)(\gamma_2 - \gamma_1)}$,

$\qquad\qquad\qquad \beta_1 = \beta_1(y, \beta_2(y))$,

$L_\sigma(s, \beta_1)$: $\quad s(y) = \dfrac{y^2(1 + f_1(y) + f_2(y))^2 + \beta_2 f_2(y)(\gamma_1 - \gamma_2)}{(1 + f_1(y) + f_2(y))(\gamma_1(y-1) - y^2)}$,

$\qquad\qquad\qquad \beta_1 = \beta_1(y, s(y))$;

$L_\sigma(s, \beta_2)$: $\quad s(y) = \dfrac{y^2(1 + f_1(y) + f_2(y))^2 + \beta_1 f_1(y)(\gamma_2 - \gamma_1)}{(1 + f_1(y) + f_2(y))(\gamma_2(y-1) - y^2)}$,

$\qquad\qquad\qquad \beta_2 = \beta_2(y, s)$.

$$(4.142)$$

The results of calculations of the parametric portraits for (4.138) on the basis of the bifurcation curves (4.141) and (4.142) are not presented here. The presence of explicit formulas for L_Δ and L_σ significantly simplifies the problem. If there is a need to plot bifurcation curves L_Δ and L_σ in other parameter planes for which explicit expressions of the type (4.141) and (4.142) cannot be obtained, it is possible to use, as above, a graphical procedure based on explicit dependencies of steady states from parameters (4.140).

Phase portrait. A phase portrait for model (4.138) can be plotted for various numbers of steady states, differing in the type of stability. An example of a phase portrait for five steady states, where two of them are unstable and three of them are stable, is shown in Fig. 4.81.

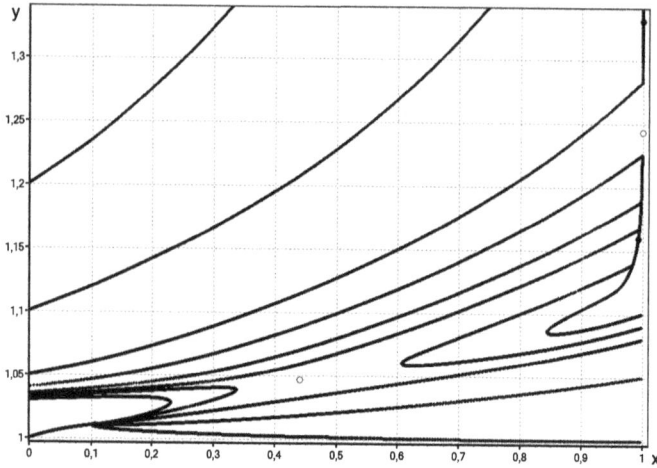

Fig. 4.81: Phase portrait for (4.138) at $\gamma_1 = 55$, $\gamma_2 = 75$, $Da_1 = 0.065$, $Da_2 = 0.001$, $s = 1.6$, $\beta_1 = 0.15$, $\beta_2 = 0.7$

4.9 Consistent scheme

A scheme of two consistent reactions is

$$1)\ A \rightarrow B, \qquad 2)\ B \rightarrow C.$$

A dimensionless mathematical model of a CSTR corresponding to the consistent scheme has the form:

$$\frac{dx_1}{d\tau} = f_1(y)(1 - x_1) - x_1,$$

$$\frac{dx_2}{d\tau} = f_2(y)(1 - x_2) - f_1(y)(1 - x_1) - x_2 + 1,$$

$$\frac{dy}{d\tau} = \beta_1 f_1(y)(1 - x_1) + \beta_2 f_2(y)(1 - x_2) - s(y - 1), \tag{4.143}$$

where x_1 and x_2 are the dimensionless concentrations of substances; y is the dimensionless temperature; $f_1(y) = Da_1 \exp(\gamma_1(1 - 1/y))$, $f_2(y) = Da_2 \exp(\gamma_2(1 - 1/y))$, β_1, β_2, Da_1, Da_2, γ_1, γ_2, s are the dimensionless parameters.

After excluding the variables x_1 and x_2 we obtain the following stationarity equation

$$\frac{\beta_1 f_1(y)}{1 + f_1(y)} + \frac{\beta_2 f_1(y)f_2(y)}{(1 + f_1(y))(1 + f_2(y))} - s(y - 1) = 0. \tag{4.144}$$

The elements of the Jacobian matrix for (4.143) are

$$a_{11} = -f_1(y) - 1, \quad a_{12} = 0, \quad a_{13} = \frac{y_1}{y^2(1 + f_1(y))},$$

$$a_{21} = f_1(y), \quad a_{22} = -f_2(y) - 1, \quad a_{23} = \frac{y_2 f_1(y)}{y^2(1 + f_1(y))(1 + f_2(y))},$$

$$a_{31} = -\beta_1 f_1(y), \quad a_{32} = -\beta_2 f_2(y),$$

$$a_{33} = \frac{y_1 \beta_1 f_1(y)}{y^2(1 + f_1(y))} + \frac{y_2 \beta_2 f_1(y) f_2(y)}{y^2(1 + f_1(y))(1 + f_2(y))} - s.$$

Parametric dependencies are obtained from the stationarity equation (4.144):

$$s(y) = \frac{\beta_1 f_1(y)(1 + f_2(y)) + \beta_2 f_1(y) f_2(y)}{(y - 1)(1 + f_1(y))(1 + f_2(y))},$$

$$\beta_1(y) = \frac{s(y - 1)(1 + f_1(y))}{f_1(y)} - \frac{\beta_2 f_2(y)}{1 + f_2(y)},$$

$$\beta_2(y) = \frac{s(y - 1)(1 + f_1(y))(1 + f_2(y))}{f_1(y) f_2(y)} - \frac{\beta_1(1 + f_2(y))}{f_2(y)},$$

$$Da_1(y) = \frac{s(y - 1)}{e_1(y)\left(\beta_1 + \frac{\beta_2 f_2(y)}{1 + f_2(y)} - s(y - 1)\right)},$$

$$Da_2(y) = \frac{s(y - 1) - \beta_1 f_1(y)/(1 + f_1(y))}{e_2(y)\left(\frac{(\beta_1 + \beta_2) f_1(y)}{1 + f_1(y)} - s(y - 1)\right)},$$

$$y_1(y) = \frac{y}{y - 1} \ln \frac{s(y - 1)}{Da_1\left(\beta_1 + \frac{\beta_2 f_2(y)}{1 + f_2(y)} - s(y - 1)\right)},$$

$$y_2(y) = \frac{y}{y - 1} \ln \frac{s(y - 1) - \beta_1 f_1(y)/(1 + f_1(y))}{Da_2\left(\frac{(\beta_1 + \beta_2) f_1(y)}{1 + f_1(y)} - s(y - 1)\right)}, \tag{4.145}$$

where $e_1(y) = \exp(y_1(1 - 1/y))$ and $e_2(y) = \exp(y_2(1 - 1/y))$.

An example of the parametric dependencies of steady states is given in Fig. 4.82. The explicit form of these dependencies (4.145) allows us to study the effect of different model parameters on the corresponding values of the stationary concentrations and temperature.

As in Section 4.8, the presence of two exponents $f_1(y)$ and $f_2(y)$ determines the possibility of the existence of up to five steady states. The difference of model (4.143) from (4.138) is that model (4.143) is a dynamic system in three dimensional phase space. This defines a much greater diversity of its dynamic behavior. For example, as in Section 4.6, the stability of steady states is determined by the roots of the characteristic equation of the third order:

$$\lambda^3 + \sigma\lambda^2 + \delta\lambda + \Delta = 0,$$

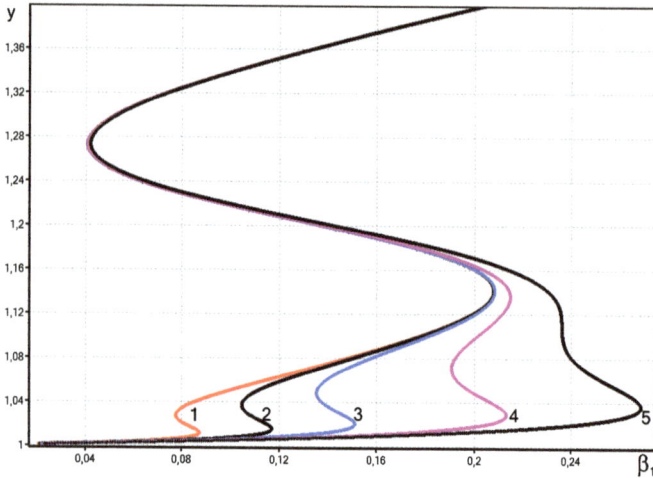

Fig. 4.82: Parametric dependencies $y(\beta_1)$ at $\gamma_2 = 40$, $Da_1 = 0.09$, $Da_2 = 0.001$, $\beta_2 = 0.6$, $s = 2$, 1) $\gamma_1 = 120$; 2) 90; 3) 70; 4) 50; 5) 40

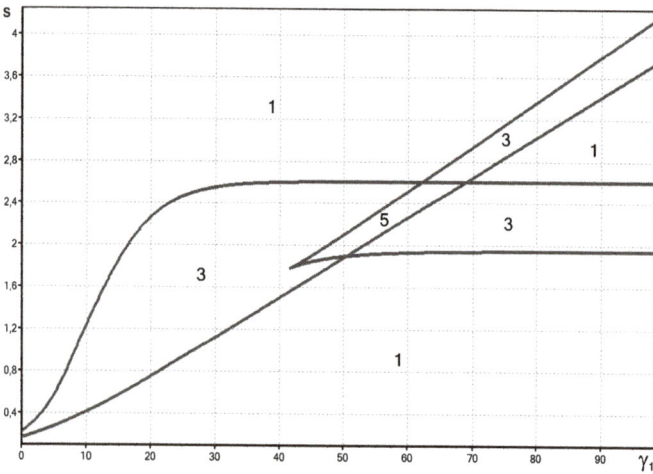

Fig. 4.83: Multiplicity curve L_Δ on the plane (γ_1, s) at $\gamma_2 = 40$, $Da_1 = 0.09$, $Da_2 = 0.001$, $\beta_1 = 0.2$, $\beta_2 = 0.6$

where the coefficients σ, δ and Δ are defined according to (4.143) through the elements of the Jacobian matrix. In addition, the dynamics of (4.143) can be characterized by chaotic behavior, as discussed in [13].

An example of the plot of a multiplicity curve is presented in Fig. 4.83. The regions with one, three, and five steady states are labeled by numbers in the figure.

The models (4.138) and (4.143) are among the important basic models of the theory of chemical reactors. Therefore, we believe that they deserve a more detailed parametric study, at least to the same degree of detail as the model of a CSTR for a single reaction of the first order (4.49).

4.10 One reversible reaction

It is well known that first-order phase transitions are frequently associated with noticeable thermal effects. For example, melting is characterized as an exothermal process, whereas liquid–solid transformation can occur with heat production. In the simplest case, the rates of these transitions can be described as a function of temperature in a standard way using the classical Arrhenius equation. The nonlinearity and inertness of thermal processes in the small neighborhood of phase transitions can lead to typical nonlinear and nonstationary effects, namely, the multiplicity of steady states and oscillations.

Here, we propose the simplest dynamic model of a phase transition and perform its parametric analysis. Conditions have been found for the existence of three and five steady states; the ranges of the parameters where auto-oscillations exist in a dynamic system are established; and characteristic parametric and phase portraits of the mathematical model have been plotted. The process dynamics in the neighborhood of a phase transition point has been shown to be rather complex. The processes observed here include the hysteresis of temperature dependences, undamped concentration and temperature oscillations, and considerable dynamic bursts during the stabilization of a steady state.

For phase transitions of the type

$$F_1 \rightleftharpoons F_2 , \tag{4.146}$$

in a system where there is heat exchange with the environment, a dimensionless spatially c6:46 model in accordance with [82] can be represented as

$$\frac{dx_1}{d\tau} = -f_1(y)x_1 + f_2(y)(1 - x_1) ,$$

$$\frac{dy}{d\tau} = \beta_1 f_1(y)x_1 + \beta_2 f_2(y)(1 - x_1) + s(y - 1) , \tag{4.147}$$

where $f_i(y) = Da_i \exp(y_i/(1 - 1/y))$, $i = 1, 2$; x_1 and y are dimensionless concentration and temperature, respectively; dimensionless parameters Da_i, y_y and β_i according to Frank-Kamenetskii, respectively, characterize the rates, activation energies, and thermal effects of $F_1 \rightarrow F_2$ and $F_2 \rightarrow F_1$ phase transitions; and s is a parameter characterizing the heat exchange intensity with the environment.

Mathematical model (4.147) is a set of two differential equations with characteristic nonlinearities $f_1(y)$ and $f_2(y)$. In combustion theory and theoretical foundations of chemical reactors, such a model is a traditional object of parametric analysis [1, 2]. Model (4.147) is specific in that it refers a non-flow-through system and contains two exponents, considerably supplementing the variety of the dynamic and nonlinear properties of the system as shown in our earlier investigations. Here, we report the parametric analysis of dynamic system (4.147), including an analysis of the number and the stability of steady states, the plotting of parametric dependences of stationary characteristics, the design of parametric and phase portraits, and calculations of temporal dependences characterized by dynamic bursts and undamped concentration and temperature oscillations.

In a steady state the equations are

$$-f_1(y)x_1 + f_2(y)(1 - x_1) = 0 , \tag{4.148}$$

$$\beta_1 f_1(y)x_1 + \beta_2 f_2(y)(1 - x_1) + s(y - 1) = 0 . \tag{4.149}$$

From this, we obtain

$$x_1 = f_2(y)/(f_1(y) + f_2(y)) , \qquad (1 - x_1) = f_1(y)/(f_1(y) + f_2(y)) .$$

The stationary temperature is determined from the heat balance equation (4.149) as follows:

$$\frac{(\beta_1 + \beta_2)f_1(y)f_2(y)}{f_1(y) + f_2(y)} = s(y - 1) . \tag{4.150}$$

If the total thermal effect of phase transitions (4.146) is positive, $\beta_1 + \beta_2 > 0$ in (4.150), and the stationary temperature y (the point of intersection of the heat production curve and the heat removal curve), exceeds unity. In this case, the presence of two exponents in (4.150) leads to the possibility of the existence of five steady states. The calculations show five steady states for a certain ratio between y_1 and y_2 within rather narrow ranges of the other parameters.

Parametric dependencies are easily obtained in an explicit form from the stationarity equation (4.150), for example,

$$s = \frac{(\beta_1 + \beta_2)f_1(y)f_2(y)}{(y - 1)(f_1(y) + f_2(y))} ,$$

$$\beta_1 + \beta_1 = \frac{s(y - 1)(f_1(y) + f_2(y))}{f_1(y)f_2(y)} ,$$

$$Da_1 = \frac{s(y - 1)f_2(y) \exp (y_1(1 - 1/y))}{(\beta_1 + \beta_2)f_2(y) - s(y - 1)} ,$$

$$y_1(y) = \frac{y}{(y - 1)} \ln \frac{s(y - 1)f_2(y)}{Da_1(\beta_1 + \beta_2)f_2(y) - s(y - 1)} . \tag{4.151}$$

An example of parametric dependencies of steady states based on (4.151) is given in Fig. 4.84. The multiplicity of steady states leads to a hysteresis in the temperature dependencies.

The presence of two exponents allows the existence of one, three, and five steady states. The stability of these steady states is determined by the roots of the characteristic equation

$$\lambda^2 + \sigma\lambda + \Delta = 0 ,$$

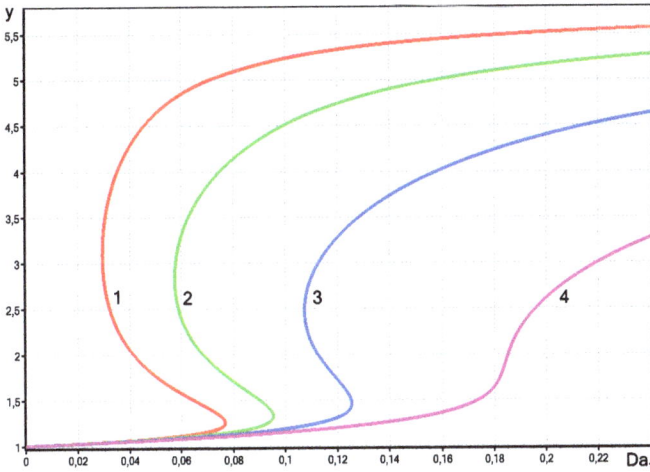

Fig. 4.84: Dimensionless stationary temperature vs. parameter Da_1 at $s = 0.2$, $\gamma_2 = 3$, $\beta_1 = 0.05$, $\beta_2 = 0.15$, $Da_2 = 0.4$, 1) $\gamma_1 = 8$; 2) 7; 3) 6; 4) 5

where the coefficients are calculated through the elements of the Jacobian matrix of the linearized system (4.147) in the neighborhood of a steady state,

$$a_{11} = -f_1(y) - f_2(y) ,$$

$$a_{12} = -\frac{\gamma_1}{y^2}f_1(y)x_1 - \frac{\gamma_2}{y^2}f_2(y)(1 - x_1) ,$$

$$a_{21} = \beta_1 f_1(y) - \beta_2 f_2(y) ,$$

$$a_{22} = \beta_1 \frac{\gamma_1}{y^2}f_1(y)x_1 + \beta_2 \frac{\gamma_2}{y^2}f_2(y)(1 - x_1) - s ,$$

as follows:

$$\sigma = -a_{11} - a_{22} , \quad \Delta = a_{11}a_{22} - a_{12}a_{21} . \tag{4.152}$$

The equality to zero of the coefficients of σ and Δ determines the values of parameters at which the number and the type of stability of steady states can change. Bifurcation curves L_Δ and L_σ, corresponding to the change in the number and stability of steady states on the basis of relations (4.151) and (4.152) can be written explicitly using the parametric analysis tools developed above. For example, the bifurcation curve L_Δ is given by the condition

$$\frac{(\beta_1 + \beta_2)f_1^2(y)f_2^2(y)}{y^2(f_1(y) + f_2(y))^2}\left(\frac{\gamma_1}{f_1(y)} + \frac{\gamma_2}{f_2(y)}\right) = s , \tag{4.153}$$

and curve L_σ by

$$\left(\beta_1 \frac{\gamma_1}{y^2}x_1 - 1\right)f_1(y) + \left(\beta_2 \frac{\gamma_2}{y^2}(1 - x_1) - 1\right)f_2(y) = s . \tag{4.154}$$

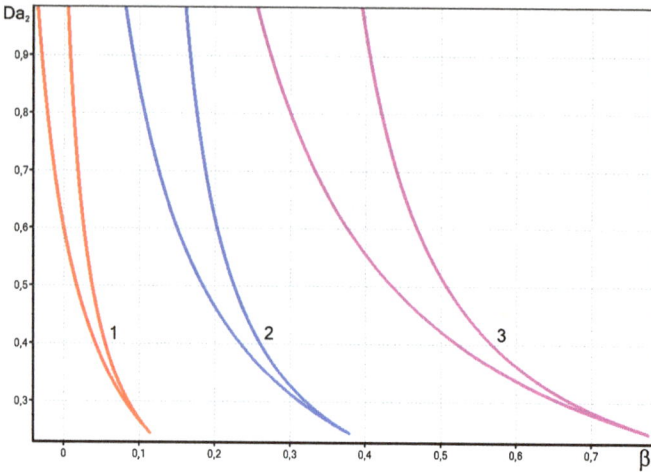

Fig. 4.85: Regions of three steady states bounded by bifurcation curves $L_\Delta(\beta_1, Da_1)$ for the parameters specified in Fig. 4.84. s varies as follows: (1) 0.2, (2) 0.4, and (3) 0.7

Together with the stationarity condition in the form (4.151), the equalities (4.153)–(4.154) give the possibility to write equations for curves L_Δ and L_σ explicitly in the plane of two selected parameters. So, to obtain $L_\Delta(\beta_1, Da_1)$, it is sufficient to consider (4.151) and (4.153) jointly, using them to express $\beta_1 = \beta_1(Da_1, y)$ and $Da_1 = Da_1(y)$, where the steady state value y is changed as a parameter. In the same way, the equations for L_σ can be written from (4.151) and (4.154) by selecting a combination of two parameters.

Figure 4.85 shows the regions of multiplicity of steady states in the parameter plane (β_1, Da_1), the boundaries of which are determined by the bifurcation curve L_Δ. Similarly, we can plot the curve L_σ which determines the stability of steady states. The combination of curves L_Δ and L_σ allows us to obtain a parametric portrait of the dynamic system. The specific type of the phase portrait for model (4.147) corresponds to each region of the two parameters in this portrait.

Calculations show that at the stabilization of a steady state the time dependencies of the concentration $x_1(t)$ and temperature $y(t)$ can be characterized by significant dynamic bursts (what we call a boomerang effect). For example, in the presence of hysteresis, the transition from one steady state to another can be accompanied by short-term temperature spikes far exceeding the stationary temperature value. A soft transition from one steady state to another can be organized at the control of the parameters determining the dynamics of this transition. "Dangerous" and "safe" conditions of transition of a system from one phase state to another and back can be determined during a detailed parametric analysis of the corresponding dynamic models of the type (4.147).

Thus, the dynamic model (4.147) can be considered to be the simplest basic model of a first-order phase transition. The parametric analysis of this model shows that it can have one, three, or five steady states. The parameter regions have been found where auto-oscillations exist in a dynamic system; characteristic parametric and phase portraits have been plotted for the mathematical model. The process dynamics in the neighborhood of a phase transition point can be rather complex and can be characterized by the hysteresis of temperature dependencies, undamped oscillations of concentrations and temperatures, as well as significant dynamic bursts during the stabilization of a system to a steady state.

4.11 Model of spontaneous combustion of brown-coal dust

In Sections 4.11 and 4.12 we presented the results of the parametric analysis of two specific mathematical models in a sufficiently compressed view. Our goal is to show that an experience of the analysis of basic models is useful for the construction and investigation of kinetic models of specific processes. Safe operation of thermopower facilities is a basis for the smooth functioning of cities and settlements. However, fires and explosions may occur in CHP.

Fires and explosions are typical of all stages of handling of coal. Spontaneous combustion is the main reason for fires at fuel stores and coal-conveying systems (50%–60%). Every sixth fire at thermal power plants or boiler houses is also caused by spontaneous combustion. It was shown in [98] that the ability of coals to ignite spontaneously strongly depends on the kinetic parameters of this process, i.e., activation energy and the pre-exponent. These parameters were found to change when coal is heated to 500–600 K; in this case, the coal becomes more reactive, i.e., its thermal activation occurs. In this section, an explosion of coal dust in a cylindrical vessel was studied experimentally. It was found that, for a certain fuel/oxidizer ratio, a second explosion can follow the first explosion as a result of thermal activation of coal at the first step. As mathematical models, equations describing the material and thermal balance in a tube reactor were used. Calculations performed for the kinetic parameters of spontaneous combustion of brown coal from the Irsha-Borodino deposit show that the models proposed are in qualitative agreement with the experiment.

Here we use three simple models that qualitatively describe the spontaneous combustion of coal dust in a cylindrical reactor under various thermal and physicochemical assumptions. First of all, this is an ideal-displacement regime in a tube. In this case, one dimensional steady profiles of the temperature and concentration distribution along the reactor were considered. Moreover, regimes of spontaneous combustion for excess oxygen were studied with allowance for a considerable change in the latter. Particular attention was given to spontaneous combustion with allowance for the so-called thermal activation of coal. In the first two cases, a one-step process occurred, whereas a three-step process was observed in the third case. The aim of modeling was to qualitatively describe the main specific features of spontaneous combustion of coal dust and to determine physicochemical and thermophysical parameters responsible for spontaneous combustion.

Reaction X → P (coal → product). In the steady case, the mathematical model of a tube reactor with a single exothermic reaction comprises the equations of material and thermal balance [97]:

$$u\frac{dX}{d\ell} = -k(T)X\,,$$

$$C_p\rho u\frac{dT}{d\ell} = (-\Delta H)k(T)X + \frac{4h}{d}(T_w - T)\,. \tag{4.155}$$

Here

$$k(T) = k^0 \exp(-E/(RT)) \tag{4.156}$$

is the reaction rate constant, k^0 is the pre-exponent, E is the activation energy, R is the universal gas constant, T is the current temperature, X is the current concentration of coal dust, T_w is the temperature of the reactor wall, ℓ is the variable length of the reactor, h is the coefficient of heat transfer from the wall of the reactor to its volume, d is the diameter of the reactor, ρ, C_p and u are the density, heat capacity, and feeding rate of the reactive mixture, respectively, and $(-\Delta H)$ is the thermal effect of the reaction. The length of the reactor varies within the limits

$$0 \le \ell \le \ell_f\,, \tag{4.157}$$

where ℓ_f is the total length of the cylindrical reactor. The conditions at the input of the reactor are written as

$$X(0) = X^0\,, \qquad T(0) = T_0\,, \tag{4.158}$$

where X^0 and T_0 are the concentration and temperature of the reactive mixture at the input of the reactor. In model (4.155), the first-order reaction is considered, and it is assumed that the process occurs under oxygen excess, which can be accepted as a first approximation if the concentration of coal dust is small.

Calculations according to model (4.155) were performed for the following parameters: $k^0 = 8.4 \cdot 10^7\,\mathrm{s}^{-1}$, $E = 53,200\,\mathrm{J\,mol}^{-1}$, $R = 8.31\,\mathrm{J\,mol}^{-1}\,\mathrm{K}^{-1}$, $(-\Delta H) = 900\,\mathrm{J\,mol}^{-1}$, $T_w = 283\,\mathrm{K}$, $T_0 = 325\,\mathrm{K}$, $X^0 = 5 \cdot 10^{-4}\,\mathrm{g\,cm}^{-3}$, $l_f = 200\,\mathrm{cm}$, $d = 10\,\mathrm{cm}$, $\rho = 5 \cdot 10^{-4}\,\mathrm{g\,cm}^{-3}$, $C_p = 1.13\,\mathrm{J\,g}^{-1}\,\mathrm{K}^{-1}$, $u = 100\,\mathrm{cm\,s}^{-1}$, $h = 0.006\,\mathrm{J\,cm}^{-2}\,\mathrm{s}^{-1}\,\mathrm{K}^{-1}$). Conditions under which coal dust is ignited spontaneously were determined for varied parameters T_0, X^0, h, T_w and u. Some calculated results are shown in Fig. 4.86. It was found that the conditions of spontaneous combustion and location and temperature of the "hot spot" are highly sensitive to the conditions at the input to the reactor and heat-exchange conditions: for a certain critical value of the parameter varied, the temperature increases abruptly, which can be characterized as a thermal explosion of coal dust in the tube. An example of this steady temperature distribution along the reactor is shown in Fig. 4.86, where the input temperature is the varied parameter.

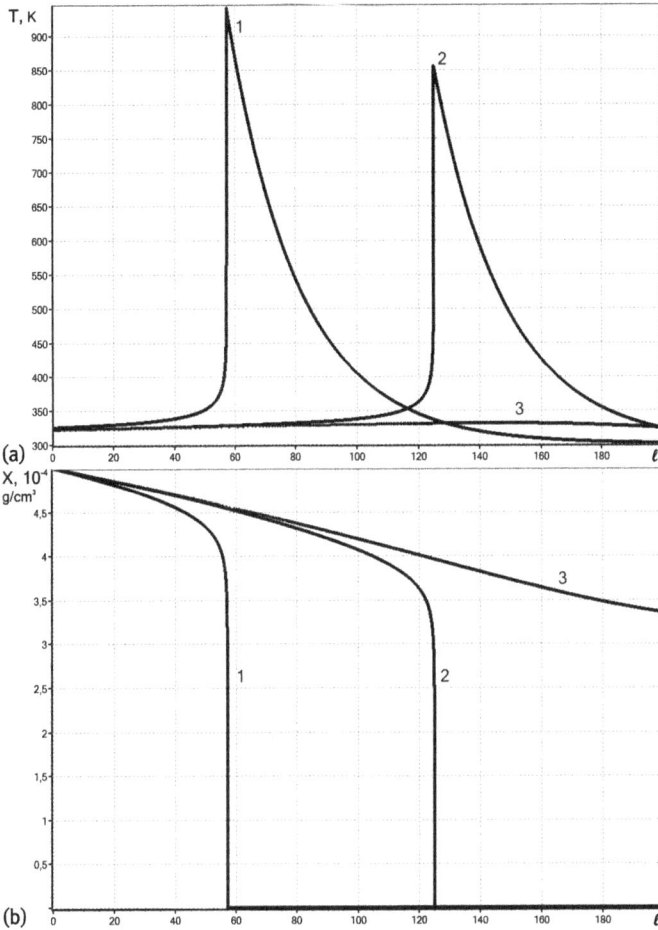

Fig. 4.86: Profiles of temperature (a) and concentration of coal dust (b) in a cylindrical reactor for varied temperature at the input of the reactor: 1) $T_0 = 324.7$ K, 2) 324.8, 3) 325

A similar effect is observed when the temperature of the reactor wall is varied. Thus, model (4.155)–(4.158) describes the phenomenon qualitatively and can be used as the simplest model of spontaneous combustion of coal dust in the tube.

Reaction $O_2 + X \rightarrow P$. To take into account the variation in the oxidizer concentration, we modify system (4.155)–(4.158):

$$u\frac{dX}{d\ell} = -k(T)X \cdot O_2 \,,$$

$$u\frac{dO_2}{d\ell} = -k(T)X \cdot O_2 \,,$$

$$C_p \rho u \frac{dT}{dl} = (-\Delta H)k(T)X \cdot O_2 + \frac{4h}{d}(T_w - T) \,. \tag{4.159}$$

Here the kinetics corresponds to the second-order reaction and O_2 is the concentration of oxygen. The notation used in system (4.159) coincides with that considered above, and conditions (4.156)–(4.158) are the same. Figure 4.87 shows the characteristic profiles of coal and oxygen concentrations and temperature for different concentrations of O_2 at the input of the reactor. These profiles were obtained for the following parameters: $k^0 = 0.84 \cdot 10^9\,\mathrm{s^{-1}}$, $E = 53,200\,\mathrm{J\,mol^{-1}}$, $R = 8.31\,\mathrm{J\,mol^{-1}\,K^{-1}}$, $(-\Delta H) = 900\,\mathrm{J\,mol^{-1}}$, $T_w = 300\,\mathrm{K}$, $T_0 = 310\,\mathrm{K}$, $X^0 = 5 \cdot 10^{-4}\,\mathrm{g\,cm^{-3}}$, $u = 100\,\mathrm{cm\,s^{-1}}$, $h = 0.006\,\mathrm{J\,cm^{-2}\,s^{-1}\,K^{-1}}$). Here the pre-exponent k^0 differs from that in the above-considered scheme. The reason is that the kinetic parameters correspond to different conditions: above, we considered the reaction under the assumption of a constant amount of oxygen, whereas its variation is taken into account in system (4.159). Calculations show that the degree of the coal burnout strongly depends on the oxidizer content. For a sufficient amount of oxygen, the temperature increases abruptly near the input to the reactor, which can be interpreted as an intense spontaneous combustion of coal dust. As the concentration of O_2 at the input to the reactor decreases, the temperature profiles become more smooth, and the characteristic time of variation in the concentration of reagents increases. Thus, the concentration of the oxidizer is one of the factors determining the spontaneous combustion of brown-coal dust.

Three-step scheme with thermal activation. A physicochemical analysis shows that coal can be activated upon heating, i.e., its ability to ignite spontaneously increases since the activation energy in (4.156) decreases. Therefore, in addition to the one-step reactions considered above, we study the three-step scheme:

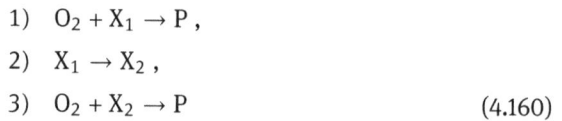

$$\begin{aligned}
&1) \quad O_2 + X_1 \rightarrow P\,, \\
&2) \quad X_1 \rightarrow X_2\,, \\
&3) \quad O_2 + X_2 \rightarrow P \qquad\qquad\qquad (4.160)
\end{aligned}$$

(X_1 and X_2 are the initial and activated coals, respectively). Step 1 corresponds to the above-considered reaction of coal oxidation, step 2 to coal activation (transition from the initial coal X_1 to the activated coal X_2), and step 3 to oxidation of the activated coal. The activation energy of this reaction is much smaller than the activation energy in the first step. Scheme (4.160) corresponds to the following thermokinetic model:

$$u\frac{dX_1}{d\ell} = -k_1(T)O_2 X_1 - k_2(T)X_1\,,$$

$$u\frac{dX_2}{d\ell} = k_2(T)X_1 - k_3(T)O_2 X_2,$$

$$u\frac{dO_2}{d\ell} = -k_1(T)O_2 X_1 - k_3(T)O_2 X_2\,, \qquad\qquad (4.161)$$

$$C_p\rho u\frac{dT}{d\ell} = (-\Delta H_1)k_1(T)O_2 X_1 +$$

$$+ (-\Delta H_2)k_2(T)X_1 + (-\Delta H_3)k_3(T)O_2 X_2 + \frac{4h}{d}(T_w - T)\,.$$

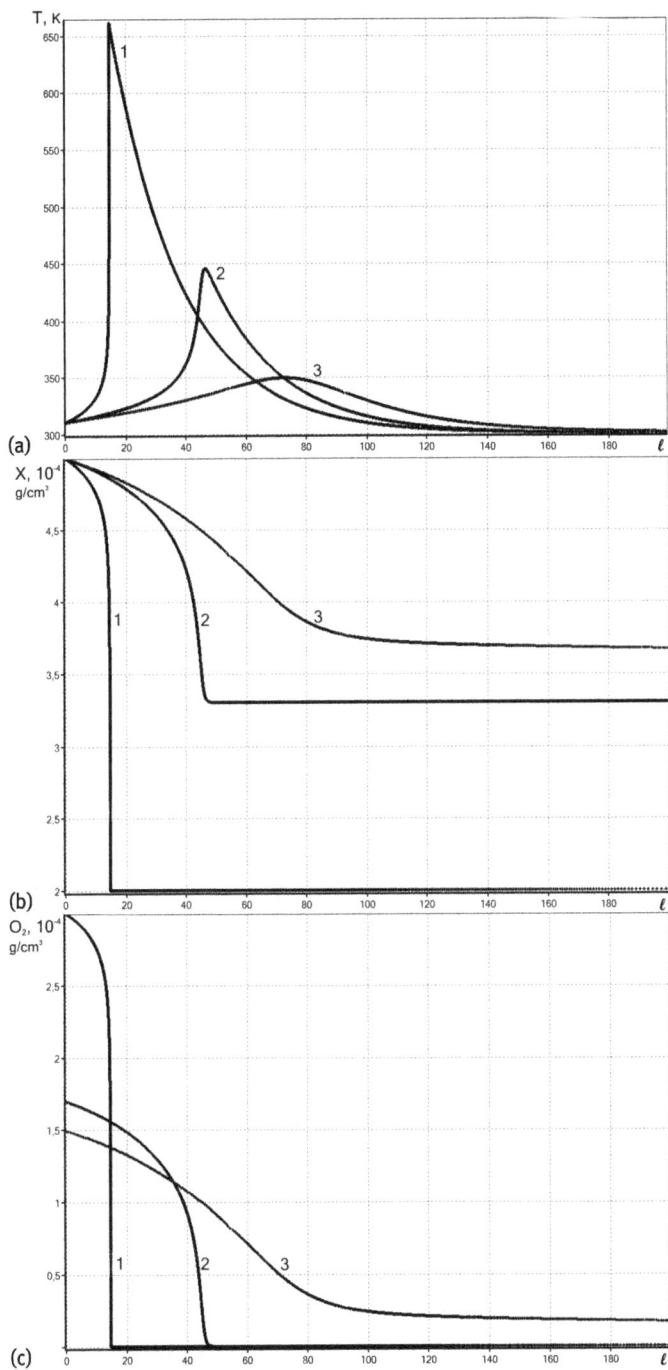

Fig. 4.87: Profiles of the coal-dust temperature (a) and concentrations of coal dust (b) and oxygen (c) with varied concentration of the oxidizer at the input to the reactor: 1) $O_2^0 = 1.5 \cdot 10^4$; 2) $1.7 \cdot 10^4$; 3) $3 \cdot 10^4$

Here $k_i(T)$ ($i = 1, 2,$ and 3) are the constants of the rates at stages 1, 2, and 3 [see (4.160)] and $(-\Delta H_i)$ ($i = 1, 2,$ and 3) are the thermal effects of these reactions. Figure 4.88 shows some profiles of the reagent concentrations and temperature. In the calculations, the following kinetic and thermophysical parameters were used: $k_1^0 = 0.84 \cdot 10^9 \, s^{-1}$, $k_2^0 = 1,200 \, s^{-1}$, $k_3^0 = 0.62 \cdot 10^9 \, s^{-1}$, $E_1 = 53,200 \, J \, mol^{-1}$, $E_2 = 20,000 \, J \, mol^{-1}$, $E_3 = 40,000 \, J \, mol^{-1}$, $(-\Delta H_1) = 900 \, J \, mol^{-1}$, $(-\Delta H_2) = 400 \, J \, mol^{-1}$, $(-\Delta H_3) = 1,100 \, J \, mol^{-1}$, $T_w = 300 \, K$, $T_0 = 310 \, K$.

The first peak of temperature in Fig. 4.88 corresponds to spontaneous combustion of the initial coal. In this case, oxygen burns out completely, and the combustion of dust is terminated. The concentration of the activated coal increases in the first interval of combustion and stabilizes at the middle of the cylindrical reactor. For $\ell \approx 60$ cm, a small-amplitude jump in the concentration of X_2 is observed. This is an indication of combustion of the activated coal, which is rapidly terminated due to the lack of the oxidizer. A new portion of the oxidizer is supplied at the middle of the reactor ($\ell = 100$ cm), which leads to a secondary burst of coal dust. In this case, the activated coal burns.

As in the case of one-step reactions, the combustion regimes and ignition conditions in model (4.161) are highly sensitive to the input conditions and conditions of heat exchange with the ambient medium. The three-step scheme (4.160) taking into account thermal activation of coal allows one to qualitatively describe the process of secondary spontaneous combustion of coal dust within the framework of the simple model (4.161).

Thus, the calculation results are in qualitative agreement with the experimental data. It was not the aim of this work to obtain a quantitative description of the processes using the models considered. Therefore, detailed statistical characteristics of the experiment are not given in this paper. For calculations, the standard thermophysical parameters were taken from [98] and the kinetic parameters E and k^0 for gas suspension of coal were borrowed from [99].

Experiments and mathematical modeling of spontaneous combustion of brown-coal dust allowed us to reveal the main factors that determine the critical conditions of combustion. The fire and explosion safety of coal dust is determined by the kinetic processes of its spontaneous combustion, ignition, and explosion propagation. This is of significant practical importance. Explosions of coal dust mixtures at heat power engineering plants are characterized by secondary explosions (puffs) in the suspended mixture. These explosions also occur when the extinguishers are used or when the direction and air velocity flows change abruptly, etc. If at a certain moment of time ignition conditions are formed in the presence of a combustible medium and an oxidizer (oxygen contained in air), an ignition source appears. Ignition of this mixture occurs in the deflagration-combustion regime. Some dangerous factors of combustion are observed: combustion products are formed, volume temperature increases, and coal dust is activated.

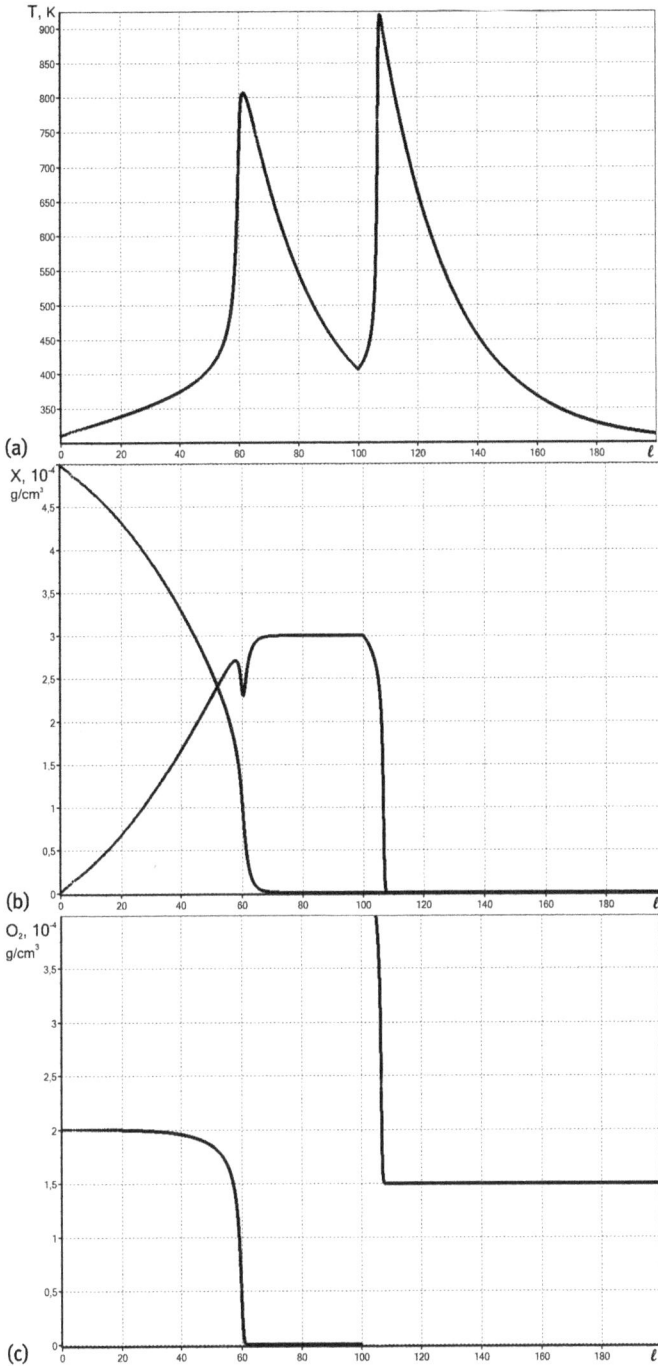

Fig. 4.88: Profiles of coal-dust temperature (a) and concentrations of coal dust (b) and oxygen (c) for the three-step scheme with thermal activation

So, to prevent the action of dangerous factors of an explosion on objects, it is necessary to study and control these technological processes. At present, however, explosion prevention and the protection of heat power engineering plants is based on passive systems (easily removable structures, explosion valves, filling of window openings, etc.), which are not effective means of the prevention of accidents in coal-feed paths. This problem can be solved with the help of mathematical modeling of the spontaneous combustion of coal dust. Detailed calculations of the macrokinetics of these processes for parameters varied within a wide range allow one to develop scientifically justified methods and specifications of estimating the fire and explosion safety of coal-feed paths at modern power plants.

4.12 Modeling of the nitration of amyl in a CSTR and a tube reactor

One major problem of conversion is the processing of propellant components, including amyl (nitrogen tetroxide). The processing of amyl, heptyl, melanges, and other components yields a number of useful products. The authors in [50] describe a model facility for processing of amyl to commercial products and give results of an experiment aimed to produce nitroalkanes by gas-phase nitration of hydrocarbons. Nitroalkanes (unique high-boiling chemical solvents) serve as the starting substances for the production of medicinal and biologically active agents. In [99], the reagents were hydrocarbon (particularly, propane), amyl, and air. The reagents fed into the reactor were heated (separately) to a definite temperature.

A difficulty in operating the facility is the possible instability of the stationary regime and transition to a high-temperature combustion regime (thermal explosion). Control of the cooling jacket temperature and the reactor inlet temperature proved to be an efficient means for stabilizing the operating regime. Based on the parametric analysis of relevant mathematical models, this section gives quantitative assessments of the effectiveness of various control parameters which can be used to develop an automatic control system for the process in the future.

The overall reaction can be written as

$$2RH + 3NO_2 \rightarrow 2RNO_2 + NO + H_2O ,$$

where RH is an alkyl radical and NO_2 is nitrogen dioxide.

In mathematical modeling we used a one-step scheme of the reaction:

$$NO_2 + \text{propane-air mixture} \rightarrow \text{reaction products} . \tag{4.162}$$

Models of a continuous stirred tank reactor and a tube reactor were considered for modeling of the nitration of amyl. By analogy with [97], we studied the operating conditions of a tube reactor and peculiarities of the processes carried out in it. The possibility of controlling the process in a tube reactor with the experimental parameters of [50] was analyzed.

4.12.1 Parametric analysis of the mathematical model of a CSTR

We carry out a parametric analysis of the model of a CSTR for scheme (4.162) using experimental dimensional parameters (see Table 4.2). The kinetic parameters correspond to the data published in the well-known review of gas-phase nitration of alkanes [99].

The pre-exponent and the activation energy of the reaction-rate constant given in [99] correspond to the temperatures at which the main experimental data array was obtained on the model setup of [50]. The thermophysical and geometrical characteristics correspond to the peculiarities of the model experimental setup for gas-phase nitration of amyl.

The mathematical model of a spherical reactor in which the reaction proceeds by scheme (4.162) can be written as

$$V\frac{dX}{dt} = -Vk(T)(X_{mix} - X)X + q(X^0 - X) = f(X, T) \, ,$$

$$C_p\rho V\frac{dT}{dt} = (-\Delta H)Vk(T)(X_{mix} - X)X +$$

$$+ qC_p\rho(T_0 - T) + hS(T_w - T) = g(X, T) \, , \tag{4.163}$$

Tab. 4.2: Parameters of the Reactor and Nitration process

Physical parameters	Parameter values
Density of mixture, ρ	$8.75 \cdot 10^{-5}$ mol cm^{-3}
Specific heat of mixture, C_p	1740.33 J mol^{-1} K^{-1}
Coefficient of heat transfer, h	0.01 J cm^{-2} s^{-1} K^{-1}
Volume of reactor, V	2.734 cm^3
Area of reactor surface, S	2.430 cm^2
Reaction tube length l	172 cm
Tube diameter, d	4.5 cm
Reaction heat, $(-\Delta H)$	$1.95 \cdot 10^6$ J mol^{-1}
Volume flow rate, q	27.3 cm^3 s^{-1}
Activation energy, E	$1.51 \cdot 10^5$ J mol^{-1}
Pre-exponent, k^0	$1.97 \cdot 10^{14}$ s^{-1}
Universal gas constant, R	8.31 J mol^{-1} K^{-1}
Reaction time, t	10 s
Inlet concentration, X^0	$5 \cdot 10^{-5}$ mol cm^{-3}
Inlet temperature, T_0	573 K
Temperature of reactor walls, T_w	573 K
Concentration of reaction mixture, X_{mix}	$30.5 \cdot 10^{-5}$ mol cm^{-3}

where X_{mix} is the concentration of the reaction mixture, X is the concentration of NO_2, $(X_{mix} - X)$ is the concentration of the propane-air mixture (in the ratio 1 : 6 for the experiment), and the notations of the other parameters are given in the Table 4.2. The kinetic dependence used in (4.163) corresponds to the reaction scheme $A + B \rightarrow C$ adopted in accordance with (4.162).

For this second-order reaction, the kinetic function has the form $k(T)(X_{mix} - X)X$. We carry out a parametric analysis of model (4.163) using the procedure described in [3].

Steady states. The stationary equations for model (4.163) are written as

$$- Vk(T)(X_{mix} - X)X + q(X^0 - X) = 0 ,$$
$$(-\Delta H)Vk(T)(X_{mix} - X)X + qC_p\rho(T_0 - T) + hS(T_w - T) = 0 . \tag{4.164}$$

We express X from the first equation:

$$X = \frac{qC_p\rho(T_0 - T) + hS(T_w - T)}{q(-\Delta H)} + X^0 \quad \text{or} \quad X = X^0 - C , \tag{4.165}$$

Here $C = [qC_p\rho(T - T_0) + hS(T - T_w)]/(q(-\Delta H))$. Finally, the stationary equation has the form

$$(-\Delta H)Vk(T)(X_{mix} - X^0 + C)(X^0 - C) = qC_p\rho(T - T_0) + hS(T - T_w) , \tag{4.166}$$

where the left expression is a function of heat production and the right expression is a function of heat removal.

Parametric dependencies. From (4.166) we obtain:

$$X^0(T) = C + \frac{X_{mix}}{2} - \sqrt{\left(C + \frac{X_{mix}}{2}\right)^2 - C\left(X_{mix} + C + \frac{q}{Vk(T)}\right)} ,$$

$$V(T) = \frac{Cq}{k(T)(X_{mix} - X^0 + C)(X^0 - C)} , \tag{4.167}$$

where C is determined in (4.165). Examples of parametric dependencies plotted using formulas (4.167) are given in Fig. 4.89.

Jacobian matrix elements. The Jacobian matrix elements for system (4.163) are as follows:

$$a_{11} = \frac{\partial f}{\partial X} = -k(T)(X_{mix} - 2X) - \frac{q}{V} ;$$

$$a_{12} = \frac{\partial f}{\partial T} = -\frac{E}{RT^2}k(T)(X_{mix} - X)X ;$$

$$a_{21} = \frac{\partial g}{\partial X} = \frac{(-\Delta H)k(T)}{\rho C_p}(X_{mix} - 2X) ;$$

$$a_{22} = \frac{\partial g}{\partial T} = \frac{E}{RT^2}\frac{(-\Delta H)k(T)}{\rho C_p}(X_{mix} - X)X - \frac{q}{V} - \frac{hS}{V\rho C_p} . \tag{4.168}$$

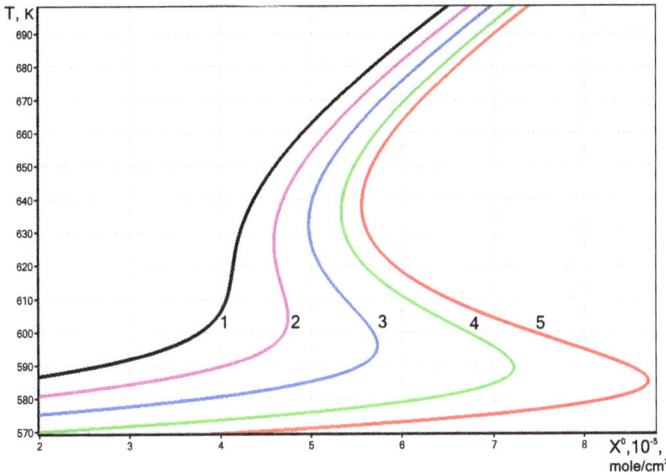

Fig. 4.89: Parametric dependencies of $T(X^0)$ at 1) $T_0 = 623$; 2) 593; 3) 563; 4) 533; 5) 513

According to the standard procedure of parametric analysis using (4.168), we obtain explicit equations for the bifurcation curves of steady states.

The multiplicity curve is plotted in accordance with (4.167) and (4.168).

$L_\Delta(X^0, V)$:

$$V(T, X^0) = \frac{Cq}{k(T)(X_{\text{mix}} - X^0 + C)(X^0 - C)} \, ,$$

$$X^0(T) = \left(\frac{X_{\text{mix}}}{2} + C - \frac{AC}{A - BC} \right) - $$

$$- \sqrt{\left(\frac{X_{\text{mix}}}{2} + C - \frac{AC}{A - BC} \right)^2 - C(X_{\text{mix}} + C) + \frac{AC(X_{\text{mix}} + 2C)}{A - BC}} \, , \qquad (4.169)$$

where

$$A = k(T)\left(q + \frac{hS}{\rho C_p} \right) , \qquad \text{and} \qquad B = \frac{E}{RT^2} \frac{q(-\Delta H)k(T)}{\rho C_p} \, ,$$

and C is determined in (4.165).

Examples of multiplicity curves (4.169) in the plane (X^0, V) with variation of the third parameter are given in Fig. 4.90.

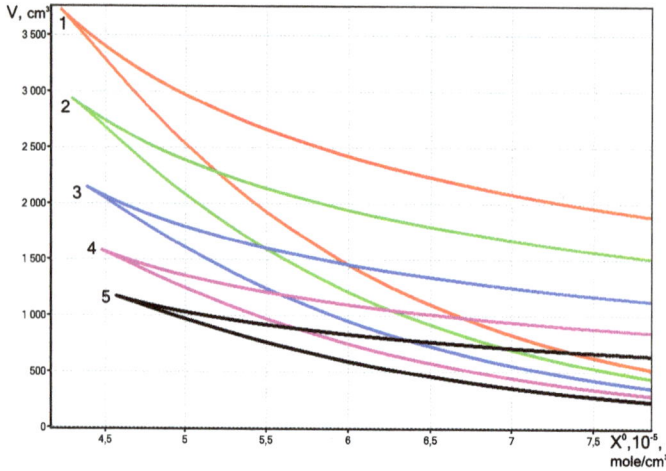

Fig. 4.90: Multiplicity curves in the (X^0, V) plane at 1) $T_0 = 573$; 2) 603; 3) 643; 4) 683; 5) 723

Neutrality Curve

$L_\sigma(X^0, V)$:

$$V(T, X^0) = \frac{Cq}{k(T)(X_{\mathrm{mix}} - X^0 + C)(X^0 - C)} ,$$

$$X^0(T) = \left(\frac{X_{\mathrm{mix}}}{2} + C - \frac{Cq}{D}\right) -$$

$$- \sqrt{\left(\frac{X_{\mathrm{mix}}}{2} + C - \frac{Cq}{D}\right)^2 - C\left(X_{\mathrm{mix}} + C - \frac{q}{D}(X_{\mathrm{mix}} + 2C)\right)} . \qquad (4.170)$$

Here

$$D = 2q + \frac{hS}{\rho C_p} - \frac{E}{RT^2}\frac{Cq(-\Delta H)}{\rho C_p} .$$

The size of the region with an unstable steady state depends appreciably on the value of the heat-transfer coefficient. A parametric portrait in the plane (X^0, V) is given in Fig. 4.91. The region of low-temperature stationary regimes can be determined by analogy with [54]. It is characterized by relatively low values of X^0 and V. Calculations show the high parametric sensitivity of the stationary dependencies on values of h. Thus, the given parameter can be recommended as a control action.

The time dependence corresponding to the presence of auto-oscillations is given in Fig. 4.92. The corresponding phase portrait is given in Fig. 4.93. The obtained parametric dependencies, multiplicity and neutrality curves, parametric portraits, and the corresponding phase portraits can be used to evaluate the dynamic characteristics of a CSTR. In addition, the calculated characteristics can be useful in designing a CSTR.

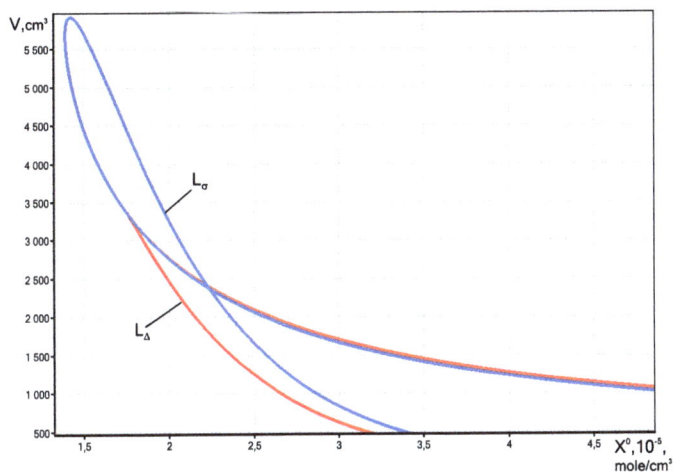

Fig. 4.91: Parametric portrait in the plane (X^0, V) at $h = 0.003$

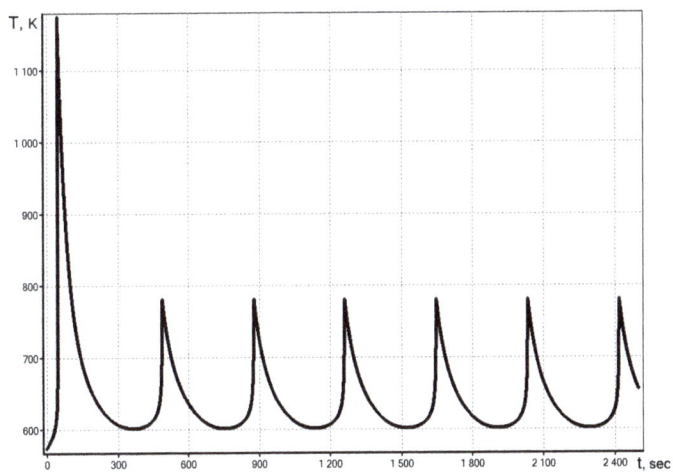

Fig. 4.92: Time dependence $T(t)$ at $V = 8,000$, $h = 0.01$, $X^0 = 4 \cdot 10^{-5}$

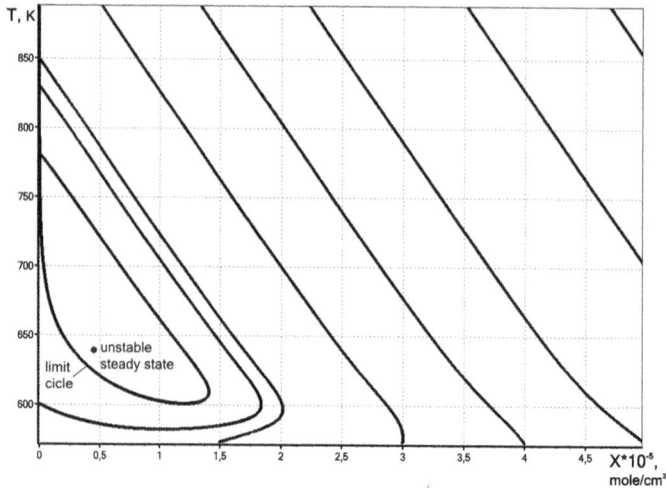

Fig. 4.93: Phase portrait for the unstable steady state for $V = 8,000$, $h = 0.01$, $X^0 = 4 \cdot 10^{-5}$

4.12.2 Model of a tube reactor

In the stationary case, the mathematical model of a tube reactor with the exothermic nitration reaction (4.162) has the form

$$u\frac{dX}{d\ell} = -k(T)(X_{\text{mix}} - X)X \,,$$

$$C_p\rho u\frac{dT}{d\ell} = (-\Delta H)k(T)(X_{\text{mix}} - X)X + \frac{4h}{d}(T_w - T) \,, \tag{4.171}$$

where ℓ is the current length of the reactor ($0 \le \ell \le L$, where L is the total length of the tubular reactor), d is the diameter of the reactor, and u is the feed speed of the reaction mixture. In (4.171), the kinetic dependences are chosen similarly to (4.163). The conditions at the reactor inlet are given by

$$X(0) = X^0 \,, \qquad T(0) = T_0 \,. \tag{4.172}$$

The values and ranges of the thermophysical parameters are taken from the model of a CSTR (see the Table 4.2).

The calculations and analysis of model (4.171) are performed similarly to 4.5.4. A special feature of model (4.171) is that its takes into account the concentration of the propane-air mixture, and this shows up as the appearance of the factor $(X_{\text{mix}} - X)$.

The diameter and length of the reactor were chosen such that the volume of the tube reactor was similar to the volume of the CSTR. A comparison of the auto-oscillating regime in the CSTR (see Fig. 4.92) and the corresponding temperature and concentration profiles in the tube reactor is given in Fig. 4.94-4.95. Auto-oscillations in the stirred reactor are characterized by a significant periodic increase and decrease in

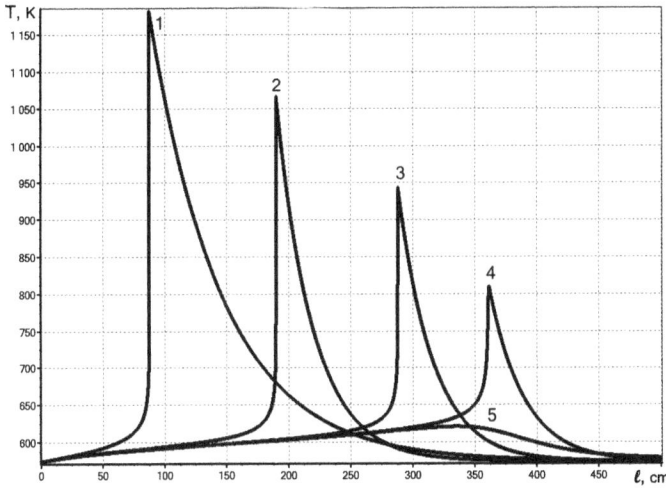

Fig. 4.94: Temperature profiles along the length of the tube reactor under varied heat-transfer conditions $u = 1.717386$, $T_0 = 573$, $X^0 = 5 \cdot 10^{-5}$:
1) $h = 0.005$; 2) 0.01; 3) 0.0106; 4) 0.010645; 5) 0.01065

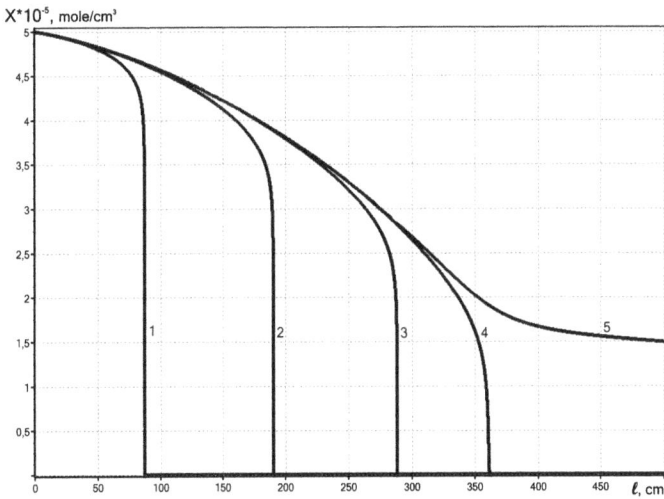

Fig. 4.95: Concentration profiles along the length of the tube reactor under varied heat-transfer conditions $u = 1.717386$, $T_0 = 573$, $X^0 = 5 \cdot 10^{-5}$:
1) $h = 0.005$; 2) 0.01; 3) 0.0106; 4) 0.010645; 5) 0.01065

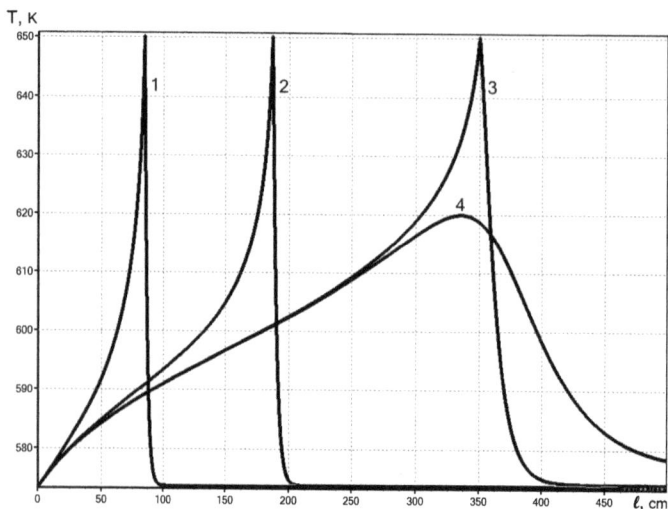

Fig. 4.96: Temperature profiles along the reactor length with varied inlet concentration of NO_2 at $u = 1.717386$, $T_0 = 573$, $X^0 = 5 \cdot 10^{-5}$:
1) $h = 0.005$; 2) 0.01; 3) 0.010645; 4) 0.01065

temperature. The corresponding temperature profile in the tube reactor also has a distinct hot spot. As in previous calculations, the position of the hot spot approximately corresponds to $\ell_* = ut_*$, where t_* is the period of auto-oscillations in the stirred reactor. Thus, the study of the dynamic characteristics of the stirred tank reactor allows one to evaluate the position of the hot spot in the tube reactor.

The high temperature sensitivity of the hot spot with variation of parameters should be noted. There is a certain bifurcation value of the parameters for which the rather smooth temperature profile becomes a profile with a sharp rise of temperature, which can be treated as a thermal explosion in the tube. An example of such transitions from smooth to sharp temperature profiles is given in Fig. 4.94, where the coefficient of heat transfer through the reactor wall and the inlet concentration of NO_2 were used as varied parameters. Calculations show that there are critical values of these parameters for which the temperature rises sharply in a definite section of the tubular reactor.

When carrying out the process in a pilot plant, it is necessary to carry out nitration in soft temperature regimes. In this case, the oxidation is incomplete and the yield of the useful product is maximal. In our calculations, we assumed the following limits: $T < T_{max}$, where T_{max} is the maximum permissible temperature. It is used as a parameter and can be varied in calculations. For example, the calculations for $T_{max} = 650$ K are given in figures 4.96–4.97. For sufficiently low values of h, a sharp rise of temperature far exceeding T_{max} can occur in a definite section of the reactor. To prevent this accidental situation, it is necessary to sharply increase the value of the heat-transfer

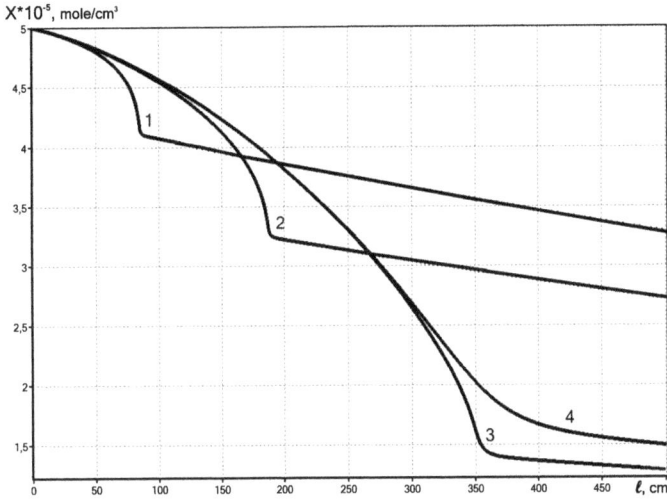

Fig. 4.97: Concentration profiles along the reactor length with varied inlet concentration of NO_2
$u = 1.717386$, $T_0 = 573$, $X^0 = 5 \cdot 10^{-5}$:
1) $h = 0.005$; 2) 0.01; 3) 0.010645; 4) 0.01065

coefficient. Exactly these regimes were used in a pilot plant to stabilize temperature profiles. This further proves that the mathematical model (4.171) is adequate to available experimental data [50]. In addition, we note that the inlet concentration X^0 can be used as a control parameter. The sensitivity of the temperature and concentration profiles to this parameter is also high.

Thus, a mathematical model for the gas-phase nitration of amyl in a CSTR and a tube reactor was developed and analyzed. The high parametric sensitivity of the dynamics of the process to heat-transfer conditions and input data was shown, and this makes it possible to propose effective methods for preventing accidents. Parametric analysis of relevant models allows one to obtain quantitative characteristics of the conditions leading to instability of the nitration process, and to define control parameters to whose variation the dynamics of the system is most sensitive. Analysis of time dependences allows one to evaluate the characteristic times of transition processes, which are required to choose a type of reactor control. Dynamic overshoots, relaxation times, and critical ignition conditions can be determined from calculations of phase portraits, time dependencies, and corresponding bifurcation curves. In this sense, parametric analysis is a necessary step in the analysis of the dynamics of a system aimed at its optimal operation.

Bibliography

[1] Abramov VG, Merzhanov AG. Thermal explosion in homogeneous flow reactors. Combust Explos Shock Waves 1968, 4(4), 548–556.

[2] Aris R, Amundson NR. An analysis of chemical reactor stability and control. Chem Eng Sci 1958, 7(3), 121–155.

[3] Bykov VI. Modeling of critical phenomena in chemical kinetics. Moscow, URSS, 2014 (in Russian).

[4] Vaganov DA, Abramov VG, Samoilenko NG. Finding the regions of existence for oscillating processes in well-stirred reactors. Dokl AN SSSR 1977, 234(3), 640–643 (in Russian).

[5] Kholodniok M, Klich A, Kubichek M, Marek M. Methods of analysis of nonlinear dynamical models. Moscow, Mir, 1991.

[6] Sheplev VS, Bryksina NA, Slinko MG. The algorithm of calculation of Lapunov coefficient by analysis of chemical self-sustained oscillations. Dokl Akad Nauk 1998, 359(6), 789–792 (in Russian).

[7] Sheplev VS, Slinko MG. Periodic regimes of ideal mixing reactor. Dokl Akad Nauk 1997, 352(6), 781–784 (in Russian).

[8] Sheplev VS, Treskov SA, Volokitin EP. Dynamics of a stirred tank reactor with first-order reaction. Chem Eng Sci 1998, 53(21), 3719–3728.

[9] Aris R. Mathematical aspects of chemical reactions. Ind Eng Chem 1969, 61(1), 17–24.

[10] Aris R. Chemical reactors and some bifurcation phenomena. Annals NY Acad Sci 1979, 316, 314–331.

[11] Aris R. Forced oscillations of chemical reactors. Preprint Univ. of Minnesota-Minneapolis. 1987.

[12] Aris R. Mathematical modeling. A chemical engineering's perspective. London, England, Academic Press, 1999. 252–281.

[13] Aris R. The mathematical theory of diffusion and reaction in permeable catalysts. Oxford, England, Clarendon Press, 1975.

[14] Aris R. Elementary chemical reactor analysis. Englewood Cliffs, NJ, USA, Prentice-Hall, 1969.

[15] Bykov VI, Yablonskii GS. On the space-time organization of catalytic reactions. Dokl AN SSSR 1980, 251(3), 616–619 (in Russian).

[16] Hlavacek V, Van Rompay P. Current problems of multiplicity, stability and sensitivity of states in chemically reacting systems. Chem Eng Sci 1981, 36(10), 1587–1597.

[17] Hlavacek V, Votruba J. Hysteresis and periodic activity behavior in catalytic reactors. Adv Catal 1979, 27(1), 59–96.

[18] Jorgensen SE, Svirezhev YM. Towards a thermodynamic theory for ecological systems. Amsterdam, NL, Elsevier, 2004.

[19] Sheintuch M. Nonlinear kinetics in catalytic oxidation reaction: periodic and aperiodic behavior and structure sensitivity. J Catal 1985, 96(2), 326–346.

[20] Sheintuch M, Nekhamkina O. Pattern formation in homogeneous and heterogeneous reactor models. Chem Eng Sci 1999, 54(20), 4535–4546.

[21] Takoudis CG, Schmidt LD, Aris R. Isothermal sustained oscillations in a very simple surface reaction. Surf Sci 1981, 105(1), 325–333.

[22] Takoudis CG, Schmidt LD, Aris R. Steady state multiplicity in surface reactions with coverage dependent parameters. Chem Eng Sci 1981, 36(11), 1795–1802.

[23] Uppal A, Ray WH, Poore AB. On the dynamic behavior of continuous stirred tank reactors. Chem Eng Sci 1974, 29(4), 967–985.

[24] Uppal A, Ray WH, Poore AB. The classification of the dynamic behavior of continuous stirred tank reactors – influence of reactor residence time. Chem Eng Sci 1976, 31(3), 205–214.

[25] Vaganov DA, Abramov VG, Samoilenko NG. Finding the regions of existence for oscillating processes in well-stirred reactors. Dokl AN SSSR 1977, 234(3), 640–643 (in Russian).

[26] Vaganov DA, Samoilenko NG, Abramov VG. Periodic regimes of continuous stirred tank reactors. Chem Eng Sci 1978, 33(8), 1131–1140.

[27] Bykov VI, Volokitin EP, Treskov SA. Parametric analysis of the mathematical model of a nonisothermal well-stirred reactor. Combust Explos Shock Waves 1997, 33(3), 294–300.

[28] Bykov VI, Tsybenova SB. Parametric analysis of the simplest model of the theory of thermal explosion – the Zel'dovich–Semenov model. Combust Explos Shock Waves 2001, 37(5), 523–534.

[29] Zel'dovich YB, Barenblatt GA, Librovich VB, Machviladze DV. Mathematical theory of flame propagation. Moscow, Nauka, 1980.

[30] Frank-Kamenetskii DA. Diffusion and heat transfer in chemical kinetics. Moscow, Nauka, 1987. (in Russian).

[31] Bykov VI, Tsybenova SB. Semenov diagram as a steady-state stability criterion. Dokl Akad Nauk 2000, 374(5), 640–643.

[32] Volter BV. Odd number of stationary regimes in chemical reactors. Theor Found Chem Eng 1968, 2(3), 472–474 (in Russian).

[33] Volter BV, Sal'nikov IE. Stability of work regimes of the chemical reactors. Moscow, Khimiya, 1982 (in Russian).

[34] Volter BV, Shatkhan FA. Diagram of operation of a chemical reactor. Theor Found Chem Eng 1972, 6(5), 756–764 (in Russian).

[35] Bykov VI, Tsybenova SB. Modified method of continuation on parameter for one equation. Comput Technol 2001, 6(4), 9–15 (in Russian).

[36] Bykov VI, Tsybenova SB. Nonlinear models of kinetics and thermokinetics. XVIII Int. Conf. on Chem. Reactors (CHEMREACTOR-18), Malta, 2008, 255.

[37] Bykov VI, Tsybenova SB. Nonstationary two-temperature model of the reactor with catalytic walls. Combust Plasma Chem (Gorenie Plazmokhimiya) 2008, 6(1), 44–49 (in Russian).

[38] Bykov VI, Tsybenova SB. The parametric analysis of Aris–Amundson's model for arbitrary order reaction. XV Int. Conf. on Chem. Reactors (CHEMREACTOR-15), Helsinki, 2001, 256–258.

[39] Bykov VI, Tsybenova S.B. The parametric analysis of Turing's two-dimensional model. XVI Int. Conf. on Chem. Reactors (CHEMREACTOR-16), Berlin, 2003, 286–289.

[40] Bykov VI, Tsybenova SB. The thermocatalytic oscillators. XVII Int. Conf. on Chem. Reactors (CHEMREACTOR-17), Athens, 2006, 625–627.

[41] Bykov VI, Valiulin FKh, Golovin YuM, et al. Modeling of carbonization processes in flowing channels of liquid propellant engine cooling systems as a result of thermal decomposition of hydrocarbon propellant. Space and Rocket Sci 2010, 3(60), 100–109 (in Russian).

[42] Bykov VI, Tsybenova SB. A model of thermokinetic oscillations on the surface of a catalyst. Russian J Phys Chem 2003, 77(9), 1402–1405.

[43] Bykov VI, Tsybenova SB. A parametric analysis of the basic models of the mechanisms of the simplest catalytic reactions. Russian J Phys Chem 2009, 83(4), 609–617.

[44] Bykov VI, Tsybenova SB. Basic models of thermal oscillations. Combust Plasma Chem (Gorenie Plazmokhimiya) 2007, 5(1–2), 120–155.

[45] Bykov VI, Tsybenova SB. Parametric analysis of the continuous stirred tank reactor model. Theor Found Chem Eng 2003, 37(1), 59–69.

[46] Bykov VI, Tsybenova SB. Parametric analysis of the models of a stirred tank reactor and a tube reactor. Combust Explos Shock Waves 2001, 37(6), 634–640.

[47] Bykov VI, Tsybenova SB. A parametric analysis of the basic nonlinear models of the catalytic reactions. Math Model Nat Phenom 2015, 10(5), 68–83.

[48] Bykov VI, Tsybenova SB. Realization of the method of continuation on parameter for two equations system. Comput Technol 2002, 7(5), 21–28 (in Russian).

[49] Bykov VI, Tsybenova SB, Serafimov LA. Emergency starting regimes of a continuous stirred tank reactor. Theor Found Chem Eng 2015, 49(4), 361–369.

[50] Bykov VI, Tsybenova SB, Kuchkin AG. Modeling of the nitration of amyl in a continuous stirred tank reactor and a tube reactor. Combust Explos Shock Waves 2002, 38(1), 30–36.

[51] Bykov VI, Tsybenova SB, Slinko MG. Imperfectly stirred continuous reactor dynamics. Dokl Chem 2001, 380(4), 298–301.

[52] Bykov VI, Tsybenova SB, Slinko MG. Modeling of a reaction on a catalyst surface. Dokl Phys Chem 2003, 388(4), 67–70.

[53] Bykov VI, Tsybenova SB, Slinko MG. Andronov–Hopf bifurcations in the Aris–Amundson model. Dokl Phys Chem 2001, 378(2), 134–137.

[54] Bykov VI, Tsybenova SB, Slinko MG. Safe and unsafe boundaries of regions of critical phenomena in the kinetics of exothermic reactions. Dokl Phys Chem 2001, 378(1), 138–140.

[55] Andronov AA, Vitt AA, Khaikin CE. The theory of oscillations. Moscow, Nauka, 1981 (in Russian).

[56] Bautin NN. Behavior of dynamical systems near the boundary of stability. Moscow, Nauka, 1984. (in Russian).

[57] Bautin NN, Leontovich EA. Methods and techniques for qualitative analysis of dynamical systems on the plane. Moscow, Nauka, 1984 (in Russian).

[58] Boreskov GK, Yablonskii GS. The evolution of ideas about the regularities of the kinetics of reactions in heterogeneous catalysis. J VKHO 1977, 22(5), 556–561 (in Russian).

[59] Slinko MG, Yablonskii GS. Nonstationary and nonequilibrium processes in heterogeneous catalysis. In: Krilov OV, Shibanova MD, eds. Problems of kinetics and catalysis–17. Moscow, Nauka, 1978, 154–169.

[60] Malinetskii GG, Potapov AB, Podlazov AV. Nonlinear dynamics. Approaches, results, expectations. Moscow, URSS, 2011 (in Russian).

[61] Bykov VI, Pushkareva TP. Modeling of combustion of a mixture of n-heptane and iso-octane in a cylindrical reactor. Combust Explos Shock Waves 1989, 25(2), 163–166.

[62] Bykov VI, Pushkareva TP, Stepanskii YY. Simulating self-excited oscillations in the cold-flame combustion of a mixture of n-heptane with iso-octane in an ideal-mixing reactor. Combust Explos Shock Waves 1987, 23(2), 137–142.

[63] Stepanskii YY, Yablonskii GS, Bykov VI. Investigation of the characteristics of cold-flame oxidation of hydrocarbon mixtures as a function of their octane numbers. Combust Explos Shock Waves 1982, 18(1), 45–48.

[64] Bernatosyan SG, Mantashyan AA. Bistability in the oscillating oxidation of propane and the effects of weak perturbations. Combust Explos Shock Waves 1986, 22(3), 337–340.

[65] Bernatosyan SG, Mantashyan AA. Laws of the appearance of oscillation in the gas-phase oxidation of hydrocarbons. Combust Explos Shock Waves 1987, 24(2), 80–83.

[66] Vardanyan IA, Nalbandyan AB. On the mechanism of thermal oxidation of methane. Int J Chem Kinet 1985, 17(8), 901–924.

[67] Vardanyan IA, Sachyan GA, Nalbandyan AB. Kinetics and mechanism of formaldehyde oxidation. Combust Flame 1971, 17(3), 315–322.

[68] Vardanyan IA, Yan S, Nalbandyan AB. Mechanism of the thermal oxidation of methane. I. Mathematical modeling of the reaction. Kinet Catal 1981, 22(4), 845–851.

[69] Vardanyan IA, Yan S, Nalbandyan AB. Mechanism of the thermal oxidation of methane. II. Effect of addition of intermediate products of methane oxidation on the kinetics. Computer calculation. Kinet Catal 1981, 22(5), 1100–1103.

[70] Griffiths JF. Thermokinetic oscillations at homogeneous gas-phase oxidation. In: Field RJ, Burger M, eds. Oscillations and traveling waves in chemical systems, New York, NY, USA, Wiley, 1985, 529–567.

[71] Mansurov ZA. Formation of soot from polycyclic aromatic hydrocarbons as well as fullerenes and carbon nanotubes in the combustion of hydrocarbon. J Eng Phys Thermophys 2011, 84(1), 125–159.

[72] Mansurov ZA. Soot formation in combustion processes (review). Combust Explos Shock Waves 2005, 41(6), 727–744.

[73] Mansurov ZA, Bobykov DU, Tashuta VN, Abil'gazinova SS. Oxidation of hexane in oscillatory conditions. Combust Explos Shock Waves 1991, 27(4), 421–424.

[74] Mansurov ZA, Matafonov AA, Nesterev VI. Oscillations in cold butane flames. Chem Phys 1988, 7(8), 1152–1154.

[75] Mansurov ZA, Bobykov DU, Tashuta VN, Abil'gazinova SS. Oxidation of hexane in oscillatory conditions. Combust Explos Shock Waves 1991, 27(4), 421–424.

[76] Mansurov ZA, Bodykov DU, Ksandopulo GI. Tracing of peroxy radicals in hexane cool flame. React Kinet Catal Lett 1988, 37(1), 31–36.

[77] Mansurov ZA, Matafonov AA. Thermokinetic oscillations in the butane cool flame. In: Korobeinichev OK, ed. Flame structure. Novosibirsk, Nauka, 1991, 33–36.

[78] Mansurov ZA, Matafonov AA, Abdikarimov MS. Thermokinetic oscillations during butane oxidation. Chem Phys 1991, 10(5), 633–637.

[79] Mansurov ZA, Matafonov AA, Konnov AA, Ksandopulo GI. Radical concentrations and temperature oscillations in cool flame oxidation of butane. React Kinet Catal Lett 1990, 41(2), 265–270.

[80] Mantashyan AA. Kinetic manifestations of low-temperature combustion of hydrocarbons and hydrogen: Cool and intermittent flames. Combust Explos Shock Waves 2016, 52(2), 125–138.

[81] Snagovskii YS, Ostrovskii GM. Modeling of kinetics of heterogeneous catalytic processes. Moscow, Khimiya, 1976 (in Russian).

[82] Azatyan VV. New laws governing branched-chain processes and some new theoretical aspects. Chem Phys 1982, 1(4), 491–508 (in Russian).

[83] Azatyan VV, Andrianova ZS, Ivanova AN. The role played by chain avalanches in the developed burning of hydrogen mixtures with oxygen and air at atmospheric pressure. Russian J Phys Chem 2006, 80(7), 1044–1049.

[84] Azatyan VV, Ayvazyan RG, Kalkanov VA, et al. Kinetic features of silane oxidation. Chem Phys 1983, 2(8), 1056–1060 (in Russian).

[85] Azatyan VV, Soroka LB. Nonstationary nature of the state of the surface of a reaction volume and laws of the combustion of phosphorus vapor. Kinet Catal 1981, 22(2), 279–285.

[86] Azatyan VV, Vagner GG, Vedeshkin GK. Effect of reactive additives on detonation in hydrogen-air mixtures. J Phys Chem A 2004, 78(6), 1036–1044.

[87] Azatyan VV, Rubtsov NM, Tsvetkov GI, Chernysh VI. The participation of preliminarily adsorbed hydrogen atoms in reaction chain propagation in the combustion of deuterium. Russ J Phys Chem 2005, 79(3), 320–331.

[88] Bykov VI, Yablonskii GS. Simplest model of catalytic oscillator. React Kinet Catal Lett 1981, 16(4), 377–384.

[89] Bykov VI, Yablonskii GS. Steady state multiplicity in heterogeneous catalytic reactions. Int Chem Eng 1981, 21(1), 142–155.

[90] Bykov VI, Yablonskii GS. Steady states and dynamic characteristics of two-center mechanisms of catalytic reactions. React Kinet Catal Lett 1981, 17(1–2), 29–34.

[91] Bykov VI, Yablonskii GS., Elokhin VI. Steady state multiplicity of the kinetic model of CO oxidation reaction. Surf Sci 1981, 107(2), L334–L338.

[92] Bykov VI, Yablonskii GS., Kumbilieva-Buda K. Influence of the number of active sites on kinetic characteristics. React Kinet Catal Lett 1979, 11(2), 97–101.

[93] Bykov VI, Yablonskii GS., Kuznetsova IV. Simple catalytic mechanism permitting a multiplicity of catalyst steady states. React Kinet Catal Lett 1979, 10(4), 307–310.

[94] Bykov VI. Simple-models of oscillating catalytic reactions. J Phys Chem 1985, 59(11), 2712–2716.

[95] Bykov VI, Gilev SE, Gorban AN, Yablonskii GS. Imitation modeling of the diffusion on the surface of a catalyst. Dokl AN SSSR 1985, 283(5), 1217–1220.

[96] Kubicek M, Marek M. Computation methods in bifurcation theory and dissipative structures. New York, Berlin, Tokyo, Springer Verlag, 1983.

[97] Amel'chugov SP, Bykov VI, Tsybenova SB. Spontaneous combustion of brown-coal dust. Experiment, determination of kinetic parameters, and numerical modeling. Combust Explos Shock Waves 2002, 38(3), 295–300.

[98] Babiy VI, Kuvaev YF. The coal dust combustion and coal-dust flame calculation. Moscow, Energoatomizdat, 1986 (in Russian).

[99] Ballod AP, Shtern VY. The gas-phase nitration of alkanes. Russ Chem Rev 1976, 45(8), 1428–1460.

5 Models of macrokinetics

The fifth chapter is devoted to models of macrokinetics. These represent a system of partial differential equations that describe the processes of heat and mass transfer and chemical transformations. Note that in a modern reprint of the book by D.A. Frank-Kamenetsky, "Heat transfer and diffusion in chemical kinetics" [1] the title was replaced with one word – macrokinetics. In this part, we also had to limit ourselves. On the one hand, the number of works and the range of tasks are too huge to give a full idea of modern macrokinetics in a small book. On the other hand, the specificity of macrokinetics models requires the use of advanced mathematical methods and algorithms. Therefore, we leave the problem of any detailed description of the models and methods of macrokinetics for later studies. Here, we give only a few results. First of all, note that the study of the number and stability of the steady states homogeneous in space fits into the framework of parametric analysis techniques, which was demonstrated in the prior chapters for dynamic systems of ODEs. However, it is necessary to clarify that with the analysis of stability of macrokinetics models, the parameters characterizing the intensity of the processes of heat and mass transfer are included in the conditions of changing the stability type. Under certain conditions, a stable steady state in a dynamic system may become unstable in the presence of diffusion. In this case, we talk about diffusion instability and the appearance of Turing structures. The investigation of stability and bifurcation of inhomogeneous steady states is rather difficult, but nonetheless an interesting task (see, e.g., work of V.A. Volpert [2–5], A.N. Ivanova [6–9], and E.S. Kurkina [10–13]).

According to our research, the geometry of the chemically active surfaces has an important value for the occurrence of nonlinearity, nonstationarity, and inhomogeneity in space. On considering the four classical geometries of surfaces: plate, sphere, cylinder, and torus, already the analysis of the homogeneous steady states shows that the geometric parameters (the radius of the sphere, the radius of the cylindrical surface, and two characteristic radii of the torus) are included in the conditions of their stability. Therefore, the local curvature of surface is an important factor for the occurrence of critical phenomena. Proof of this is the presence of the spin regimes of combustion of cylindrical samples (including the well-known Bengali candles).

A powerful tool for simulation of processes at the active surface is the Monte-Carlo method in combination with the method of probabilistic finite automata [14–17]. However, even a brief consideration of these approaches is not included in our plans. Otherwise, it would lead us away from our main line – parametric analysis of the basic models of chemical kinetics.

https://doi.org/10.1515/9783110464948-005

5.1 Homogeneous–heterogeneous reaction

With modeling of heterogeneous catalytic processes of the type "gas–solid" the mathematical model of a plug-flow reactor is often used, for which it is assumed that the temperatures of gas and catalyst are the same. This approximation works well in the stationary case. However, it is known [18, 19] that many catalytic reactions in the kinetic region can be characterized by critical effects, such as a multiplicity of steady states, the hysteresis of the stationary rate reaction on variation of parameters, and significantly nonstationary modes (undamped in time oscillations of the rate reaction).

Nonstationarity, arising at critical phenomena, can lead to the disequilibrium between gas and catalyst surface. This case requires the construction of a nonstationary two-temperature model. It should be noted that tubular reactors are not only distributed in chemical technology. For example, in modern aircraft engines hydrocarbon fuel is often used as a chiller for heat-stressed structural elements. The feature, which greatly complicates the adequate simulation of the thermohydrodynamic processes that occur in the flow channels, is that the fuel is under supercritical pressure [20]. With the experimental study of such processes, devices that include metal tubes are used. The inner surface of the metal tubes has catalytic properties. In [20], a mathematical model is considered, which allows prediction of the conditions of the loss of thermal stability of hydrocarbon fuels when they flow in channels under supercritical conditions as a result of coking.

In this section, the two-temperature model of a nonstationary flow reactor with catalytic walls in the approximation of ideal plug-flow is proposed. As an example, we consider the catalytic oxidation mechanism, which admits the multiplicity of steady states and auto-oscillations in the kinetic region. The specifics of the interaction between temperature and kinetic nonlinearities and nonstationarity is highlighted. It is shown that these specifics can lead to quite complicated dynamical behavior of reactors with catalytic walls.

In accordance with the accepted assumptions of nonstationarity and nonisothermality, the two-temperature model of a flow catalytic reactor has the form:
the equations of material and heat balances for the gas phase:,

$$V\left(\frac{\partial \mathbf{c}}{\partial t} + \frac{\partial \mathbf{c}}{\partial \ell}\right) = S\sum_s \gamma_s^\Gamma w_s(\mathbf{c}, \mathbf{x}, T_k)\,, \tag{5.1}$$

$$C_p V\left(\frac{\partial T}{\partial t} + \frac{\partial T}{\partial \ell}\right) = \alpha S(T_k - T)\,; \tag{5.2}$$

for the catalytic surface,

$$\rho_k C_{pk}\frac{\partial T_k}{\partial t} = \sum_s \Delta H_s w_s(\mathbf{c}, \mathbf{x}, T_k) + \alpha(T - T_k) + \beta(T_w - T_k)\,, \tag{5.3}$$

$$\frac{\partial x}{\partial t} = \sum_s \gamma_s w_s(\mathbf{c}, \mathbf{x}, T_k)\,, \tag{5.4}$$

where **c** and **x** are the vectors of reactant concentrations in the gas phase and the degrees of catalyst surface coverage with adsorbed substances, respectively; T and T_k are the temperatures of gas and catalyst, respectively; S, V and v are the area of the catalytic surface, the volume, and the flow rate in the reactor, respectively; ρ, ρ_k, C_p and C_{pk} are the densities and the heat capacities of the gas and catalyst; y_s^Γ and y_s are the stoichiometric vectors for reactants in the gas and intermediates; T_w is the temperature of the reactor walls; ΔH_s and w_s are the thermal reaction effect and the rate of the s-th reaction; α and β are the coefficients of heat transfer of the "gas – solid" and the catalyst wall, respectively; t is the time, and ℓ is the current length of the reactor.

The assumption of the absence of reactions in the gas phase was taken into account in the equations (5.1) and (5.2). The change of the reactant concentrations and temperature depends only on the interaction of the gas phase with the active surface. In addition, it is assumed that the reaction takes place without any significant volume changes, so that v = const. Nonstationarity and nonisothermality of the catalyst (5.3) and (5.4) induce nonstationarity and nonisothermality of the gas phase (5.1) and (5.2). The system (5.1)–(5.4) must be supplemented with appropriate initial and input data.

If the catalytic subsystem (5.4) allows for multiple steady states or oscillations, the behavior of the whole system (5.1)–(5.4) can have quite a complex character, because the interaction "gas – solid" is already nonlinear [18].

As an example, consider the catalytic oxidation mechanism:

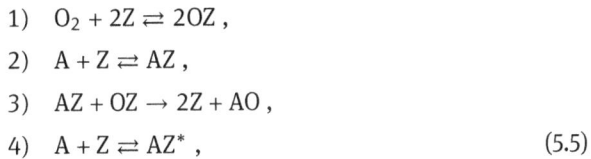

$$
\begin{aligned}
&1) \quad O_2 + 2Z \rightleftarrows 2OZ\,, \\
&2) \quad A + Z \rightleftarrows AZ\,, \\
&3) \quad AZ + OZ \rightarrow 2Z + AO\,, \\
&4) \quad A + Z \rightleftarrows AZ^*\,,
\end{aligned}
\tag{5.5}
$$

where Z is the catalyst. AZ is the substance A adsorbed on the catalyst surface Z. AZ^* is the second form of the adsorbed substance.

Schemes of type (5.5) are often used in the modeling of catalytic oxidation, such as hydrogen or carbon monoxide on platinum:

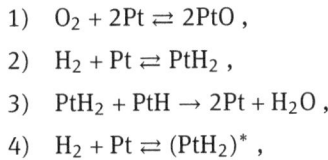

$$
\begin{aligned}
&1) \quad O_2 + 2Pt \rightleftarrows 2PtO\,, \\
&2) \quad H_2 + Pt \rightleftarrows PtH_2\,, \\
&3) \quad PtH_2 + PtH \rightarrow 2Pt + H_2O\,, \\
&4) \quad H_2 + Pt \rightleftarrows (PtH_2)^*\,,
\end{aligned}
$$

or

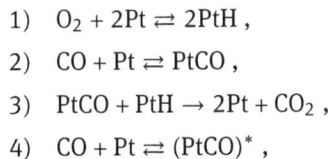

$$
\begin{aligned}
&1) \quad O_2 + 2Pt \rightleftarrows 2PtH\,, \\
&2) \quad CO + Pt \rightleftarrows PtCO\,, \\
&3) \quad PtCO + PtH \rightarrow 2Pt + CO_2\,, \\
&4) \quad CO + Pt \rightleftarrows (PtCO)^*\,,
\end{aligned}
$$

For scheme (5.5), model (5.1)–(5.4) can be rewritten as:

for the gas phase,

$$\frac{\partial c_o}{\partial t} + v\frac{\partial c_o}{\partial \ell} = -\varepsilon w_1 \,,$$

$$\frac{\partial c_a}{\partial t} + v\frac{\partial c_a}{\partial \ell} = -\varepsilon(w_2 + w_4) \,,$$

$$\frac{\partial c_{ao}}{\partial t} + v\frac{\partial c_{ao}}{\partial \ell} = \varepsilon w_3 \,,$$

$$\frac{\partial T}{\partial t} + v\frac{\partial T}{\partial \ell} = \alpha^*(T_k - T) \,, \qquad (5.6)$$

for the catalyst surface,

$$\frac{\partial x_1}{\partial t} = 2w_1 - w_3 \,,$$

$$\frac{\partial x_2}{\partial t} = w_2 - w_3 \,,$$

$$\frac{\partial x_3}{\partial t} = w_4 \,,$$

$$\frac{\partial T_k}{\partial t} = \sum_i h_i w_i + \alpha_k(T - T_k) + \beta_k(T_w - T_k) \,, \qquad (5.7)$$

where

$$w_1 = k_1 c_o z^2 - k_{-1} x_1^2 \,,$$

$$w_2 = k_2 c_a z - k_{-2} x_2 \,,$$

$$w_3 = k_3 x_1 x_2 \,,$$

$$w_4 = k_4 c_a z - k_{-4} x_3 \,, \qquad (5.8)$$

$$z = 1 - x_1 - x_2 - x_3 \,,$$

$$\varepsilon = \frac{S}{V} \,, \qquad \alpha^* = \frac{\alpha}{\rho C_p} \,, \qquad \alpha_k = \frac{\alpha}{\rho_k C_{pk}} \,,$$

$$\beta_k = \frac{\beta}{\rho_k C_{pk}} \,, \qquad h_s = \frac{\Delta H_s}{\rho_k C_{pk}} \,,$$

x_1, x_2 and x_2 are the degrees of coverage of the intermediates OZ, AZ, AZ*, respectively; z is the fraction of free catalyst surface; c_o, c_a and c_{ao} are the concentrations of O_2, A, and AO; k_i are the reaction rates of the stages (5.5).

The system (5.6)–(5.7) must be supplemented with initial and input conditions:

$$t = 0: \qquad x_i = x_i^0(\ell) \,, \quad i = 1, 2, 3 \,,$$

$$T_k = T_k^0(\ell) \,, \qquad T = T^0(\ell) \,, \qquad c_o = c_o^0(\ell) \,, \qquad c_a = c_a^0(\ell) \,,$$

$$\ell = 0: \qquad c_o = c_o^{in}(t) \,, \qquad x_a = c_a^{in} \,, \qquad T = T^{in} \,, \qquad (5.9)$$

where x_i^0, T_k^0, T^0, c_o^0, c_a^0, c_o^{in}, c_a^{in}, and T^{in} are the given functions of their arguments; $0 \le t \le t_k$; $0 \le \ell \le \ell_k$; ℓ_k is the given value of the reactor length and t_k is the time of the reactor work.

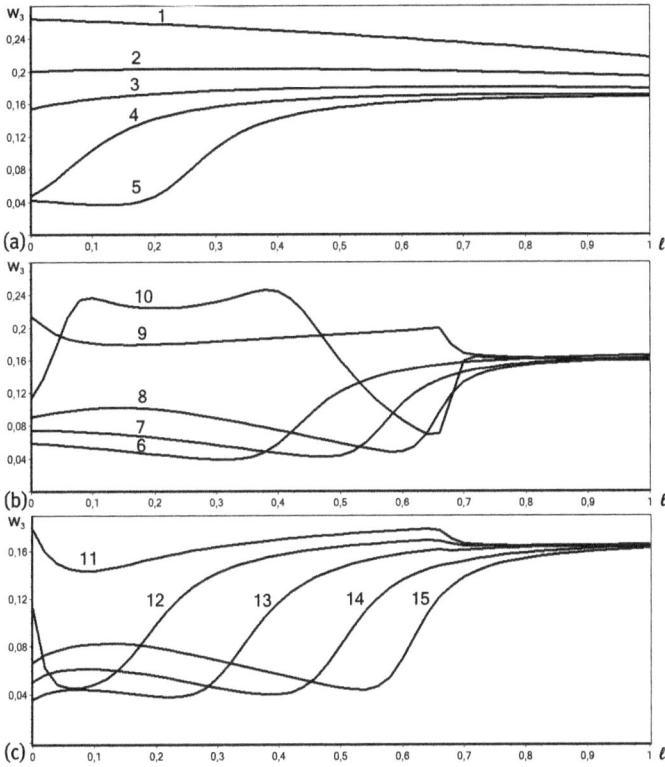

Fig. 5.1: Dynamics of change of the profile of a catalytic reaction rate (w_3).
(a) curve $1 - t = 20$, $2 - 40$, $3 - 60$, $4 - 80$, $5 - 100$;
(b) curve $6 - t = 120$, $7 - 140$, $8 - 160$, $9 - 180$, $10 - 200$;
(c) curve $11 - t = 220$, $12 - 240$, $13 - 260$, $14 - 280$, $15 - 300$

It is known [18, 19] that the kinetic subsystem (5.7) allows three steady states and oscillations when T_k, c_o, c_a are constant. This means that the dynamics of the whole system (5.6)–(5.8) depends on the initial data and inlet conditions. For fixed external conditions, the catalytic reaction rate in a stationary mode will be determined by the fact that the profiles $x_i(\ell)$ in the initial time were in the region of attraction of a stable steady state. If the steady state in the catalytic subsystem is single and unstable, then there is no stabilization to a stationary mode in the full system "gas–solid" at all. Moreover, the system is distributed, so it can have phenomena associated with the stratification of the phase space. At different points of the reactor length the system will be stabilized to different steady states [18, 19].

The calculations were performed at values of the kinetic and thermophysical parameters, providing a multiplicity of steady states and oscillations in the kinetic region: $k_1 = 2.5$, $k_{-1} = 1$, $k_2 = 1$, $k_{-2} = 0.1$, $k_3 = 10$, $k_4 = 0.0675$, $k_{-4} = 0.022$, $\varepsilon = 0.1$, $v = 0.1$, and $T_w = 500$. The process in the reactor is carried out in oscillation

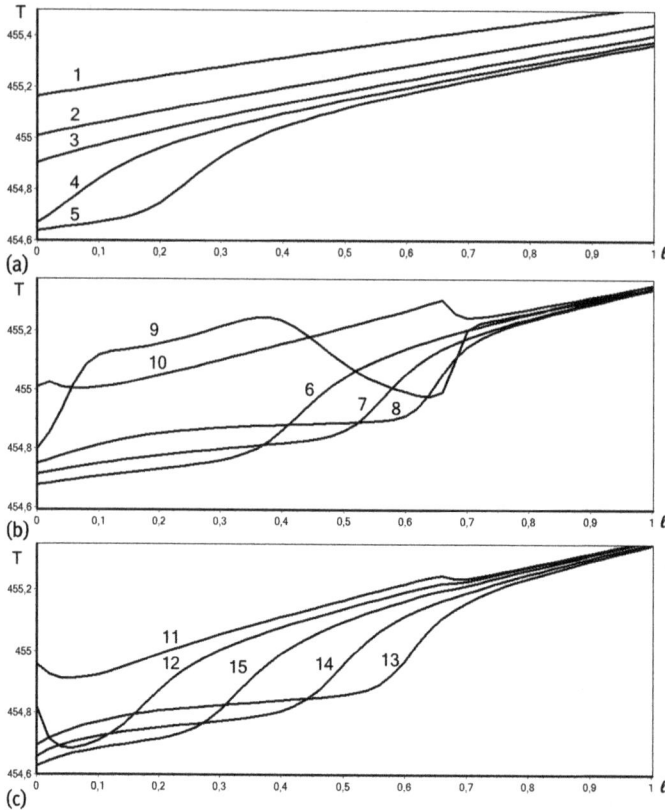

Fig. 5.2: Dynamics of change of the profile of a temperature (T) in the gas phase.
(a) curve $1 - t = 20, 2 - 40, 3 - 60, 4 - 80, 5 - 100$;
(b) curve $6 - t = 120, 7 - 140, 8 - 160, 9 - 180, 10 - 200$;
(c) curve $11 - t = 220, 12 - 240, 13 - 260, 14 - 280, 15 - 300$

mode. Pulsations of temperature and reagent concentrations in time and along the reactor length significantly depend on the thermal inertia of the "gas–solid" system.

The dynamics of the change of the profiles of the oxidation reaction rate (w_3) on the catalytic surface of the tubular reactor is shown in Fig. 5.1. After the transient process (Fig. 5.1a), the system enters the regime of oscillations in time (Fig. 5.1b). Along the reactor length the reaction wave runs, then fades out at the exit of the reactor. The variation of the rate in the gas phase (v) leads to the displacement of the point of disappearance of the oscillations on catalytic surfaces.

The results of calculations of temperature profiles of the gas phase at different points in time are presented in Fig. 5.2. In the oscillation mode corresponding to the conditions in Fig. 5.1, periodic changes of the temperature profiles are observed. Kinetic nonstationarity also leads to instability of the temperature of the reactor walls, which in turn determines the change in time of the concentration and temperature profiles in the gas phase. The amplitude and period of oscillations of the profile of the catalytic reaction rate are largely determined by kinetic parameters. The character of change of the profile along the reactor length and the region of pulsations is determined by the corresponding thermophysical characteristics of the "gas–solid" system and flow conditions.

Thus, the nonstationarity of the catalytic reaction induces nonstationarity in the gas phase. Also, the conditions of the flow organization in a tubular reactor significantly influence the stability of the work modes of its catalytic walls. The coordination in the "gas–solid" system is nonlinear and at a certain combination of thermokinetic and physicochemical parameters can lead to instability of the tubular reactor with catalytic walls.

5.2 Model of an imperfectly stirred continuous reactor

It is well known [21] that the mathematical model of a perfectly stirred continuous reactor in which an exothermic reaction is performed allows for self-oscillations and multiplicity of steady states. Under real conditions, however, the assumption that the reaction mixture in the reactor is stirred perfectly, i.e., that the spatial distributions of its parameters are uniform, is most often invalid [21]. Therefore, in modeling the dynamics of an imperfectly stirred continuous reactor, one should take into account the heat and mass transfer processes; i.e., one should use a reaction and diffusion model. In this model, a distributed dynamic system is considered. Its parametric analysis requires the application of special mathematical and computer techniques [18, 22, 23]. Moreover, this model involves new parameters that determine the intensities of the transfer processes (convection and diffusion). The interplay between nonlinearity, unsteadiness, and spatial heterogeneity leads to great diversity in the dynamic behavior of distributed systems [21, 24]. In addition, knowledge of the features of the behavior of the corresponding lumped system may also prove useful in analyzing its distributed analogue.

In this section, through an example of the basic model in the chemical reactor theory, specifically, a model of an imperfectly stirred continuous reactor, we demonstrate the synchronizing role of the transfer processes (diffusion and heat transfer) in the dynamics of the reactor, where an exothermic reaction is performed such that auto-oscillations and multiplicity of steady states can occur in the kinetic region. At certain ratios between the kinetic parameters and the heat and mass-transfer coefficients, the calculated temperature and concentration profiles prove to be oscillating. Such autowave processes in the distributed system are analogous to auto-oscillations in the corresponding lumped system. Diffusion and heat transfer synchronize these auto-oscillations in different regions of the reactor.

The basic model in chemical reactor theory that allows for auto-oscillations and multiplicity of steady states is the following system of two differential equations [21, 25]:

$$\frac{dx}{dt} = e(y)f(x) - x = P(x, y),$$

$$\frac{dy}{dt} = \beta e(y)f(x) - s(y - 1) = Q(x, y), \tag{5.10}$$

where x and y are the dimensionless concentration and temperature, respectively; t is the dimensionless time; $e(y) = Da \exp(y(1 - 1/y))$ is the reaction rate as a function of the dimensionless temperature; $f(x)$ is the kinetic function that describes the reaction stoichiometry (for example, for a first-order reaction, $f(x) = 1 - x$); and Da, β, y, and s are dimensionless parameters. The dimensionless parameters of model (5.10) characterize the heat and mass-transfer conditions in the reactor and the process kinetics [21, 25]. System (5.10) describes the dynamics of the continuous reactor under the assumption that the stirring is perfect. If the stirring is imperfect, and if one needs to take into account the nonuniformity of the spatial distributions of the parameters of the reaction mixture, then the equations of set (5.10) should be supplemented with terms characterizing the reactant diffusion and the heat transfer:

$$\frac{\partial x}{\partial t} + v\frac{\partial x}{\partial \ell} = D_x \frac{\partial^2 x}{\partial \ell^2} + P(x, y),$$

$$\frac{\partial y}{\partial t} + v\frac{\partial y}{\partial \ell} = D_y \frac{\partial^2 y}{\partial \ell^2} + Q(x, y), \tag{5.11}$$

where v is the convective transfer rate; D_x and D_y are the mass and heat-transfer intensities, respectively; and ℓ is the current reactor length. The system (5.11) should be complemented with initial and boundary conditions:

$$t = 0: \quad x = x^0(\ell), \quad y = y^0(\ell), \tag{5.12}$$

$$\ell = 0: \quad x = x_0(t), \quad y = y_0(t), \tag{5.13}$$

$$\ell = \ell_k: \quad \frac{\partial x}{\partial \ell} = 0, \quad \frac{\partial y}{\partial \ell} = 0, \tag{5.14}$$

where $x^0(\ell)$ and $y^0(\ell)$ are the initial concentration and temperature profiles, respectively; $x_0(t)$ and $y_0(t)$ are the concentration and temperature, respectively, at the reactor inlet; ℓ_k is the total reactor length; and conditions (5.14) are impermeability conditions. Boundary value problem (5.12)–(5.14) can also be formulated differently. For example, one could assume that the reactor wall temperature depends on ℓ or that the heat and mass-transfer conditions at the reactor inlet and outlet are other than conditions (5.14). However, for our purposes of studying the synchronization of auto-oscillations in a distributed system described by set (5.11), we consider the problem (5.11)–(5.14).

Problem (5.11)–(5.14) was solved for such values of the dimensionless parameters Da, β, y and s that there are auto-oscillations in the corresponding lumped system described by set (5.10), namely, $Da = 0.14, \beta = 0.4, y = 33, s = 3$ [25]. These parameter

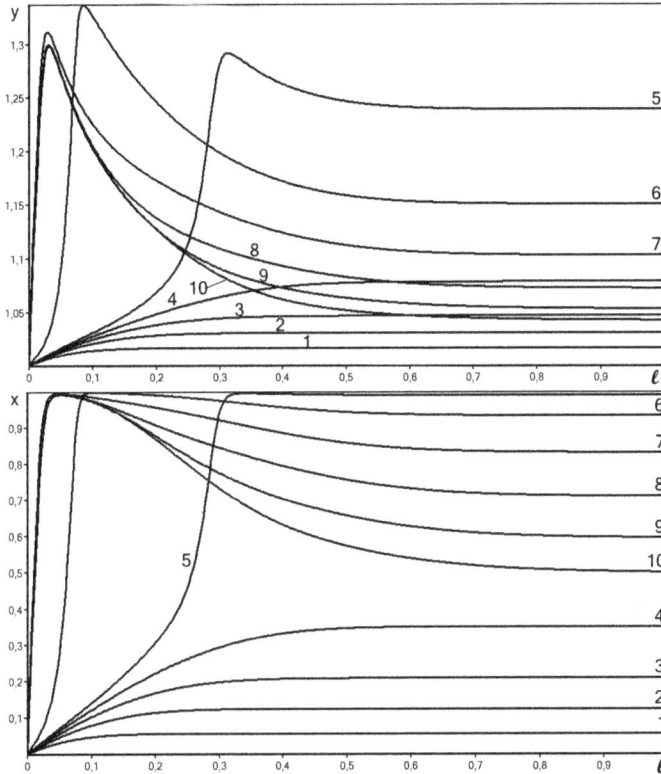

Fig. 5.3: Changes in the profiles of the concentration x and the temperature y along the length ℓ at $D_x = 0.01$, $D_y = 0.01$, $v = 0.1$, and $t_k = 3.6$. Curves 1–10 are plotted at a step of $\Delta t = 0.36$

values correspond to real reactor characteristics in accordance with the universally accepted method of bringing parameters into a dimensionless form [21, 22, 25]. In model (5.11)–(5.14), the initial and boundary conditions were fixed: $x^0 \equiv 0$, $y^0 \equiv 1$, $x_0 \equiv 0$, $y_0 \equiv 1$. The parameters D_x, D_y and v were varied. The results of one of the variants of calculations are presented in Figs. 5.1 and 5.2

Fig. 5.3 shows the dynamics of the transient process of attaining an oscillating mode. At the beginning of this induction period, the concentration and temperature profiles change comparatively slowly (curves 1–4); then, the temperature abruptly increases almost throughout the reactor length; the reactor "ignites" (curve 5); and then a temperature profile with a hot spot near the reactor inlet is established (curves 6–10).

Fig. 5.4 demonstrates the development of the process over time and the attainment of an auto-oscillating mode. The temperature maximum near the reactor inlet continues to be sufficiently stable (curves 1–4), but it is followed by a train of heat waves traveling along the reactor length (curves 5–7). Then, the temperature profile (curves 8, 9) stabilizes to the initial profile (curve 1) and oscillations in the reactor recur. The temperature oscillations determine the auto-oscillating character of changes

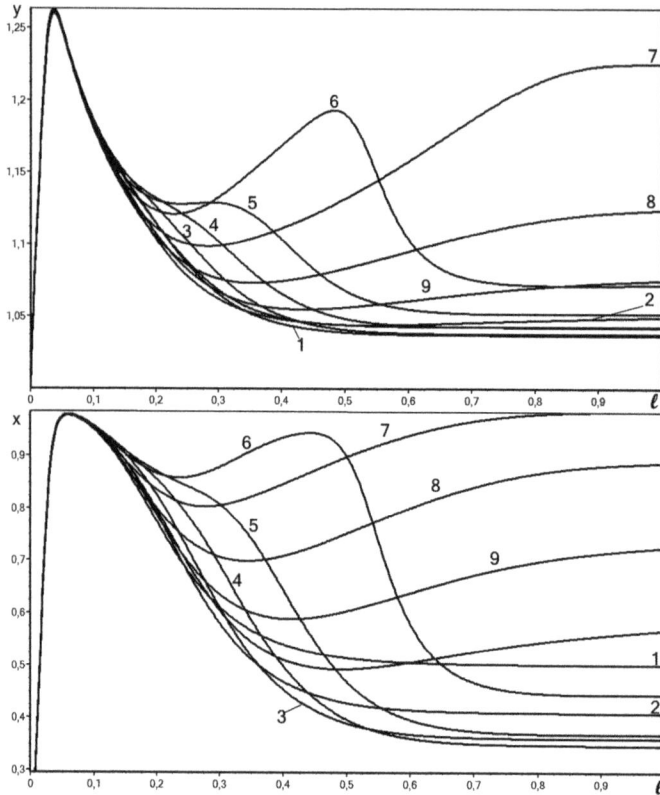

Fig. 5.4: One period of the autowave process in an imperfectly stirred continuous reactor at $t_k = 8$ (the continuation of Fig. 5.3 in time). Curves 1–9 are plotted at a step of $\Delta t = 0.5$ beginning with $t = 3.6$

in the concentration. The oscillating operating modes of the reactor are substantially dependent on the convective transfer rate. With an increase in v, the hot spot in the reactor shifts to the right, the oscillations disappear, and stable concentration and temperature profiles are established.

The period and the amplitude of the oscillations at each point along the reactor length are synchronized by the mass and heat transfer processes, whose intensities are D_x and D_y, respectively. Calculations showed that with an increase in these parameters (which can be attained through sufficiently intense stirring), the instability of the reactor operation disappears and a stable mode is attained. At small D_x and D_y, the synchronizing role of diffusion weakens and the probability of emergence of unstable modes in the reactor increases. If there is instability in kinetic region, then it can also emerge in the stirred reactor, e.g., in the starting period of the reactor operation, when the stirring intensity is still insufficient. Varying D_x and D_y over wide ranges allows us to reveal the synchronization band in different regions of the reaction mixture. In so doing, we can find the parameter ranges within which the imperfect stirring model should be used. Calculations in terms of model (5.11) of the distributed system upon varying its parameters enables us to determine, with given accuracy, the applicability of model (5.10) of the corresponding perfectly stirred system.

Note that in the general case, for nonlinear reaction and diffusion systems, the role of diffusion is not as unambiguous as in our model. It is known [26, 27] that when the coefficients D_x and D_y are essentially different and there is autocatalysis, diffusion alone can cause a loss of stability of a uniform steady state of a distributed system described by a model of type (5.11) and can give rise to Turing patterns. However, in our case, when $D_x \approx D_y$ and there is no autocatalysis, one can observe a stabilizing role of the heat and mass transfer in the stirred continuous reactor where an exothermic reaction is performed, such that auto-oscillations and multiplicity of steady states can occur in the kinetic region.

Thus, the calculation of the dynamics of the imperfectly stirred continuous reactor in terms of model (5.11)–(5.14) have demonstrated the important role of stirring in the stable reactor operation. The stabilizing role of diffusion and heat transfer consists in the fact that the heat and mass transfer processes synchronize possible local oscillations of the reactant concentrations and the reactor temperature that may occur in kinetic region. At certain ratios between the kinetic and macrokinetic parameters, a chemical reactor can operate in oscillating modes, which are most often technologically undesirable or emergency modes. A detailed parametric analysis of models (5.10) and (5.11)–(5.14) enables us to reveal the parameter ranges within which the operation of the stirred reactor is stable.

5.3 Dissipative structures on the active surface

In the region of temperatures when the model of an ideal adsorbed layer ("high" temperatures) and the model of localized adsorption ("low" temperatures) do not work, we can use the following simplified macrokinetic model for processes on the catalytic filament:

$$\frac{dc_i}{dt} = f_i(\mathbf{c}, \bar{\mathbf{x}}) + v_0 c_i^0 - v c_i , \quad i = 1, \ldots, n , \tag{5.15}$$

$$\frac{\partial x_j}{\partial t} = g_j(\mathbf{c}, \mathbf{x}) + D_j \frac{\partial^2 x_j}{\partial \ell^2} , \quad j = 1, \ldots, m , \tag{5.16}$$

where $\mathbf{c} = (c_1, \ldots, c_n)$ and $\mathbf{x} = (x_1, \ldots, x_m)$ are the vectors of reagent concentrations in the gas phase and on the catalyst surface, respectively; $0 \le \ell \le 1$, ℓ is the nondimensional current length of the catalytic filament; v_0 and v are the rates of the flow reactants at the inlet and outlet of the reactor (in the simplest case $v_0 = v$); c_i^0 are the reactant concentrations in the gas phase at the inlet of the reactor; D_j are diffusion coefficients of the reactants on the catalyst surface; and f_i and g_j are the functions of the kinetic dependencies specified in accordance with the specified mechanism of chemical transformations;

$$\bar{x}_j = \int_0^1 x d\ell .$$

The model (5.15) and (5.16) is written down in the assumption of one-dimensional diffusion and full mixing in the gas phase. Boundary and initial conditions can be specified in the form:

$$\frac{\partial x_j}{\partial \ell}(0, t) = \frac{\partial x_j}{\partial \ell}(1, t) = 0 , \tag{5.17}$$

$$c_i(0) = c_i^0 , \qquad x_j(\ell, 0) = x_j^0(\ell) , \tag{5.18}$$

where $x_j^0(\ell)$ is the initial distribution profile of intermediates on the catalyst surface. Conditions (5.17) have the sense of impermeability conditions.

Consider model (5.15) and (5.16), under the assumption that the flow of the reactants in the gas phase is arranged so that $c_i \equiv \text{const}$. Then the basic model discussed in this section has the form [19, 28]:

$$\dot{x}_j = \sum_s \gamma_{js} w_s(x) + D_j \Delta x_j , \quad j = 1, \dots, n , \tag{5.19}$$

where γ_{js} are the stoichiometric coefficients, w_s is the rate of the s-th reaction and Δ is Laplace operator. The Laplace operator is one-dimensional for catalytic filament: $\Delta = \partial^2/\partial \ell^2$.

The stoichiometric matrix $\mathbf{\Gamma} = ((\gamma_{js}))$ has the property that it corresponds to the law of conservation of atoms of this type in each elementary reaction. This means that there is a vector m (for simplicity, we consider one conservation law), the components of which are proportional to molecular weight of intermediates

$$\mathbf{\Gamma m} = 0 . \tag{5.20}$$

The equation (5.20) for the lumped system entails the execution of the laws of conservation of mass. In the case of the distributed system (5.19), this law can be obtained as follows. We integrate (5.19) on the space variable, taking into account (5.17) and obtain

$$\frac{\partial}{\partial t} \int_0^1 x d\ell = \int_0^1 \mathbf{\Gamma w(x)} d\ell , \tag{5.21}$$

where $\mathbf{w(x)}$ is the rate vector of stages. After multiplying (5.21) on $\mathbf{m}$ taking into account (5.20), we have:

$$\frac{\partial}{\partial t}(\mathbf{m}, \bar{x}) = 0 .$$

So, it means

$$(\mathbf{m}, \bar{x}) \equiv \sum m_i \bar{x}_i^0 = \text{const} , \qquad (5.22)$$

where $\bar{x}_i^0 = \int_0^1 x_i^0(\ell) d\ell$.

The identity (5.22) is the integral law of conservation of intermediates on the catalyst surface. Note that in the case of equal diffusion coefficients $D_j = D$, there is a law of conservation in each point of the space ℓ

$$\sum m_i x_i(\ell, t) \equiv 1 , \qquad (5.23)$$

provided that

$$\sum m_i x_i^0(\ell) = 1 .$$

Indeed, for the total mass

$$\bar{x} = \sum m_i x_i$$

of (5.23) we obtain one equation

$$\frac{\partial \bar{x}}{\partial t} = D \frac{\partial^2 \bar{x}}{\partial \ell^2}$$

with boundary conditions (5.17), which with the initial data

$$\bar{x}(\ell, 0) = \text{const} .$$

has a solution

$$\bar{x}(\ell, t) \equiv \text{const} .$$

However, we are interested in the case of unequal diffusion coefficients. Then the basic model (5.19) represents a system for which the integral conservation law (5.22) must be executed. At the numerical calculations of this model, the ratio (5.22) should be performed. This ratio can serve as one of the criteria for the correct operation of the computational algorithm.

The ratio is executed in the steady state

$$\sum_i D_i m_i x_i \equiv \text{const} . \qquad (5.24)$$

It is known that in a lumped and closed system the conservation law

$$\sum_i m_i x_i(t) = \text{const} .$$

is always valid. In a stationary distributed system it is converted into the ratio (5.24). As a simple example, consider the catalytic reaction going according to the mechanism that we call a catalytic trigger:

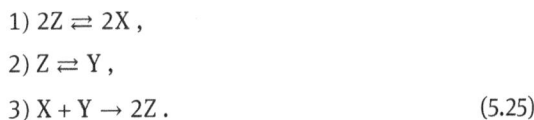

$$1) \ 2Z \rightleftarrows 2X ,$$
$$2) \ Z \rightleftarrows Y ,$$
$$3) \ X + Y \rightarrow 2Z . \qquad (5.25)$$

We assume that the substances X and Y diffuse on the catalyst surface Z, and then the stationary and distributed model corresponds to scheme (5.25)

$$D_x \frac{d^2 x}{d\ell^2} + 2k_1 z^2 - 2k_{-1} x^2 - k_3 xy = 0 , \tag{5.26}$$

$$D_y \frac{d^2 y}{d\ell^2} + k_2 z - k_{-2} y - k_3 xy = 0 , \tag{5.27}$$

$$- 2k_1 z^2 + 2k_{-1} x^2 - k_2 z + k_{-2} y + 2k_3 xy = 0 . \tag{5.28}$$

For system (5.26)–(5.28) the ratio (5.24) means

$$D_x x + D_y y \equiv \alpha , \tag{5.29}$$

where α is a constant defined by initial data. From (5.28), taking into account (5.29), we obtain

$$x(y) = \alpha - \beta y ,$$

$$z(y) = \left(-k_2 + \sqrt{k_2^2 + 8k(y)} \right) / (4k_1) , \tag{5.30}$$

where $\beta = D_y / D_x$, $k(y) = 2k_{-1} x^2(y) + k_{-2} y + 2k_3 x(y) y$.

Taking into account (5.30), the equation (5.27) is converted to the form

$$D_y \frac{d^2 y}{d\ell^2} + k_2 z(y) - (k_{-2} + k_3 x(y)) y = 0 . \tag{5.31}$$

It is easy to verify that the equation (5.31) is integrable in quadratures. So, for $\varphi(y) = dy/d\ell$ we have

$$\varphi(y) = A - \int_0^y f(u) du ,$$

where A is an arbitrary constant,

$$f(y) = (k_2 z(y) - (k_{-2} + k_3 x(y)) y) / D_y .$$

The function $\varphi(y)$ for (5.31) has the meaning of the motion rate of a particle of unit mass with an energy A in a field with potential energy

$$U(y) = \int_0^y f(u) du .$$

As shown above, the lumped system corresponding to the mechanism (5.25) can have three steady states. In this case, the function $U(y)$ has three extrema, two of which satisfy the stable steady states (y_1, y_3), and one the unstable steady state (y_2). Obviously, if $A_2 < A < \min\{A_1, A_3\}$, where $A_i = U(y_i)$, then (5.31) has a periodic solution on ℓ. This is the desired dissipative structure.

The constant α in (5.29) can be determined from the condition of normalization. So we can accept that

$$\int_0^1 (z + x + y)d\ell = 1 .$$

In addition, note that if D_x or D_y is zero, there are only flat steady states. So, in our case, the presence of two diffusing substances is substantial. The question of the stability of emerging heterogeneous distributed solutions is not considered in this section.

The selected properties of the basic model of the macrokinetics of the systems of type (5.19) and the example (5.26)–(5.28) show that during a nonlinear catalytic reaction complicated by surface diffusion, the appearance of periodic spatial structures is possible. In this formulation of the task, these dissipative structures are of purely macroscopic origin. The specificity of the catalytic systems discussed is that the presence of these effects is possible with diffusion of at least two adsorbed substances.

Consider the question of the stability of the homogeneous steady state for models of the type (5.19). The analysis will be done on the example of an autocatalytic oscillator:

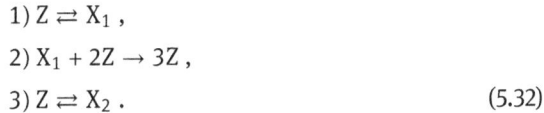

$$1)\, Z \rightleftarrows X_1 ,$$
$$2)\, X_1 + 2Z \rightarrow 3Z ,$$
$$3)\, Z \rightleftarrows X_2 . \tag{5.32}$$

The diffusion model corresponds to scheme (5.32)

$$\frac{\partial x_1}{\partial t} = D_1 \frac{\partial^2 x_1}{\partial \ell^2} + f_1(x_1, x_2) ,$$

$$\frac{\partial x_2}{\partial t} = D_2 \frac{\partial^2 x_2}{\partial \ell^2} + f_2(x_1, x_2) , \tag{5.33}$$

where

$$f_1 = k_1 z - k_{-1} x_1 - k_2 x_1 z^2 ,$$
$$f_2 = k_3 z - k_{-3} x_2 ,$$
$$z = 1 - x_1 - x_2 . \tag{5.34}$$

The homogeneous steady states x_1^*, x_2^* are defined as the solutions of a system of equations $f_1 = f_2 = 0$.

Linearization of system (5.33) in a neighborhood x_1^*, x_2^* leads to the characteristic equation

$$\lambda^2 - \sigma\lambda\delta = 0 ,$$

where

$$\sigma = f_{11}^* + f_{22}^* - (D_1 + D_2)\omega^2 ,$$
$$\delta = f_{11}^* f_{22}^* - f_{12}^* f_{21}^* - \omega^2 (f_{22}^* D_1 + f_{11}^* D_2) + D_1 D_2 w^4 .$$

It is easy to see that σ and δ can be converted to the form

$$\sigma = \sigma_0 - (D_1 + D_2)\mu ,$$
$$\delta = \delta_0 - (f_{22}^* D_1 + f_{11}^* D_2)\mu + D_1 D_2 \mu^2 ,$$

where σ_0 and δ_0 are the coefficients of the characteristic equation for the corresponding lumped system; $\mu = \omega^2$; ω is the frequency of spatial perturbations of the homogeneous steady state. Bifurcation curves satisfying $\sigma = 0$ and $\delta = 0$ can be written similarly to previously in explicit form in different planes of parameters, including diffusion coefficients D_1, D_2.

The two-dimensional diffusion model of the type (5.19) is

$$\frac{\partial x_1}{\partial t} = D_1 \left(\frac{\partial^2 x_1}{\partial \ell_1^2} + \frac{\partial^2 x_1}{\partial \ell_2^2} \right) + f_1(x_1, x_2),$$

$$\frac{\partial x_2}{\partial t} = D_2 \left(\frac{\partial^2 x_2}{\partial \ell_1^2} + \frac{\partial^2 x_2}{\partial \ell_2^2} \right) + f_2(x_1, x_2), \tag{5.35}$$

for the considered scheme (5.32). The stability of the homogeneous steady state is determined by the solutions of the linearized task

$$e^{\lambda t} e^{i w_1 \ell_1} e^{i w_2 \ell_2},$$

where w_1 and w_2 are the frequencies of the perturbations in the spatial variables ℓ_1 and ℓ_2; i is the imaginary unit; λ are the roots of the characteristic equation with the coefficients

$$\sigma = \sigma_0 - (D_1 + D_2)(w_1^2 + w_2^2),$$

$$\delta = \delta_0 - (f_{22}^* D_1 + f_{11}^* D_2)(w_1^2 + w_2^2) + D_1 D_2 (w_1^2 + w_2^2)^2,$$

where

$$f_{11} = -k_1 - k_{-1} - k_2 z^2 + 2 k_2 x_1 z, \quad f_{12} = 2 k_3 x_1 z - k_1,$$

$$f_{21} = -k_3, \qquad\qquad\qquad f_{22} = -k_2 - k_{-3}.$$

From the conditions $\sigma = 0$ and $\delta = 0$, as above, the equations of the bifurcation curves can be obtained. From the conditions of stationarity of the homogeneous steady state we obtain

$$x_2 = \frac{k_3}{k_3 + k_{-3}} (1 - x_1), \quad z = (1 - x_1) \frac{k_{-3}}{k_3 + k_{-3}},$$

$$k_1 = k_2 x_1 z + k_{-1} x_1 / z, \quad \text{or} \quad k_2 = \frac{k_1}{x_1 z} - \frac{k_{-1}}{z^2}.$$

Then, for example, for the plane of parameters (k_1, k_2), we can define

$$L_\sigma(k_1, k_2): \quad k_2 = \frac{k_{-1} A + k_3 + k_{-3} + (D_1 + D_2)(w_1^2 + w_2^2)}{x_1 z - z^2},$$

$$k_1 = k_2 x_1 z + \frac{k_{-1} x_1}{z}, \quad \text{where} \quad A = 1 + \frac{x_1}{z};$$

$$L_\delta(k_1, k_2): \quad k_2 = \frac{k_{-1} x_1 (\alpha + \beta D_2 - k_3)/z + (\alpha + \beta D_2)(k_{-1} + \beta D_1)}{x_1 (\alpha + \beta D_2 - k_3)/z + \alpha + \beta D_2},$$

$$k_1 = k_2 x_1 z + \frac{k_{-1} x_1}{z}.$$

Note that D_1 and D_2 as parameters are included in the expressions for L_σ and L_δ. This defines the dependence of the stability conditions of steady states on the values of diffusion coefficients explicitly. In addition, the perturbation frequencies w_1 and w_2 also are included. So, for different perturbation frequencies the homogeneous steady state can have a different type of stability.

The results of the calculations of model (5.35) with kinetics (5.34) are shown in Fig. 5.5. The values of the parameters k_i have been taken [29] such that the undamped oscillating modes exist in the corresponding lumped system (5.35) ($k_1 = 0.12, k_{-1} = 0.01, k_2 = 1, k_3 = 0.0032, k_{-3} = 0.002, D_1 = D_2 = 10^{-7}$). Figure 5.5 shows the results of calculations of a system of differential equations (5.35) at (a) $t_1 = 2,000$, (b) $t_2 = 2,050$, (c) $t_3 = 2,100$, (d) $t_4 = 2,150$, (e) $t_5 = 2,200$, (f) $t_6 = 2,250$, (g) $t_7 = 2,300$, and (h) $t_8 = 2,350$). The dark color shows the degree of coverage of the substance X_1 (the black tone corresponds to $x_1 = 1$ and the white tone to $x_1 = 0$). Figure 5.5a–h corresponds to the different moments in time of the periodic change of the total degree of coverage of the active surface by adsorbed substance X_1. The catalytic surface is characterized by substantial heterogeneity, despite the presence of diffusion. Visually, Fig. 5.5a–h reminds us of the distribution of intermediates on the active surface in the modeling of similar processes using the Monte Carlo method.

The nonlinear nature of the kinetics leads to the fact that stable homogeneous steady state in a lumped system in the presence of diffusion in the distributed system can lose stability, which in (5.35) is characterized by the presence of stable inhomogeneous solutions (dissipative structures). These structures can be stable in time or can change periodically, which corresponds to the autowave modes on the catalyst surface. Such a "flicker" of intermediates on the catalyst surface leads to a complex dynamics of the total rate of the catalytic reaction. These macrocharacteristics essentially depend on the ratio of parameters of nonlinear interaction of intermediates and their diffusion.

Thus, the procedure of parametric analysis of the number and the stability of the homogeneous steady states of the diffusion model can also be used in this case. The corresponding bifurcation curves can also be obtained in an explicit form. The specificity of a distributed system consists in that a steady state can be stable regarding some of the perturbation frequencies and unstable regarding others. At present, there are many research techniques for studying heterogeneous steady states [6–13]. However, the demonstration of this rather time-consuming procedure in this context is not included in our plans. Moreover, besides a steady state a distributed system can have nonstationary solutions, which correspond to complex spatiotemporal organization of macrokinetics systems up to diffusion chaos [30].

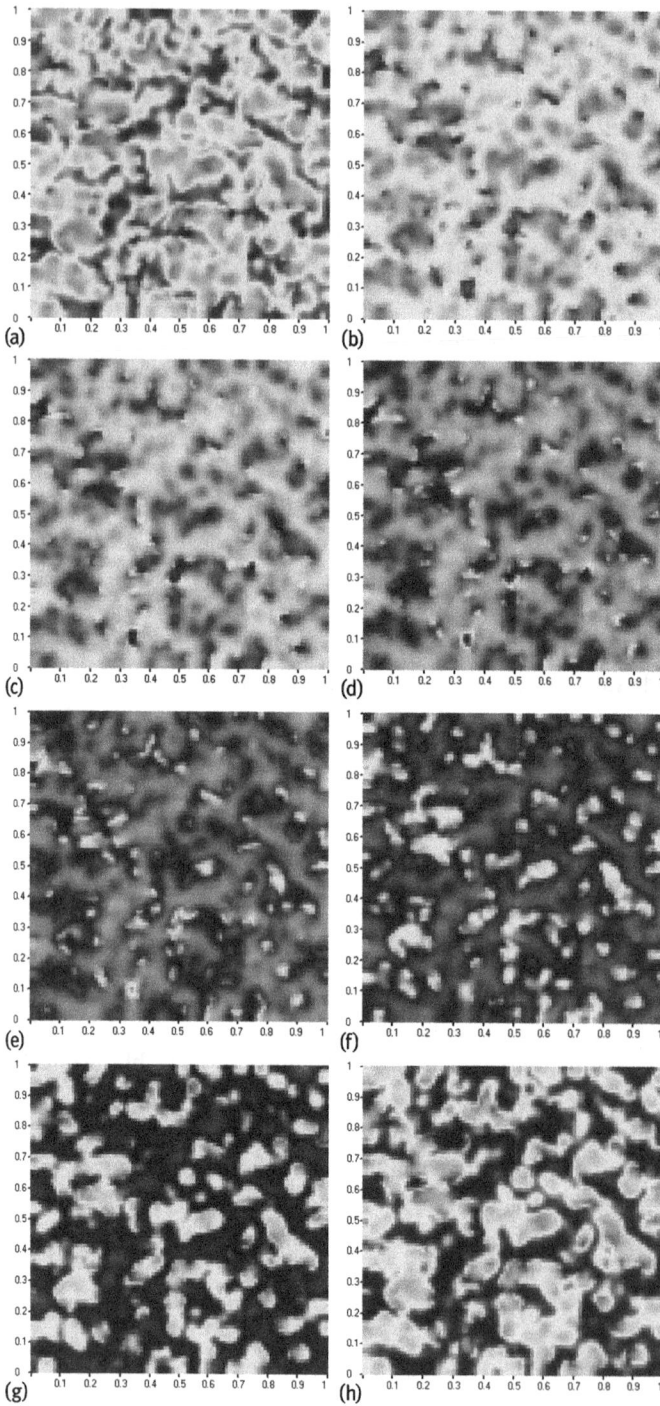

Fig. 5.5: Dynamics of changes in the coverage of the surface by the adsorbed substance X_1

5.4 The model of sorption–reaction–diffusion

At present, a popular method for modeling the processes on a catalyst surface is the Monte Carlo method [31–41]. In the framework of this method, one can take into account a large number of physicochemical factors and construct a sufficiently detailed imitation model of the catalyst surface. However, an essential drawback of this modeling is the absence of developed methods for making a priori estimates and for performing a parametric analysis of the models. From this standpoint, classical macrokinetic models are preferable. Classical macrokinetic models are sets of nonlinear partial differential equations, for which there is a rich arsenal of qualitative and numerical methods of a study [2–5].

In this section, in the localized adsorption approximation, we propose a two-dimensional model of the sorption, reaction, and diffusion of intermediates on the surface of a solid catalyst. This model takes into account that adsorbed molecules can jump from an occupied active site to a neighboring free active site. Diffusion coefficients depend on the degree of coverage of the catalyst surface by adsorbed molecules and characterize the diffusion interaction of intermediates on the catalyst surface. The kinetics of their chemical conversion is described by the right sides of the macrokinetic equations. For the two-dimensional diffusion operator for the simplest nonlinear autocatalytic reaction mechanism, calculations are performed whose results demonstrate the possibility of diffusion instability and essential inhomogeneity of processes on the catalyst surface.

In the simplest case, diffusion and reaction on the catalyst surface is mathematically described in terms of the lattice model. The catalyst surface is modeled by a square lattice of active sites Z. Each active site has four nearest neighbors. Farther neighbors are not considered. Each intermediate X is assumed to occupy a single active site Z. A diffusion event is modeled as a jump of a molecule X from an occupied active site to any neighboring free, active site. The probability of this event depends on the degree of coverage of the active surface by adsorbed molecules.

In the one-dimensional (in space) case, the change in the occupancy of the catalyst surface by molecules X in a section ξ is written as [42]

$$- p_0 x(\xi) z(\xi + \varepsilon) - p_0 x(\xi) z(\xi - \varepsilon) + p_0 x(\xi - \varepsilon) z(\xi) + p_0 x(\xi + \varepsilon) z(\xi) , \tag{5.36}$$

where x and z are the degrees of coverage for X and Z, respectively; p_0 is the microscopic constant or the probability of a jump of a molecule X from an occupied active site to a free, active site Z; and ε is the characteristic lattice size. Since the following expansions,

$$x(\xi \pm \varepsilon) = x(\xi) \pm \varepsilon \frac{\partial x}{\partial \xi} + \frac{1}{2} \varepsilon^2 \frac{\partial^2 x}{\partial \xi^2} + o(\varepsilon^2) ,$$

$$z(\xi \pm \varepsilon) = z(\xi) \pm \varepsilon \frac{\partial z}{\partial \xi} + \frac{1}{2} \varepsilon^2 \frac{\partial^2 z}{\partial \xi^2} + o(\varepsilon^2) ,$$

are valid, the discrete expression (5.36) can be generalized to the continuous model

$$D\left(z(\xi)\frac{\partial^2 x}{\partial \xi^2} - x(\xi)\frac{\partial^2 z}{\partial \xi^2}\right),$$

where $D = p_0\varepsilon^2$ is diffusion coefficient. Thus, the model of diffusion and reaction on the catalyst surface can be written as

$$\frac{\partial x_i}{\partial t} = f_i(x_1, \ldots, x_n) + D_i\left(z\frac{\partial^2 x_i}{\partial \xi^2} - x_i\frac{\partial^2 z}{\partial \xi^2}\right), \quad i = 1, 2, \ldots, n, \tag{5.37}$$

where x_i are the degrees of coverage of the catalyst surface by mobile molecules, z is the fraction of free active sites, D_i are diffusion coefficients, f_i are the kinetic functions of surface reactions, and t is time. In each section ξ, the conservation law is satisfied:

$$z = 1 - \sum_{i=1}^{n} x_i. \tag{5.38}$$

The set of equations (5.37) and (5.38) must be complemented with boundary conditions. For example, at the ends of a catalyst filament ($0 \le \xi \le \xi_k$), under the impermeability conditions, we have

$$\frac{\partial x_i}{\partial \xi} = 0, \quad i = 1, 2, \ldots, n, \quad \text{at} \quad \xi = 0, \quad \xi = \xi_k.$$

In the two-dimensional approximation, diffusion at the microscopic level is considered along two spatial coordinates, (ξ_1 and ξ_2). In this case, similarly to expression (5.36), the change in the degree of coverage of the catalyst surface by molecules X has the form

$$- p_0x(\xi_1, \xi_2)z(\xi_1 + \varepsilon, \xi_2) - p_0x(\xi_1, \xi_2)z(\xi_1 - \varepsilon, \xi_2) -$$
$$- p_0x(\xi_1, \xi_2)z(\xi_1, \xi_2 + \varepsilon) - p_0x(\xi_1, \xi_2)z(\xi_1, \xi_2 - \varepsilon) +$$
$$+ p_0x(\xi_1 - \varepsilon, \xi_2)z(\xi_1, \xi_2) + p_0x(\xi_1 + \varepsilon, \xi_2)z(\xi_1, \xi_2) +$$
$$+ p_0x(\xi_1, \xi_2 - \varepsilon)z(\xi_1, \xi_2) + p_0x(\xi_1, \xi_2 + \varepsilon)z(\xi_1, \xi_2). \tag{5.39}$$

Generalizing the discrete expression (5.39) to a continuous model at the macroscopic level, similarly to the derivation of equations (5.37), we obtain the equations describing diffusion and reaction on the catalyst surface:

$$\frac{\partial x_i}{\partial t} = f_i(\mathbf{x}) + D_i(z\Delta x_i - x_i\Delta z), \quad i = 1, 2, \ldots, n, \tag{5.40}$$

where $\mathbf{x}$ is the vector of the intermediates concentrations, $\mathbf{x} = (x_1, \ldots, x_n)$; the fraction z of free, active sites is found from the conservation law (5.38); and Δ is the two-dimensional Laplace operator,

$$\Delta = \frac{\partial^2}{\partial \xi_1^2} + \frac{\partial^2}{\partial \xi_2^2}.$$

Equations (5.37) and (5.40) involve the kinetic functions f_i, which correspond to the mechanism of the chemical interaction of the intermediates. In analysis of the dynamics of such distributed systems, of particular interest is the case where there is a kinetic subsystem involving a trigger or an oscillator. Then, the combination of the spatial distribution and the nonlinear kinetics results in a great variety of the properties of the system as a whole [18].

As an example of the application of the macrokinetic model (5.40), we consider the equations describing the reaction and diffusion for the simplest reaction mechanism allowing multiplicity of steady states and auto-oscillations in the kinetic region [18]:

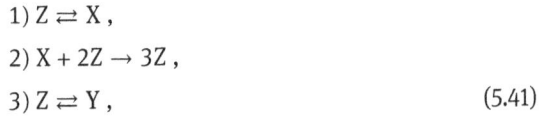

$$1)\ Z \rightleftarrows X,$$

$$2)\ X + 2Z \rightarrow 3Z,$$

$$3)\ Z \rightleftarrows Y, \tag{5.41}$$

where Z is a free, active site on the catalyst surface and X and Y are intermediates. The formal mechanism (5.41) can be interpreted, e.g., as the mechanism of the adsorption and desorption of some substance A on metal Me:

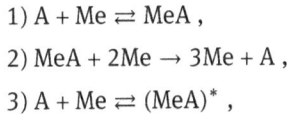

$$1)\ A + Me \rightleftarrows MeA,$$

$$2)\ MeA + 2Me \rightarrow 3Me + A,$$

$$3)\ A + Me \rightleftarrows (MeA)^*,$$

where MeA and $(MeA)^*$ are weakly and strongly bound adsorption complexes, respectively, of the substance A on the metal Me, and the second step characterizes the possibility of desorption of the substance A if there are two neighboring free, active sites. Assuming that the concentrations of intermediates in the gas phase are quasi-steady state, we have the hypothetical reaction mechanism (5.41).

Under the assumption that mechanism (5.41) involves diffusion of only the substance X, from the set (5.40), we obtain the macrokinetic model

$$\frac{\partial x}{\partial t} = k_1 z - k_{-1} x - k_2 x z^2 + D(z\Delta x - x\Delta z),$$

$$\frac{\partial y}{\partial t} = k_3 z - k_{-3} y, \tag{5.42}$$

where $z = 1 - x - y$; k_i are the reaction rate constants in mechanism (5.41); D is diffusion coefficient of the intermediate X, and Δ is the two-dimensional differential diffusion operator:

$$\Delta x = \frac{\partial^2 x}{\partial \xi_1^2} + \frac{\partial^2 x}{\partial \xi_2^2},$$

$$\Delta z = \frac{\partial^2 z}{\partial \xi_1^2} + \frac{\partial^2 z}{\partial \xi_2^2}. \tag{5.43}$$

In the absence of diffusion, system (5.42) at certain parameter values describes self-sustained oscillations, which take place in a single and unstable steady state

(x^*, y^*):

$$k_1 z^* - k_{-1} x^* - k_2 x^* (z^*)^2 = 0,$$
$$k_3 z^* - k_{-3} y^*,$$
$$z^* = 1 - x^* - y^*. \tag{5.44}$$

The stability of the homogeneous state (5.44) of the distributed system (5.42) can be analyzed according to a standard procedure. The linearization in the vicinity of the state (x^*, y^*) leads to a characteristic equation, whose roots depend on the frequencies ω_1 and ω_2 of perturbations along the spatial variables ξ_1, ξ_2:

$$u(t, \xi_1, \xi_2) = u(0, \xi_1, \xi_2) e^{\lambda t} e^{i\omega_1 \xi_1} e^{i\omega_2 \xi_2}. \tag{5.45}$$

By analogy with Turing's problem [26, 27], for (5.45), one can write expressions for the critical values of ω_1 and ω_2 at which the type of stability of the homogeneous state (x^*, y^*) changes.

Since model (5.41) can describe auto-oscillations, the stable homogeneous state in the lumped system in the presence of diffusion in the distributed system may become unstable. In this case, system (5.42) may have inhomogeneous solutions (dissipative structures). These structures can be stable in time or can vary periodically, thereby representing autowave modes on the catalyst surface. In the latter case, adsorbed molecules are self-organized into clusters, and these clusters periodically grow, decline, and vanish. Such fluctuations of intermediates on the catalyst surface result in a complicated dynamics of the overall rate of the catalytic reaction. This macroscopic characteristic essentially depends on the ratio between the parameters of the nonlinear interaction of intermediates and their diffusion at the microscopic level.

Figure 5.6 presents the results of calculations for model (5.42)–(5.43). The values of the parameters k_i were taken so that the lumped system that corresponds to system (5.42) can describe undamped oscillatory modes. A detailed parametric analysis of model (5.42) deserves separate consideration. Here, we only note that parametric analysis methods enable construction of bifurcation curves in the planes of different parameters. These bifurcation curves separate the planes into regions differing in the type of the dynamic behavior of the system. Knowing these regions, one can predict features of the nonlinear and unsteady-state properties of reaction–diffusion systems [18].

The differently shaded areas of Fig. 5.6a–l represent differently occupied regions of the catalyst surface (at the black and white pixels, the degrees of coverage of the catalyst surface by molecules X are $x = 1$ and 0, respectively). Figure 5.6a–l illustrates the periodic change in the total degree of coverage of the active surface by the adsorbed substance X at different moments of time. Although there is diffusion, the catalyst surface is substantially inhomogeneous. Visually, Fig. 5.6a–l is reminiscent of the distributions of intermediates on the active surface at the simulation of similar processes by the Monte Carlo method. However, Fig. 5.6a–l corresponds to a classical macrokinetic model, for which there are qualitative methods of analysis.

Fig. 5.6: Dynamics of the occupancy of the catalyst surface by the adsorbed substance X. The values k_i at $D = 10^{-7}$: $k_1 = 0.12$, $k_{-1} = 0.01$, $k_2 = 1$, $k_3 = 0.0032$, $k_{-3} = 0.002$; and at t (a) 0, (b) 50, (c) 100, (d) 150, (e) 200, (f) 250, (g) 300, (h) 350, (i) 400, (j) 450, (k) 500, (l) 550

Model (5.40) was obtained under the assumption that "jumps" occur only in the free sites of Z. In the typical case, when one may let all possible transitions of the type $X_i \rightleftarrows X_j$, the basic diffusion model by analogy with (5.40) can be written as

$$\dot{x}_i = f_i(x) + \sum_{j=1}^{n} D_{ij}(x_j \Delta x_i - x_i \Delta x_j), \quad i = 1, \ldots, n, \tag{5.46}$$

where $D_{ij} \geq 0$, $D_{ij} = D_{ji}$ and $x_1 + \cdots + x_n = 1$.

Because of the problem of dissipative structures and autowave processes the question of the stability of x^* and the existence of stable and periodic in space solutions of system (5.46) is important.

Consider a case of a system of two equations:

$$\dot{x}_1 = f_1(x_1, x_2) + (D_{12}x_2 + D_{13}(1 - x_2))\Delta x_1 + (D_{13} - D_{12})x_1 \Delta x_2,$$
$$\dot{x}_2 = f_2(x_1, x_2) + (D_{23} - D_{21})x_2 \Delta x_1 + (D_{23}(1 - x_1) + D_{21}x_1)\Delta x_2, \tag{5.47}$$

where Δ is the one-dimensional Laplace operator $\Delta = \partial^2/\partial \ell^2$.

For deviations $u_i = x_i - x_i^*$ the linearized system corresponding to (5.47) is written as

$$\dot{u}_1 = f_{11}^* u_1 + f_{12} u_2 + D_{11}^* \Delta u_1 + D_{12}^* \Delta u_2,$$
$$\dot{u}_2 = f_{21}^* u_1 + f_{22} u_2 + D_{21}^* \Delta u_1 + D_{22}^* \Delta u_2, \tag{5.48}$$

where $f_{ij}^* = \partial f_i/\partial x_j$, calculated in the steady state x_1^*, x_2^*,

$$D_{11}^* = D_{12}x_2^* + D_{13}(1 - x_2^*), \quad D_{12}^* = (D_{13} - D_{12})x_1^*,$$
$$D_{21}^* = (D_{23} - D_{21})x_2^*, \quad D_{22}^* = D_{23}(1 - x_1^*) + D_{21}x_1^*.$$

In accordance with the general procedure of stability analysis of steady states we will seek the solution to (5.48) in the form

$$u_1 = u_1^0 e^{\lambda t} e^{i\omega \ell}, \quad u_2 = u_2^0 e^{\lambda t} e^{i\omega \ell}. \tag{5.49}$$

In our case, $D_{11}^*, D_{22}^* < 0$, so the sign of the roots of the characteristic equation $\lambda^2 + \sigma\lambda + \delta = 0$, where

$$\sigma = (D_{11}^* + D_{22}^*)\omega^2 - f_{11}^* - f_{22}^*,$$
$$\delta = (D_{11}^* \omega^2 - f_{11}^*)(D_{22}^* \omega^2 - f_{22}^*) - (D_{12}^* \omega^2 - f_{12}^*)(D_{21}^* \omega^2 - f_{21}^*),$$

at $f_{11}^*, f_{22}^* < 0$ (the absence of autocatalysis) is determined by the sign δ. The equality $\delta = 0$ is considered as an equation for the squares of the wavenumbers $\mu = \omega^2$:

$$a_0 \mu^2 + a_1 \mu + a_2 = 0,$$

where

$$a_0 = D_{13}D_{23}(1 - x_1^* - x_2^*) + D_{21}D_{13}x_1^* + D_{12}D_{23}x_2^*,$$
$$a_1 = D_{13}(f_{21}^* x_1^* - f_{22}^*(1 - x_2^*)) + D_{23}(f_{12}^* x_2^* - f_{11}^*(1 - x_1^*)),$$
$$a_2 = f_{11}^* f_{22}^* - f_{12}^* f_{21}^*.$$

The values of these coefficients depend on the structure of the reaction mechanism. For example, for the linear catalytic cycle

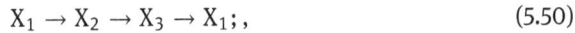

$$X_1 \to X_2 \to X_3 \to X_1; ,$$ (5.50)

we have

$$a_0 = (D_{12}D_{23}k_1k_3 + D_{12}D_{13}k_2k_3 + D_{13}D_{23}k_1k_2)/a_2 ,$$

$$a_1 = D_{12}k_3 + D_{13}k_2 + D_{23}k_1 ,$$

$$a_2 = k_2k_3 + k_1(k_2 + k_3) .$$

In this case, a_0, a_1, $a_2 > 0$, and then the equation on μ has no positive roots. Therefore, diffusion cannot change the character of the stability of the homogeneous steady state in space.

The same conclusion is valid for the simplest catalytic trigger,

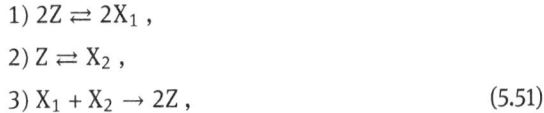

1) $2Z \rightleftarrows 2X_1 ,$

2) $Z \rightleftarrows X_2 ,$

3) $X_1 + X_2 \to 2Z ,$ (5.51)

which allows three steady states in the kinetic region. One can show that for (5.51) $a_1 > 0$. Therefore, the character of the stability of the homogeneous steady state (x_1^*, x_2^*) in the distributed system (5.47) is determined by the sign of a_2. If the steady state is stable in the lumped system, $(a_2 > 0)$, then it is stable and in a distributed system. If the steady state is unstable $(a_2 < 0)$, then it remains unstable and in the presence of diffusion. In the last case, there are such frequency perturbations w_{cr} (the solution of the equation $\delta = 0$) that when $w > w_{cr}$ the harmonics (5.49) decay $(\lambda < 0)$. Moreover, if $w < w_{cr}$ the harmonics grow indefinitely $(\lambda > 0)$. For simplicity, we consider the case when one substance diffuses, for example, X_2. Then, the following model corresponds to (5.51):

$$\dot{x}_1 = 2k_1z^2 - 2k_{-1}x_1^2 - k_3x_1x_2 ,$$

$$\dot{x}_2 = k_2z - k_{-2}x_2 - k_3x_1x_2 + D(z\Delta x_2 - x_2\Delta z) ,$$ (5.52)

where $z = 1 - x_1 - x_2$. For (5.52), we have

$$w_{cr}^2 = \frac{\det(f_{ij}^*)}{D(f_{11}^*(1 - x_1^*) - f_{12}^*x_2^*)} .$$ (5.53)

It is possible to show that the character of the stability of the homogeneous steady state is determined by the value of a_2, namely, if $a_2 > 0$, then the steady state is stable; if $a_2 < 0$, then the steady state is unstable. It is easy to check that the denominator in (5.53) is negative. So, the unstable steady state remains unstable and in a distributed system. This case is favorable for the occurrence of periodic solutions on ℓ.

Dissipative structures for system (5.52) can be found similarly to the problem (5.47)–(5.49). The homogeneous and inhomogeneous steady states for (5.52) are determined from the equation

$$D((1 - x_1)x_{2\ell\ell}'' + x_2x_{1\ell\ell}'') + k_2z - k_{-2}x_2 - k_3x_1x_2 = 0 ,$$ (5.54)

where $x_1 = x_1(x_2)$ is expressed from the first equation (5.52). The equation (5.54) is integrable in quadratures. For $\varphi(x_2) = x'_{2\ell}$ we obtain:

$$\varphi(x_2) = (A - U(x_2)) \exp\left(-2 \int\limits_0^{x_2} r(u)du \right),$$

$$U(x_2) = 2 \int\limits_0^{x_2} s(u) \exp\left(2 \int\limits_0^u r(v)dv \right) du,$$

where A is the arbitrary constant. If there are three steady states $(x_2^{*(1)} < x_2^{*(2)} < x_2^{*(3)})$ in the appropriate lumped system and if

$$U(x_2^{*(2)}) < A < \min\{U(x_2^{*(1)}), U(x_2^{*(3)})\},$$

then the equation (5.54) along with homogeneous steady states has periodic solutions. In the limiting case where $A = \min\{U(x_2^{(*(1))}), U(x_2^{(*(3))})\}$, the periodic solution degenerates to the solitary wave.

So, on the one hand, diffusion may lead to the existence of inhomogeneous steady states if there are critical effects on the kinetic level. On the other hand, the specificity of the considered kinetic nonlinearity (no autocatalysis) is that it cannot change the character of the stability of a steady state, i.e., the stable steady state in the kinetic region remains so and in the presence of diffusion. As is known, diffusion can change the type of stability of a homogeneous steady state in the presence of autocatalysis. Moreover, if there is a single and unstable steady state in the kinetic region, the observed surface state is inhomogeneous.

Thus, in the localized adsorption approximation, we proposed the reaction–diffusion model of type (5.40), in which diffusion operator is quasilinear, since the diffusion coefficients depend on the degree of coverage of the catalyst surface by adsorbed molecules. The proposed approach can also be extended to the case where there is diffusion of further intermediates. Then, the corresponding continuous description at the macroscopic level involves diffusion operators of higher order, e.g., $\partial^4/\partial \ell^4$. A formal substantiation of proceeding from the discrete description (5.10) to the continuous model (5.11) (from the microscopic to the macroscopic level) is a difficult problem. However, our experience of modeling at the microscopic and macroscopic levels showed the adequacy of reaction–diffusion models of type (5.40) at medium degrees of coverage. The advantage of macrokinetic models of type (5.40) is the possibility of their qualitative analysis.

5.5 Macrokinetics of catalytic reactions on surfaces of various geometries

A large body of experimental and theoretical data on dissipative structures and autowaves on the surfaces of metal catalysts has been accumulated [43, 44]. As a rule, the macrokinetics of nonlinear catalytic reactions has been mathematically described using models of mass transfer on a plate surface [3, 6]. Spin modes of solid flame combustion of cylindrical samples have been studied [45–49]. At the same time, the conditions for the loss of stability of uniform steady states largely depend on the geometry of a surface on which there are diffusion and chemical transformations of adsorbed reactants.

In this section, bifurcations and the dynamics of reaction–diffusion systems are numerically and qualitatively studied for four classical geometries of the active surface: plate, sphere, cylinder, and torus. It is shown that the conditions for diffusion instability are essentially dependent on the geometry and local curvature of the surface. The dynamics of emerging dissipative structures – nonstationary clusters and autowaves on the catalyst surface – is numerically studied.

Under the assumption that diffusion of an adsorbate on the catalyst surface occurs when it jumps to a neighboring free site, a mathematical model of the reaction–diffusion type can be written as [18, 50]:

$$\frac{\partial x_i}{\partial t} = f_i(\mathbf{x}, z) + D_i(z\Delta x_i - x_i\Delta z) , \qquad i = 1, \ldots, n , \tag{5.55}$$

where x_i is the concentration of the i-th reactant; $\mathbf{x} = (x_1, x_2, \ldots, x_n)$ is the vector of reactant concentrations; $f_i(\mathbf{x}, z)$ are kinetic functions; and z is the fraction of free catalyst surface:

$$z = 1 - \sum_{i=1}^{n} x_i . \tag{5.56}$$

In the macrokinetic model (5.55) and (5.56), the functions f_i are written according to a given mechanism of catalytic reactions. Diffusion coefficients D_i in (5.55) are assumed constant for surfaces of constant curvature (plate, sphere). In the general case, the differential operator of diffusion has a more complex form:

$$\Delta x = \text{div}(D(x, \ell) \,\text{grad}\, x)$$

For a plate (Cartesian coordinates ℓ_1, ℓ_2), the Laplace operator has a traditional form:

$$\Delta = \frac{\partial^2}{\partial \ell_1^2} + \frac{\partial^2}{\partial \ell_2^2} . \tag{5.57}$$

For the lateral surface of a cylinder (angular coordinate $\varphi \in [0, 2\pi]; \ell \in [0, h], h$ is the height of the cylinder of radius R), we have [51]:

$$\Delta = \frac{1}{R^2} \frac{\partial^2}{\partial \varphi^2} + \frac{\partial^2}{\partial \ell^2} . \tag{5.58}$$

The Laplace operator for the surface of a sphere of radius R can be written as:

$$\Delta = \frac{1}{R^2}\frac{\partial^2}{\partial\varphi^2} + \frac{\cot\varphi}{R^2}\frac{\partial}{\partial\varphi} + \frac{1}{R^2\sin^2\varphi}\frac{\partial^2}{\partial\psi^2}, \tag{5.59}$$

where $\varphi \in [0, 2\pi)$ is the polar angle and $\psi \in [0, 2\pi]$ is the azimuth angle (longitude). The Laplace operator for the surface of a torus can be written as

$$\Delta = \frac{1}{r^2}\frac{\partial^2}{\partial\varphi} - \frac{\cos\varphi}{r(R - r\sin\varphi)}\frac{\partial}{\partial\varphi} + \frac{1}{(R - r\sin\varphi)^2}\frac{\partial^2}{\partial\psi^2}, \tag{5.60}$$

where r is the radius of the circle the rotation of which in space about an axis generates the torus; R is the radius of rotation of this circle ($R > r$); $\varphi, \psi \in [0, 2\pi]$ are the corresponding angular coordinates, which are related to the Cartesian coordinates by the expressions

$$\ell_1 = (R - r\sin\varphi)\cos\varphi,$$
$$\ell_2 = (R - r\sin\varphi)\sin\varphi,$$
$$\ell_3 = r\cos\varphi.$$

Important geometrical characteristics of curvilinear surfaces are the principal curvatures κ_1 and κ_2 of a surface at each fixed point of it, as well as the Gaussian curvature $K = \kappa_1\kappa_2$ and the average curvature $G = (\kappa_1 + \kappa_2)/2$. For a plane, $\kappa_1 = \kappa_2 = 0$; for the lateral surface of a cylinder, the principal curvature along the cylinder axis, $\varphi_1 = 0, K = 0, G = 1/(2R)$; for the entire surface, $\kappa_1 = \kappa_2 = 1/R, K = 1/R^2, G = 1/R$. For the surface of a torus,

$$\overline{K} = \frac{1}{2\pi}\int_0^{2\pi}\frac{\sin\varphi d\varphi}{r(R - r\sin\varphi)} = \frac{1}{r^2}.$$

System (5.55) should be supplemented with some boundary conditions. For example, for a finite plate and cylinder, impermeability conditions at the boundaries were imposed. For surfaces without a boundary (sphere, torus), there are no such conditions. Laplace operators (5.58)–(5.60) contain geometrical parameters R, r. Calculations and analytical studies showed that these parameters are significant for the conditions for loss of stability of the uniform steady state for system (5.55).

The spatially uniform steady state for system (5.55) satisfies the equations

$$f_i(x_1^*, x_2^*, \dots, x_n^*, z^*) = 0, \quad i = 1, 2, \dots, n,$$
$$z^* = 1 - x_1^* - \dots - x_n^*. \tag{5.61}$$

The stability of the state x^* is determined by the solutions of a linearized system of equations, which, e.g., for a plate, are sought in the form

$$u_j = u_j(0, \ell_1, \ell_2)e^{\lambda t}e^{i\omega_1\ell_1}e^{i\omega_2\ell_2}, \quad j = 1, 2, \dots, n,$$

Here, $u_j = x_j - x_j^0$ are the deviations from the steady state, λ are the roots of the characteristic equation, and ω_1 and ω_2 are wavenumbers. The stationary equation (5.61) involves neither geometrical parameters nor diffusion coefficients; therefore, x^* is independent of them. However, the stability of the uniform states x^* may be significantly dependent both on D_i and on the geometry of the surface (in this case, on R and r). At certain (bifurcation) values of these parameters, x^* may lose stability, and nonuniform steady states corresponding to certain values of the wavenumbers ω_1 and ω_2 become stable. Analysis of the stability of nonuniform steady states of reaction–diffusion systems requires a special mathematical technique [6–9], which is beyond the scope of this book.

As an example, we consider a three-step mechanism:

$$1)\ A + Z \rightleftarrows X_1 ,$$
$$2)\ X_1 + 2Z \rightarrow 3Z + A ,$$
$$3)\ B + Z \rightleftarrows X_2 , \tag{5.62}$$

where A and B are substances in the gas phase, Z is a catalyst, and X_1 and X_2 are intermediates (substances adsorbed on the catalyst surface). The kinetic model corresponding to mechanism (5.62) is the simplest autocatalytic oscillator; the model has a single unstable steady state, which guarantees auto-oscillations [18]. Mechanism (5.62) is hypothetical and can be interpreted as a mechanism of the reversible adsorption of substance A on the surface of the metal catalyst Z. Buffer step 3) in mechanism (5.62) can be interpreted, e.g., as the exchange of active sites on the surface and in the bulk of the catalyst, as the reversible poisoning of the catalyst with substance B, etc.

Mechanism (5.62) corresponds to the following mathematical model of type (5.55):

$$\frac{\partial x_1}{\partial t} = D_1(z\Delta x_1 - x_1\Delta z) + w_1 - w_2 ,$$
$$\frac{\partial x_2}{\partial t} = D_2(z\Delta x_2 - x_2\Delta z) + w_3 , \tag{5.63}$$

where

$$w_1 = k_1 z - k_{-1} x_1 , \quad w_2 = k_1 x_1 z^2 ,$$
$$w_3 = k_3 z - k_{-3} x_2 , \quad z = 1 - x_1 - x_2 ,$$

k_i are the step rate constants, some of which contain the partial pressures of substances in the gas phase as factors; x_1 and x_2 are the degrees of surface occupation by substances X_1 and X_2, respectively; and z is the fraction of free surface.

The developed procedure of the parametric analysis of steady states in models of type (5.55) for various geometries of surfaces (5.57)–(5.60) allows one to write conditions for the loss of stability of the uniform steady state of system (5.63) and to study the dynamics of processes on the catalyst surface numerically. These conditions contain

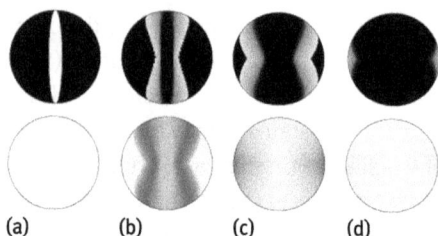

(a)   (b)   (c)   (d)

Fig. 5.7: Traveling wave on the visible side of a sphere: the dynamics of the reactant concentration fields x_1 (top) and x_2 (bottom) at $k_1 = 0.09$, $k_{-1} = 0.01$, $k_3 = 0.0032$, $k_{-3} = 0.002$, $k_2 = 1$, $D_1 = 10^{-6}$, $D_2 = 10^{-7}$ in time $t =$ (a) 0, (b) 600, (c) 1,200, and (d) 1,800

all the basic kinetic, diffusion, and geometrical parameters and also the frequencies of harmonics of perturbations of the uniform steady state.

Calculations using model (5.63) were performed with the following values of the kinetic parameters: $k_1 = 1$, $k_{-1} = 0.0256$, $k_2 = 1$, $k_3 = 0.0032$, $k_{-3} = 0.005$, which ensure oscillations in the kinetic subsystem of system (5.63). At $D_1 = 0$ (only X_2 diffuses) and $D_2 = 10^{-5}$ on the surface of a plate (unit square), model (5.63) has spatially nonuniform solutions that oscillate in time [51]. In this case, the adsorbates self-organize to form clusters, which periodically emerge, grow, reduce, and vanish. Such a "flicker" of the catalyst surface leads to a complex dynamics of the overall rate of the catalytic reaction.

For the surface of a sphere of unit radius at $D_1 = 10^{-6}$, $D_2 = 10^{-7}$, and the same kinetic parameters as for the plate, nonstationary modes of the type of traveling waves were found. Figure 5.7 presents the concentrations of two intermediates on the visible side of the sphere. Two fronts of a wave move away from the meridian sector to the reverse side of the sphere and mutually annihilate there.

Calculations showed that on the cylindrical surface after loss of stability of a uniform reaction front, a spin mode emerges; a structure (spot) moves at a constant velocity along a helix on the lateral surface of the cylinder. On the surface of a torus, spin modes are characterized by periodically varying spot velocity.

Numerical analysis and analytical studies have demonstrated that the geometry of the surface plays an important role in the manifestation of critical phenomena in reaction–diffusion systems. The local curvature of the surface on which there are mass transfer and chemical reactions significantly affects the conditions for loss of stability of spatially uniform steady states. On the real surface with local defects, it is in regions with abruptly changing curvature that dissipative structures and oscillations may emerge. Such defects may act as a sort of nuclei of microclusters and autowaves on the active surface.

In this section, bifurcations and the dynamics of reaction–diffusion processes have been investigated for four classical geometries of the active surface: plate, sphere, cylinder, and torus. It has been shown that the conditions for diffusion instability are essentially dependent on the local curvature of the surface. Other conditions being the same, the probability of loss of stability of a spatially uniform steady state increases with an increase in the curvature of the surface.

5.6 Nonlinear interaction between the active surface and bulk of a solid

The interaction between the active surface and bulk of a catalyst can make a significant contribution to the observed pattern of heterogeneous catalytic reactions [52–54]. The complex dynamics of such processes is often described under the assumption that the interaction of intermediates on the surface with the bulk of the solid can be taken into account by introducing additional stages to the reaction mechanism. However, this approach can be limited if the interaction involves not only one or two subsurface monolayers and diffusion transfer into the depth of the solid is intense enough. For example, hydrogen transfer within some metals may be characterized by considerable gradients. In this case, it is necessary to use classical mass-transfer models in solids and take into account the chemical transformations on the active surface.

It has been shown [18, 54] that reactions on the catalyst surface can have a multiplicity of steady states and auto-oscillations. In this section, we reveal the specific features of the nonlinear interaction between the active surface and bulk of a catalyst. In particular, we found that an oscillatory reaction on the surface of a solid induces an oscillatory mode of mass transfer of intermediates into the bulk of the solid. In this case, because of the nonlinearity of the interaction between the surface and bulk, the mass-transfer coefficients averaged over the oscillation period can significantly exceed those in a steady state. Thus, oscillations of the chemical reaction rate on the surface can carry out a type of work; namely, they can act as a chemical pump that injects necessary intermediates into the bulk of the solid or, conversely, extracts reactants involved in the surface reactions.

Sorption, diffusion, and chemical reactions on the surface of a solid can be described by a model of the form [50, 55]:

$$\frac{\partial x_i}{\partial t} = D_i(z\Delta x_i - x_i\Delta z) + f_i(x_1, x_2, \ldots, x_n), \quad i = 1, 2, \ldots, n, \quad (5.64)$$

where z is the fraction of free surface, $z = 1 - x_1 - \ldots - x_n$; $x_i(t, \xi_1, \xi_2)$ are the degrees of coverage of the catalyst surface with adsorbed substances (intermediates); D_i is diffusion coefficient on the surface of the i-th intermediate; n is the total number of intermediates; f_i are kinetic functions describing the rates of surface reactions; t is time; and Δ is the Laplacian for two spatial variables ξ_1 and ξ_2, which has the classical

form

$$\Delta x_i = \frac{\partial^2 x_i}{\partial \xi_1^2} + \frac{\partial^2 x_i}{\partial \xi_2^2} ,$$

$$\Delta z = \frac{\partial^2 z}{\partial \xi_1^2} + \frac{\partial^2 z}{\partial \xi_2^2} .$$

In the general case, diffusion of reactants should be taken into account not only on the surface but also in the bulk of the solid. Here we consider the case where only a single substance (with subscript i_0) diffuses into the bulk:

$$\frac{\partial x_{i_0}}{\partial t} = D_V \left(\frac{\partial^2 x_{i_0}}{\partial \xi_1^2} + \frac{\partial^2 x_{i_0}}{\partial \xi_2^2} + \frac{\partial^2 x_{i_0}}{\partial \xi_3^2} \right) , \tag{5.65}$$

where $x_{i_0}(t, \xi_1, \xi_2, \xi_3)$ is the dimensionless concentration of the i_0-th substance in the bulk; D_V is the diffusion coefficient of this substance in the bulk; and ξ_1, ξ_2, ξ_3 are spatial coordinates.

The domain of existence of the solution to (5.65) is a unit cube. We assume that the system of equations (5.64) is valid on one of the cube faces; i.e., for (5.65) on a single selected face, the Dirichlet conditions of coupling with system (5.64) are met and, on the other faces, the impermeability conditions ($\partial x_i / \partial \xi_j = 0$) are satisfied. In our case, the reaction–diffusion system (5.64) is one of the boundary conditions for the three-dimensional model (5.65) of diffusion in the bulk of the solid. Note that the simultaneous solution of (5.64) and (5.65) and the visualization of the two and three-dimensional functions obtained is a quite complex problem. However, we do not discuss the technical aspects here.

As a particular model, we consider the Langmuir–Hinshelwood mechanism of hydrogen oxidation on a metal catalyst:

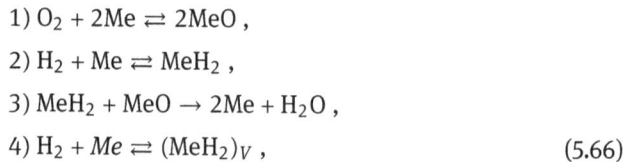

1) $O_2 + 2Me \rightleftarrows 2MeO$,

2) $H_2 + Me \rightleftarrows MeH_2$,

3) $MeH_2 + MeO \rightarrow 2Me + H_2O$,

4) $H_2 + Me \rightleftarrows (MeH_2)_V$, $\tag{5.66}$

where Me is the catalyst, MeO and MeH_2 are intermediates on the catalyst surface, and $(MeH_2)_V$ is the hydrogen that diffuses into the bulk of the catalyst. Scheme (5.66) is described by the following kinetic equations:

$$\frac{\partial x}{\partial t} = 2p_0 k_2 z^2 - 2k_{-1}x - k_3 xy ,$$

$$\frac{\partial y}{\partial t} = p_H k_2 z - k_{-2}y - k_3 xy + D(z\Delta_2 y - y\Delta_2 z) ,$$

$$\frac{\partial y_V}{\partial t} = p_H k_2 z - k_{-4}y_V , \tag{5.67}$$

where $z = 1 - x - y - y_V$ is the fraction of free surface; x and y are the degrees of surface coverage with MeO and MeH_2, respectively; y_V is the fraction of surface hydrogen diffusing into the bulk of the solid; D is the surface diffusion coefficient; p_H and p_0 are the partial pressures of the reactants in the gas phase; k_i are the step rate constants; and Δ_2 is the surface Laplacian:

$$\Delta_2 = \frac{\partial^2}{\partial \xi_1^2} + \frac{\partial^2}{\partial \xi_2^2} .$$

Model (5.67) assumes that only hydrogen diffuses over the surface and dissolves in the bulk of the solid. The equation for mass-transfer to bulk has the classical form

$$\frac{\partial y_V}{\partial t} = D_V \Delta_3 y_V , \tag{5.68}$$

where

$$\Delta_3 = \frac{\partial^2}{\partial \xi_1^2} + \frac{\partial^2}{\partial \xi_2^2} + \frac{\partial^2}{\partial \xi_3^2} .$$

The boundary conditions (on the solid surface) for (5.68) are given in the form of system (5.67). On the other faces of the unit cube, the impermeability conditions are met.

It is well known that the Langmuir–Hinshelwood mechanism (the first three stages in scheme (5.66)) together with the buffer step 4) is a catalytic oscillator, or a system in which there are auto-oscillations of the catalytic reaction rate [18]. At a certain set of parameters k_i, p_H and p_O, there are undamped oscillations of the reactant concentrations on the catalyst surface, which also persist while intermediates diffuse over the catalyst surface. Moreover, the set of the first three stages is the simplest catalytic trigger – a system with two stable steady states [18].

Calculations showed that, at a certain set of kinetic parameters, the corresponding kinetic model has multiple steady states, two of which are stable and one of which is unstable. As a rule, the stable steady states differ significantly in degree of surface coverage with adsorbed hydrogen. Therefore, the intensities of exchange between the surface and bulk also differ. In a steady state in which the degree of surface coverage with adsorbed hydrogen is low, the surface acts so that the hydrogen dissolved in the bulk of the solid is extracted from it, whereas, in a steady state in which the degree is high, the hydrogen concentration gradient is directed into the bulk of the solid.

Thus, for gas dissolution in the bulk of the solid, the active surface of the solid can act as a sort of valve that lets the gas in or out, depending on the steady state of the active surface. It seems that the described possibility of controlling the gas dissolution in the bulk of a solid by its active surface can be of practical use. The state of the surface can be controlled (brought from one state to another) by a brief change in the external parameters, namely, temperature or the partial pressures of the reactants in the gas phase.

Along with scheme (5.66), there is a very simple nonlinear mechanism of chemical oscillations, the so-called autocatalytic oscillator, which can also be interpreted as a hypothetical mechanism of adsorption and desorption of hydrogen and its diffusion

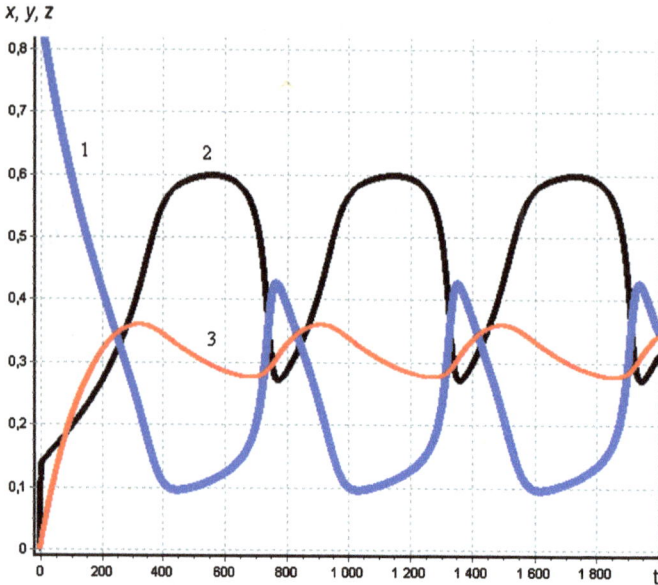

Fig. 5.8: Auto-oscillations of the average degrees of surface coverage with intermediates: (1) the fraction of free surface, (2) the average degree of surface coverage with adsorbed hydrogen, and (3) the fraction of hydrogen diffusing into the bulk

into the bulk of a metal:

$$1)\ H_2 + Z \rightleftarrows ZH_2 ,$$

$$2)\ ZH_2 + 2Z \rightarrow 3Z + H_2 ,$$

$$3)\ H_2 + Z \rightleftarrows (ZH_2)_V , \tag{5.69}$$

where ZH_2 is the hydrogen adsorbed on the metal surface Z and $(ZH_2)_V$ is the hydrogen diffusing into the bulk of the metal.

The autocatalytic step 2) characterizes the second route of H_2 desorption, provided that there are two neighboring free adsorption sites. Scheme (5.69) is described by the kinetic model

$$\frac{\partial x}{\partial t} = p_H k_1 z - k_{-1} x - k_2 x z^2 + D(z \Delta_2 x - x \Delta_2 z) ,$$

$$\frac{\partial y}{\partial t} = p_H k_3 z - k_{-3} y , \tag{5.70}$$

where $z = 1 - x - y$, x is the degree of surface coverage with ZH_2, and y is the fraction of hydrogen diffusing into the bulk.

The nonstationary mass-transfer equation into bulk is similar to (5.68). Without diffusion, system (5.70) allows auto-oscillations [18]. At the kinetic parameters $k_1 p_H = 0.12$, $k_{-1} = 0.01$, $k_2 = 1$, $k_3 p_H = 0.0032$ and $k_{-3} = 0.002$, the uniform steady state of dynamic system (5.70) may be unstable, and its solutions tend to a limit cycle starting with any initial conditions.

Specific calculations illustrating the nonlinearity of the interaction between the active surface of a metal and its bulk were performed for such oscillatory modes on the surface. In this case, on the surface, there exist nonstationary structures – hydrogen clusters. The average surface coverage with hydrogen is characterized by undamped oscillations (Fig. 5.8). These oscillations induce H_2 concentration oscillations in the bulk of the solid. The concentrations averaged over the period exceed the bulk concentrations of hydrogen dissolved in the metal in a steady state. This fact confirms that the interaction between the active surface and bulk of a solid is essentially nonlinear.

We have showed that the active surface can significantly determine the modes of gas dissolution in the bulk of a metal. Multiplicity of steady states and auto-oscillations on the metal surface can be factors that control the nonlinear interactions between the gas and the solid. By varying external control parameters (such as, e.g., temperature and pressure), one can control the state of the surface, which, in turn, determines the intensity of the gas dissolution in the bulk of the solid.

5.7 Models of wave propagation reactions

Recent studies show that the kinetics of combustion processes has a significant impact on the characteristics of flame propagation [56–66]. A combination of temperature and kinetic nonlinearities may considerably complicate the physicochemical pattern of the process. For example, if the kinetic subsystem has several steady states, which determine a hysteretic shape of the temperature dependence of the reaction rate, then there may be combustion modes with variable flame propagation rates. If the kinetics is oscillatory, then, in the entire system, there may be complex aperiodic modes (deterministic chaos).

In this section, we present a basic model of flame propagation, which consists of a temperature subsystem and a kinetic subsystem. Possible flame propagation modes are classified depending on specific features of the kinetic subsystem (multiplicity of stages, multiplicity of steady states, nonstationarity, oscillations, etc.). In particular, it is shown that for a heat release function of hysteretic shape, there exists a stable combustion wave with zero propagation rate (a standing front).

Consider a simplified one-dimensional model of combustion theory:

$$\frac{\partial T}{\partial t} = \lambda \frac{\partial^2 T}{\partial x^2} + W(T, \mathbf{c}) + \alpha(T_0 - T), \tag{5.71}$$

$$\frac{\partial \mathbf{c}}{\partial t} = F(T, \mathbf{c}), \tag{5.72}$$

where T is temperature, t is time, $\mathbf{c}$ is the reactant concentration vector, λ is the heat conduction parameter, α is the heat transfer parameter, x is length, W is the heat release function, T_0 is the ambient temperature, and F is the kinetic function.

The kinetic function $F(T, \mathbf{c})$ is constructed according to a given scheme of transformations of reactants. In the general case, it has the form

$$F(T, \mathbf{c}) = \sum_{s=1}^{m} \gamma_s w_s(T, \mathbf{c}),$$

where γ_s is the stoichiometric vector of the s-th step and w_s the rate of the s-th step. For example, for the mass action law,

$$w_s(T, \mathbf{c}) = k_s(T) \prod_{i=1}^{n} c_i^{\alpha_{is}},$$

where α_{is} are the stoichiometric coefficients of the s-th step.

The function $W(T, \mathbf{c})$ in (5.71) is given as the algebraic sum of the heat release functions of individual stages:

$$W_s(T, \mathbf{c}) = \sum_{s=1}^{m} h_s w_s(T, \mathbf{c}),$$

where h_s is the heat of the s-th step ($h_s > 0$ for exothermic reactions and $h_s < 0$ for endothermic reactions).

If kinetic subsystem (5.72) has several steady states, i.e., if the system of equations

$$F(T, \mathbf{c}) = 0: \qquad \mathbf{c} = \mathbf{c}(T)$$

has several solutions $\mathbf{c}(T)$ for the concentration vector $\mathbf{c}$, then the function $W(T, \mathbf{c}(T))$ is hysteretic in T. Let us show the possibility of such an essentially nonlinear function by a simple example. Let hydrogen be oxidized on a certain metal Me according to the scheme

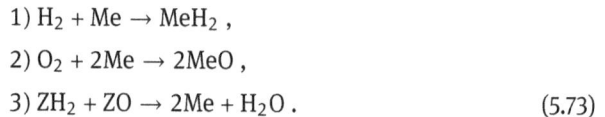

1) $H_2 + Me \rightarrow MeH_2$,

2) $O_2 + 2Me \rightarrow 2MeO$,

3) $ZH_2 + ZO \rightarrow 2Me + H_2O$. $\qquad\qquad$ (5.73)

At constant partial pressures of the reactants O_2, H_2, and H_2O in the gas phase, scheme (5.73) corresponds to the kinetic model

$$\dot{c}_H = k_1 p_H z - k_3 c_H c_O,$$
$$\dot{c}_O = 2k_2 p_O z^2 - k_3 c_H c_O, \qquad\qquad (5.74)$$

where c_H and c_O are the dimensionless concentrations of hydrogen and oxygen, respectively, adsorbed on the metal surface; $z = 1 - c_H - c_O$ is the fraction of free surface of catalyst Me; and k_i are the rate constants for reactions (5.73), characterized by the Arrhenius dependence on temperature: $k_i(T) = k_i^0 \exp(-E_i/RT)$. Steady states of kinetic subsystem (5.74) are found from the system of equations

$$k_1 p_H (1 - c_H - c_O) - k_3 c_H c_O = 0,$$
$$2k_2 p_O (1 - c_H - c_O)^2 - k_3 c_H c_O. \qquad\qquad (5.75)$$

It is easy to see that system (5.75) has two boundary steady states:

$$1)\ c_H = 0, \quad c_O = 1, \quad z = 0,$$
$$2)\ c_H = 1, \quad c_O = 0, \quad z = 0, \tag{5.76}$$

at which the hydrogen oxidation rate

$$w = k_3 c_H c_O \tag{5.77}$$

is zero. Along with steady states (5.76), under the condition

$$k_3(2k_2 p_O - k_1 p_H)^2 \geq 8k_2 p_O (k_2 p_H)^2, \tag{5.78}$$

there are two more steady states with nonzero reaction rate (5.77),

$$w = \frac{k_1^2 p_H^2}{2k_2 p_O}. \tag{5.79}$$

The critical temperature value T_*, at which a branch of nonzero reaction rate emerges, is found from the equality in expression (5.78). Of the two boundary steady states, one is stable and the other is unstable. Similarly, of the two inner steady states, one is stable and the other is unstable. Calculations show that if the reversibility of adsorption stages in scheme (5.73) is low, then the function $w(T)$ is hysteretic. Within a certain temperature range (T_*, T_{**}), there are two branches of reaction rate $w(T)$. A jump from one branch to the other occurs once the temperature value crosses the boundary of the above interval.

The specificity of the hysteretic dependence of the heat release function $W(T)$ on temperature in (5.71) consists in the fact that the combustion wave velocity on the strength of (5.71) may have an interval of zero values. The length of this interval is determined by the size of the hysteresis. It was shown [67] that in systems with hysteresis, frontal phenomena always occur even at arbitrarily large α values (at corresponding values $T_0 \in [T_*, T_{**}]$). A steady combustion wave with zero propagation velocity (a standing front) is a stable structure.

As a rule, kinetic subsystem (5.72) is quasistationary with respect to (5.71). Therefore, expressing the reactant concentrations in terms of temperature from the quasistationarity conditions, we obtain the reaction rate $W(T, c(T))$ as a function of temperature only. In this case, for modeling flame propagation, it is sufficient to use classical heat balance equation (5.71). However, if the kinetic subsystem is essentially unsteady, e.g., when there are auto-oscillations in the kinetically controlled region, then it is necessary to simultaneously consider the heat balance equation (5.71) and the material balance equation (5.72). As an example of an oscillating kinetic subsystem, let us analyze scheme (5.73) supplemented with the so-called buffer step

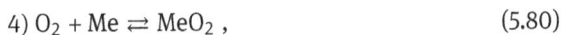

$$4)\ O_2 + Me \rightleftharpoons MeO_2, \tag{5.80}$$

where MeO_2 is a second product of oxygen adsorption, which is unreactive, unlike MeO.

The mathematical model of the kinetic subsystem corresponding to scheme (5.73) + (5.80) has the form

$$\dot{c}_H = k_1 p_H z - k_{-1} c_H - k_3 c_H c_O \, ,$$
$$\dot{c}_O = 2 k_2 p_O z^2 - k_{-2} c_O^2 - k_3 c_H c_O \, ,$$
$$\dot{c}_{(O)} = k_4 p_O z - k_{-4} c_{(O)} \, , \tag{5.81}$$

where $z = 1 - c_H - c_O - c_{(O)}$ and $c_{(O)}$ is the concentration of intermediate substance MeO_2 in the buffer step. System (5.81) is the simplest catalytic oscillator, i.e., a system that has an essentially unsteady mode – undamped oscillations of reactant concentrations and, thus, the reaction rate $W(T, c(T))$. Oscillations arise at a certain ratio of the parameters p_H, p_O, T, k_i, E_i, etc. Their existence conditions were derived previously [18] at given partial pressures of reactants in the gas phase and in an isothermal case. However, in entire system (5.71)+(5.72), such unsteady modes are also quite feasible.

Kinetic nonstationarity can be described not only by mechanisms of the type in (5.73) and (5.80), but also using simpler autocatalytic schemes. For example, the simplest of such schemes is a reaction containing an autocatalytic step:

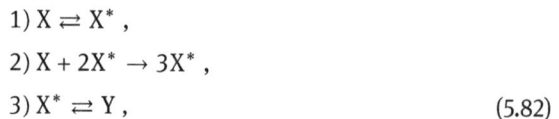

$$1)\, X \rightleftarrows X^* \, ,$$
$$2)\, X + 2X^* \rightarrow 3X^* \, ,$$
$$3)\, X^* \rightleftarrows Y \, , \tag{5.82}$$

where X^* is the excited form of particle X, step 1) is the excitation of X, step 2) is multiplication of active particles X^*, and step 3) is their reversible interaction with inert substance Y. As was shown above, the kinetic model corresponding to scheme (5.82),

$$\dot{x} = -k_1 x + k_{-1} x_* - k_2 x x_*^2 \, ,$$
$$\dot{y} = k_3 x_* - k_{-3} y \, , \tag{5.83}$$

where $x_* = 1 - x - y$, is the simplest kinetic oscillator. At any initial data, the solutions $x(t)$ and $y(t)$ of the kinetic model (5.83) tend to a stable limit cycle, which is characterized by a certain oscillation period and amplitude.

If the kinetic subsystem is characterized by a unique dependence of the reaction rate on temperature, but the reaction occurs in several stages, including endothermic processes, then the function $W(T)$ may either be monotonic, or have intervals of nonmonotonic change. In the presence of endothermic nonlinear stages that can lead to multiplicity of steady states, a hysteresis can be observed in the dependence $W(T)$. Moreover, the kinetic subsystem can have an oscillating character. In each of these qualitatively different cases, the flame propagation dynamics will have its own specificity. Details of this specificity can be studied by modern methods of parametric analysis using the basic model (5.71) and (5.72), which is one of the reaction–diffusion systems.

5.8 Macroclusters on the catalyst surface at the CO oxidation on Pt

In this section, we will analyze one simplified model of the catalytic reaction of the CO oxidation, taking into account diffusion of intermediates on the catalyst surface. Suppose that on the surface of the platinum only CO diffuses. Then the scheme of transformations will be

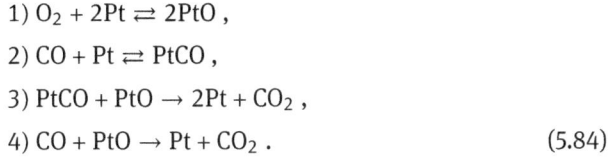

$$1)\ O_2 + 2Pt \rightleftharpoons 2PtO ,$$
$$2)\ CO + Pt \rightleftharpoons PtCO ,$$
$$3)\ PtCO + PtO \rightarrow 2Pt + CO_2 ,$$
$$4)\ CO + PtO \rightarrow Pt + CO_2 . \tag{5.84}$$

The macrokinetic model for the (5.84) takes the form

$$\dot{x}_1 = 2k_1 p_1 z^2 - 2k_{-1} x_1^2 * - k_3 x_1 x_2 - k_4 p_2 x_1 ,$$
$$\dot{x}_2 = k_2 p_2 z - k_{-2} x_2 - k_3 x_1 x_2 + D(z \Delta x_2 - x_2 \Delta z) , \tag{5.85}$$

where $z = 1 - x_1 - x_2$, x_1, and x_2 are the degrees of coverage of PtO and PtCO, respectively; Δ is the one-dimensional Laplace operator $\partial^2 / \partial \ell^2$; p_1 and p_2 are the partial pressures of O_2 and CO in the gas phase (considered constants), and k_i is the rate constant of stages.

As above, the desired dissipative structures can be obtained from the stationary equation [55]

$$D(1 - x_1) \frac{d^2 x_2}{d\ell^2} + x_2 \frac{d^2 x_1}{d\ell^2} + k_2 p_2 z - k_{-2} x_2 - k_3 x_1 x_2 , \tag{5.86}$$

where $x_1(x_2)$ is found from the condition $\dot{x}_1 = 0$. The equation is integrable by quadratures. For $\varphi(x_2) = dx_2/d\ell$, we have

$$\varphi(x_2) = (A - U(x_2)) \exp\left(-2 \int_0^{x_2} r(u)du\right) ,$$

$$U(x_2) = 2 \int_0^{x_2} s(u) \exp\left(2 \int_0^u r(v)dv\right) du ,$$

where the functions $r(u)$ and $s(u)$ are written according to (5.86):

$$s(x_2) = a_0(x_2)/a_2(x_2) , \quad r(x_2) = a_1(x_2)/a_2(x_2) ,$$
$$a_2(x_2) = D\left(1 - x_1(x_2) - x_2 \frac{dx_1}{dx_2}\right) , \quad a_1(x_2) = Dx_2 \frac{d^2 x_1}{dx_2^2} ,$$
$$a_0(x_2) = k_2 p_2(1 - x_1(x_2) - x_2) - k_{-2} - k_3 x_1(x_2) x_2 ,$$
$$x_1(x_2) = \frac{1}{2a}(b - \sqrt{b^2 - 4ac}), \quad a = 2k_1 p_1 - 2k_{-1} ,$$
$$b = 4k_1 p_1(1 - x_2) + k_3 x_2 + k_4 p_2 , \quad c = 2k_1 p_1(1 - x_2)^2 .$$

Note that if x_1, $x_2 \geq 0$ and $x_1 + x_2 \leq 1$, the function $x_1(x_2)$ is uniquely defined and is differentiable. If the arbitrary constant A is chosen so that

$$A_2 < A < \min\{A_1, A_3\},$$

where $A_i = U(x_2^{*(i)})$, $i = 1, 2, 3$, $x_2^{*(i)}$ is one of the three steady states of the appropriate lumped system, then equation (5.86) (the second initial condition for it is chosen so that $x_2(0) = x_2^{*(2)}$) along with the homogeneous $x_2^{*(1,2,3)}$ has periodic solutions in space. These solutions are the desired dissipative structures.

It is especially important to note the limiting case when $A = \min\{A_1, A_3\}$. Here, the periodic solution degenerates to the solitary wave. Such decision corresponds to the fact that one of the stable homogeneous steady states is implemented at $\xi \to \pm\infty$ on the catalyst surface. But in some region the steady state approaches another stable state (but does not reach it). This heterogeneity can be interpreted as a macrocluster on the catalyst surface (e.g., a spot "CO on O_2" or on the contrary, a spot "O_2 on CO").

The calculations were carried out at varying temperatures and partial pressures in the range of: $400\,K \leq T \leq 500\,K$ and $10^{-8} \leq p_1, p_2 \leq 10^{-3}$, 10^{-6} Torr, which correspond to the real conditions of the experiment and where the multiplicity of steady states is possible. The values k_i^0 and E_i are given Section 5.6. The dependencies of the homogeneous steady states $x_{1,2}^*$ on parameters p_1, p_2, and T in the given region of changes can be characterized by the hysteresis. Here, the multiplicity of steady states is observed in the temperature range of $[T_1 = 417.5\,K$ and $T_2 = 452\,K]$. The values of T_1 and T_2 are the bifurcation values of temperature. At $T < T_1$ there is a single steady state with almost zero reaction rate $W = k_3 x_1 x_2 + k_4 p_2 x_1$. At $T \in (T_1, T_2)$ there are three values of the stationary rate. At $T > T_2$ the single steady state with the already significant value of W remains once more.

The limiting heterogeneous solutions (5.86) (of the type of solitary wave), as well as the corresponding distributions of x_1 and W along a length ξ correspond to the single spots CO and O_2 on the catalyst surface. Note that in the particular conditions of heterogeneity there will be the spots of two types. The first type is the spot "CO on O_2" (almost all of the surface is occupied by adsorbed molecules with a larger share of O_2, and in some finite region an increase in the share of O_2 is observed). The second type is the spot "O_2 on CO". This case corresponds to the specifics of the lumped system that the steady states in it are characterized by almost complete filling of the platinum surface by the adsorbed CO and O_2.

The nature of heterogeneous steady states depends on the temperature. It is interesting that if T changes in the interval (T_1, T_2), a "revolution" of spots occurs. For example, if $T < T_P = 422.5$ K the heterogeneous state has the nature of spot "CO on O_2" and if $T > T_P$ the state has the spot "O_2 on CO". In the first case, the situation is characterized by a high activity of the catalyst surface (by the significant reaction rate W) with separate "cold" spots. In the second case, the surface is inactive (W is small), but separate "hot" spots are observed. The size of these macroclusters changes with varying parameters and at $D = 10^{-7}$ cm^2 s^{-1} is equal to the value of 10^{-3} cm.

Write the critical frequency in the form:

$$\omega_{cr}^2 = \frac{f_{11}^* f_{22}^* - f_{12}^* f_{21}^*}{D(f_{11}^*(1 - x_1^*) - f_{12}^* x_2^*)} . \tag{5.87}$$

Disturbances with frequencies $\omega > \omega_{cr}$ in a small neighborhood of $\mathbf{x}^*$ fade out over time. If $\omega < \omega_{cr}$, the disturbances grow indefinitely. In (5.87) $f_{i,j}^*$ are elements of the Jacobian matrix of the lumped system (5.85), which are calculated in the steady state $\mathbf{x}^*$. It is easy to show that in (5.87) the denominator is negative in any $\mathbf{x}^*$, and the numerator is negative in unstable $\mathbf{x}^*$ and positive in stable $\mathbf{x}^*$. Thus, ω_{cr} exists only for unstable $\mathbf{x}^*$, i.e., the stable $\mathbf{x}^*$ remains so and in the presence of diffusion, and the unstable $\mathbf{x}^*$ in the distributed system remains unstable at relatively low-frequency disturbances.

In the calculations conformance of the spot half-width ξ_P and the value $\lambda_{cr} = (2\pi/3)/\omega_{cr}$, which corresponds to the half-width of the critical disturbances, was observed. It is difficult to establish a direct relationship between ξ_P and λ_{cr}, but the value λ_{cr}, as confirmed by the calculations, can serve as a good a priori estimate of the characteristic size of the possible macroclusters on the catalyst surface. The value λ_{cr} can be computed for the given D with the knowledge of only the values of the unstable stationary homogeneous covering of $\mathbf{x}^*$.

Similarly to the temperature dependencies, revolutions of spots on the catalyst surface also occur with varying the parameters p_1 and p_2 in the specified range. The calculations show that the largest part of the allocated region is characterized by a structure of type a spot "O_2 on CO".

Let us describe the evolution of heterogeneities with, for example, varying temperature. Here, we have not analyzed the stability of the periodic solutions (5.86). It seems most probable that periodic solutions are unstable because in numerical computations of the nonstationary model (5.85) they were observed. At a small disturbance a solitary spot disappears or generates a traveling wave, after the passage of which one of the stable homogeneous steady states is established. If the periodic solutions are unstable, then the size of solitary spots is a characteristic size of the local instability. If it is small enough that fluctuations can lead to the fact that instead of a multiplicity of steady states (the hysteresis on the parameters), only a jump in the values of the parameters corresponding to the revolution of the spots is observed.

Hence, by varying, for example, a temperature, the following sequence of change of the steady state is possible. When $T < T_1$ a single and stable homogeneous steady state with a small reaction rate (a surface is cold) exists. When varying T in the range (T_1, T_P) there are the stationary heterogeneities along with the two stable homogeneous states. These stationary heterogeneities correspond to the specific "cold" spots on the catalyst surface and characterize the size of the local instability. Further, when $T_P < T < T_2$ the local instability already corresponds to the "hot" spots. And when $T > T_2$ again remains the only steady and homogeneous steady state, which is responsible for the active condition of the catalyst. Note that with increasing T the size of the spots "CO on O_2" grows, and the size of the spots "O_2 on CO" is decreased.

Similar evolution of the character and sizes of macroclusters on the catalyst surface, induced by diffusion, is possible and by varying the partial pressures p_1 and p_2. If the characteristic size of the local instability is comparable to the size of fluctuations, the length of the hysteresis of the stationary reaction rate in the presence of diffusion, which is observed in the kinetic region, can be much smaller. In this case we can talk about the diffusional narrowing of the hysteresis of the stationary rate.

Note that a typical way of obtaining in models of stable dissipative structures is a proof of the instability of all homogeneous steady states. In our case, a diffusion cannot change the character of stability of such states. A single and unstable homogeneous steady state exists in the certain parameter region for the considered mechanism of the CO oxidation (5.86), added by the "buffer" step: 5) $CO+Pt \rightleftarrows (PtCO)$, where $(PtCO)$ is a nonreactive form CO on the catalyst surface. For the mentioned region of parameters the observed state of the catalyst surface will always be heterogeneous.

Here, the macroclusters on the catalyst surface, induced by a diffusion, were analyzed. In terms of average values of the degrees of coverage and concentrations it was shown the possibility of the existence of macrostructures in the systems of the "reaction–diffusion". On the other hand, last time a development of the subtle physical methods of investigation of the surface allows us to talk about the need for modeling microstructures formed on the catalyst surface by adsorbed substances. A review of the method of the direct imitation modeling of such processes of sorption and diffusion at the microlevel is given in[31].

5.9 Model of coking the feed channels of the fuel

In modern aircraft engines hydrocarbon fuel is often used as a cooler for heat-stressed construction elements. The fuel is under supercritical pressure. This feature greatly complicates the adequate modeling of the thermohydrodynamic processes occurring in the flow channels. This region of thermodynamic parameters differs by specifics, the consideration of which is usually difficult.

As is known, the increase in pressure in the cooling system of the engine leads to an increase in heat flow. The cooling system for hydrocarbon fuels keeps working until the wall temperature reaches the temperature of thermal decomposition of the fuel, or the so-called coking temperature. When this occurs, the precipitation of the solid phase on the wall of the flow channel, which causes an increase in its full thermal resistance, and a quick rise of temperature, and leads to its destruction. To date there is no complete understanding, and accordingly no adequate quantitative description of the whole totality of the thermophysical and chemical processes flowing in the system cooling channels of rocket engines. From general considerations it is clear that in supercritical conditions, the process of coking of channels includes the precipitation of the source of hydrocarbon fuel, precipitation of free radicals on the surface of metals, and coking. Sometimes, additional cleaning of fuel and its modification improve its thermal stability and reduce the formation of the solid phase. On the other hand, the thermal reactions of decomposition can be accelerated in the presence of a metal catalyst surface, for example copper.

The characteristics of heat exchange and precipitation rates in the flow of rocket fuel RP-1, technically pure propane, industrial propane, and a mixture of fuel RP-1 and propane were shown in [19]. The flow occurred in a copper tube with an inner diameter of 1.96 mm and a section length 38.1 mm, heated by an alternating electric current. The rates in the experiments ranged up to $30\,\mathrm{m\,s^{-1}}$. The fuel pressure at the inlet was 13.8 MPa. The initial wall temperatures ranged up to 700 K. In the experiments, the presence of a maximum precipitation rate of carbon was recorded. The maximum shifted towards high temperature while the rate of flow of fuel increased. Earlier data were obtained on the maximal growth rate of the thermal resistance of the tube wall due to the precipitation of carbon, which at the rate of flow of fuel RP-1 equalling $20\,\mathrm{m\,s^{-1}}$ was about $2\,\mathrm{K\,m^2\,J^{-1}}$ (the dimension value is taken from [19]).

The results of the study of heat exchange at the flow of kerosene of mark RT at supercritical parameters in tubes with a length of 1 m with an inner diameter ranging between 1 and 4 mm are described in [70]. The main attention was directed on the study of the intensifying action of knurling, i.e., the way comprehensively studied by the authors. A relative deterioration of heat exchange under certain combinations of operating parameters was discovered in the experiments.

The experiments conducted at the Keldysh Research Center, gave results that qualitatively confirmed the proposed method. The experiments were conducted by flowing hydrocarbon fuel at supercritical pressure in a copper tube. The heating of the middle section of the tube was carried out by the direct passage of electric current. Temperature control was carried out with a thermal imager. At a first glance, the longitudinal temperature distributions of the tube obtained gave paradoxical results. In the middle section of the tube at some time there is a "failure" temperature. The explanation of this fact, apparently, should be sought in the fact that with the beginning of thermal decomposition of the liquid, intense "flow" of heat and, in addition, precipitation of carbon, starts, that is, the formation of a heat-insulating layer.

This section is dedicated to consideration of a fairly simple mathematical model of the process of precipitation of a solid phase on the walls of the channels of engine cooling systems. For hydrocarbon fuels in conditions close to the conditions of standard techniques, it is a model of heat and mass transfer with phase transformations, chemical destruction, and coking of the walls of the channels of flow. Calculations for the process of decomposition of hydrocarbon fuel in a single copper tube with specific kinetics and phenomenological kinetics of deposition of the solid phase on the channel walls were conducted.

Consider the processes of heat and mass transfer in a tube with metal walls, the inner surface of which has catalytic properties. As the idealization of the processes in the liquid phase the approximation of ideal displacement is used. Equations of material and heat balances for the movement and decomposition of the fuel in the channel have the following form:

$$\frac{\partial c}{\partial t} + v\frac{\partial c}{\partial \ell} = \sum y_s w_s + \sum y_s^* w_s \,,$$

$$\frac{\partial T}{\partial t} + v\frac{\partial T}{\partial \ell} = \sum (\Delta H_s)w_s + \alpha(T_* - T) \,, \tag{5.88}$$

where c is the vector of concentrations of reagents in the liquid phase; v is the flow rate; t is the time; ℓ is the current reactor length; y_s and y_s^* are the stoichiometric vectors of the s-th reactions in the liquid phase and on the catalyst surface; w_s is the rate of the s-th reaction; T is the temperature of flow in the channel; T_* is the temperature of the channel walls; α is the heat transfer coefficient between the flow and the wall of the reactor; and (ΔH_s) is the thermal effect of the s-th reaction.

The stoichiometric vectors (y_s and y_s^*) correspond to given mechanisms of chemical transformations in the liquid phase and the process of interaction of intermediate substances with the wall. Density, the heat capacity of the reacting mixture, the rate constants of the stages and their temperature dependencies, geometric characteristics of a channel, etc., are included into the system (5.88) as the parameters.

For the catalytic surface we have:

$$\frac{\partial T_*}{\partial t} = \sum (\Delta H_s)w_s + \beta(T_* - T) + \delta(T_0 - T_*) \,,$$

$$\frac{\partial x}{\partial t} = \sum y_s^* w_s \,, \tag{5.89}$$

where x is the vector of concentrations of reagents on the inner surface of the channel wall; β is the coefficient of heat exchange between the channel wall and the liquid phase; δ is the coefficient of heat exchange between the channel wall and the environment; and T_0 is the environment temperature. Contributions of thermal and chemical processes on the channel wall and its heating by the electric current are taken into account in the right sides of system of differential equations (5.89).

The member $D_T\Delta T_*$, which corresponds to the finite thermal conductivity of the material of the channel wall, can be included into the heat balance equation of the system (5.89). Here, D_T is the coefficient of thermal conductivity of the channel wall and ΔT_* is the Laplace operator. The character of the dependencies of the reaction rates $w_s(c, x, T)$ and $w_s(c, x, T_*)$ on the substance concentrations in the stream and on the catalyst surface is set in accordance with the accepted kinetics of the processes. It is assumed that the reaction takes place without significant volume changes, so the rate $v = $ const. The model (5.88) and (5.89) is nonstationary and two-temperature, because the nonstationarity and the nonisothermality of the tube are caused by the nonstationarity and the nonisothermality of the processes in a liquid phase. Taking into account the precipitation of hydrocarbons on a metal surface and their subsequent coking, the coefficients of heat exchange α and β are significantly dependent on the concentration of the solid phase on the catalyst surface. The equation of heat balance for the channel wall can contain the source member $Q(T_*)$, corresponding to the Joule heating of the metal tube by an electric current.

The system of differential equations (5.88) and (5.89) should be supplemented by appropriate initial and boundary conditions. For example,

$$
\begin{aligned}
t = 0: \quad & x = x^0(\ell), \quad T_* = T_*(\ell), \\
c = c^0(\ell), \quad & T = T^0(\ell), \quad 0 \le \ell \le \ell_k; \\
\ell = 0: \quad & c = c_{in}(t), \quad T = T_{in}, \quad 0 \le t \le t_k.
\end{aligned}
\tag{5.90}
$$

where c^0 and x^0 are the initial distributions of reagent concentrations in the liquid phase and on the channel wall; T^0 is the initial profile of the temperature in the liquid phase; ℓ_k is the finite length of the channel; c_{in} and T_{in} are the concentrations of reactants and the flow temperature on the inlet of the channel; and t_k is the operating time of the reactor.

The conditions (5.90) reflect the possibility of variation of the initial profiles of concentrations and temperature, and the input modes for the model (5.88) and (5.89). The method of characteristics in combination with one of the algorithms of the integration of stiff systems of ordinary differential equations can be proposed as a method of the numerical solution of the model (5.88)–(5.90) (for example, at $D_T = 0$).

The profiles of concentrations and temperatures along the tube's length can be obtained as the result of the calculation. The dynamics of their changes over time under different regimes of the process can be studied. The variation of geometrical and thermophysical parameters and the study of conditions of thermal stability allow us find optimal modes, determine limiting factors, and to match the physical ideas about the process with available experimental data.

Specific calculations were carried out for the process of thermal decomposition of hydrocarbon fuel in the copper tube. The brutto-scheme diagram of the transformations was taken as a sequential scheme:

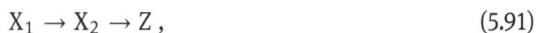

$$
X_1 \rightarrow X_2 \rightarrow Z,
\tag{5.91}
$$

where X_1, X_2 and Z are the substances of the reagents in the gas phase and on the surface of the channel wall.

The first reaction in (5.91) corresponds to the thermal decomposition of the hydrocarbon fuel and the second one to the precipitation of the solid phase on the surface of copper. Kinetic parameters of the decomposition of the hydrocarbon fuel on copper were defined previously. Here, the first-order kinetics was taken as the second step.

The general model (5.88)–(5.90) in our case is specified as follows:

- the equations of heat and mass transfer for the liquid phase are written down in accordance with (5.91)

$$\frac{\partial x_1}{\partial t} + v\frac{\partial x_1}{\partial \ell} = -w_1 \,,$$

$$\frac{\partial x_2}{\partial t} + v\frac{\partial x_2}{\partial \ell} = w_1 - w_2 \,,$$

$$\frac{\partial T}{\partial t} + v\frac{\partial T}{\partial \ell} = Q_1 w_1 + Q_2 w_2 + \alpha(z)(T_* - T) \,; \tag{5.92}$$

- heat and material balances for the catalytic surface

$$\frac{\partial T_*}{\partial t} = D_T \Delta T_* + Q(T_*) + \beta(z)(T - T_*) \,,$$

$$\frac{\partial z}{\partial t} = w_2 \,; \tag{5.93}$$

- the relative reaction rates

$$w_1 = k_1 x_1^n (a + (1 - x_1)^m) \,,$$

$$w_2 = k_2 x_2 \,; \tag{5.94}$$

where rate constants of stages (5.91) are

$$k_i = k_i^0 \exp(-E_i/RT) \,, \quad i = 1, 2 \,.$$

The heat production function due to Joule heating of the copper tube is

$$Q(T_*) = a_1(a_2 + a_3 T_*) \,.$$

The reduced coefficient of heat exchange between the liquid phase and the copper wall of the channel is

$$\alpha(z) = \alpha_0/(1 + z/\varepsilon) \,, \quad \beta(z) = \beta_0 \alpha(z) \,.$$

The initial and inlet data of type (5.90) are

$$z(0, \ell) = 0 \,, \quad x_1(t, 0) = 1 \,, \quad x_2(t, 0) = 0 \,,$$

$$T(t, 0) = T^0 \,, \quad T_*(0, \ell) = T_*^0 \,. \tag{5.95}$$

In model (5.92)–(5.95), the following basic notations were taken: x_1 and x_2 are dimensionless concentrations of the reactants in the liquid phase; z is the concentration of coke on the channel wall; Q_1 and Q_2 are the thermal effects of reaction; $Q(T_*)$ is the function of the Joule heating of the reactor walls by electric current; k_i are the rate constants of stages. The specific kinetics of the decomposition of the hydrocarbon fuel, obtained from the processing of a specific experiment, were used in (5.94). Here

a and m are the model parameters and E_i is the activation energy. The dependencies $Q(T_*)$, $\alpha(z)$ and $\beta(z)$ in the first approximation were taken as a nonlinear functions with parameters a_i, α_0, β_0 and ε.

The solution of the system of differential equations (5.92)–(5.95) allows us to obtain the profiles of temperatures and concentrations and to study their behavior in time. The parameter values correspond to the conditions of the experiment. The profiles of the thermal fields were obtained at the parameter values, corresponding to the conditions of the experiment. The calculated characteristics the thermal fields agree with the experimental data.

The dynamics of the process is next. At the beginning, a temperature profile with a characteristic deflection in the middle of the channel is formed. Further, during the accumulation of solid phase on the copper wall the process loses its thermal stability, and the temperature begins to increase sharply.

The deflection of the temperature in the beginning of the process is determined by the absorption of heat at the decomposition of the hydrocarbon fuel. The position of the temperature minimum along the length significantly depends on the rate of fluid flow. In the experimental equipment the thermal, hydrodynamic, and chemical parameters are such that the temperature minimum is observed in the middle of the copper tube. This means that the maximum rate of thermal decomposition of hydrocarbon fuel falls out for half the length of the channel. The same behavior is observed in the calculations, which is an indirect confirmation of the adequacy of the proposed model (5.92)–(5.95).

The calculations by the proposed model are presented in Fig. 5.9a–c. At the initial moment of time the temperature profile is formed with a characteristic minimum in the middle part of the channel (Fig. 5.9a). Its position corresponds to the region of intense decomposition of the fuel. On further accumulation on the channel wall of solid hydrocarbons, the system loses its stability, and the wall temperature begins to rise sharply. With an increasing flow rate the temperature minimum is shifted to the right end of the channel, because with increasing v, the region of intensive decomposition of the fuel is shifted to the right.

The process of stability loss of the system "component fuel–metal wall" in the cooling device of engines is quite complex and it is not studied in detail here. It includes phase transitions, chemical decomposition of the fuel, adsorption of free radicals on the metal surface, which often has catalytic properties, the precipitation of the solid phase on the surface of the channel, ultimately leading to loss of stability. In the present section we proposed a quite simple mathematical model that takes into account these components of the process. At this stage of the study, the analysis was limited to the use of phenomenological dependencies without consideration of the detailed mechanisms of fuel decomposition and precipitation on the channel walls of the solid phase. For hydrocarbon fuels in conditions close to the conditions of standard techniques, we have a model of heat and mass transfer with phase transformations, chemical decomposition, and coking of the channel walls. The calculations are made

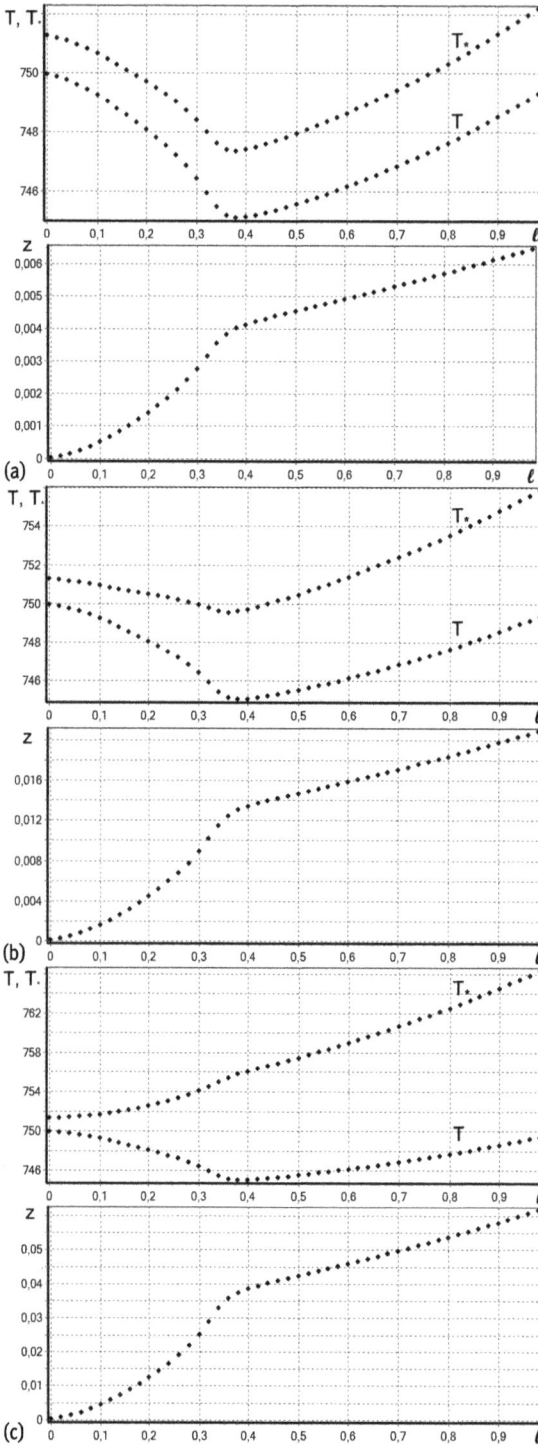

Fig. 5.9: Profiles of temperatures and concentrations at the different time moments:
(a) − $t = 0.5$; (b) − 3; (c) − 10; $v = 3$

for the process of the decomposition of hydrocarbon fuel in a copper tube with specific kinetics and phenomenological kinetics of precipitation of the solid phase on the channel walls.

To conclude the fifth chapter, we note that we considered some basic models of macrokinetics. In the simplest case, they describe the processes of heat and/or mass transfer in chemically active environments. The main attention was focused on processes of diffusion on the catalyst surface. The procedure of the parametric analysis of such models of the type "reaction–diffusion" we described briefly here. A more detailed implementation that is similar to models of isothermal kinetics is a task for the future.

Here, we would like to highlight two main results. First, the local curvature of an active surface plays an important role in the appearance of critical effects on the surface. The study of four classical geometries: plate, sphere, cylinder, and torus, shows that the corresponding radii affect the conditions of the stability loss of the homogeneous steady state. This means that at the simulation of diffusion processes on the active surface of the variable curvature (which is the first step for real surfaces) it is necessary to take into account the dependence of diffusion coefficients on the curvature of the surface. Regions with significant local heterogeneity can become the embryos of the formation of complex spatiotemporal structures. Secondly, the active surface itself can sometimes be described as a separate phase, which separates the gas phase from the volume of the solid body [69]. Therefore, at significant nonlinearity and nonstationarity of the processes on the catalyst surface it is necessary to separate the interaction "gas–surface" and "surface–volume of the solid body". In this case, the three-temperature model "gas–surface–volume" may be adequate.

Finally, we note that currently methods that imitate the modeling of processes on the catalyst surface [41, 68], including methods for probabilistic finite automata [14–17] are very popular. A combination of these approaches with classical methods macrokinetics, using models of the type "reaction–diffusion" can significantly advance our understanding and control of the processes of heat and mass transfer on chemically active surfaces that have a real structure.

Bibliography

[1] Frank-Kamenetskii DA. Diffusion and heat transfer in chemical kinetics. Moscow, Nauka, 1987. (in Russian).
[2] Vol'pert AI, Khudyaev SI. Analysis of classes of discontinuous functions and mathematical physics equations. Moscow, Nauka, 1975.
[3] Vol'pert AI, Ivanova AN. Mathematical models in chemical kinetics. In: Samarskii AA, Kurdyumov SP, Mazhukin VI, eds. Mathematical modeling: nonlinear differential equations of mathematical physics. Moscow, Nauka, 1987, 57–102.

[4] Volpert AI, Volpert VA, Volpert VA. Traveling wave solutions of parabolic systems. Trans of Math Monographs. V.140. Amer Math Society, USA, Providence, 1994.

[5] Vol'pert AI. Differential equations on graphs. Mathematics of the USSR-Sbornik 1972, 17(4), 571–582.

[6] Ivanova AN. Conditions for uniqueness of the stationary states of kinetic systems, connected with the structures of their reaction-mechanisms (I, II). Kinet Catal 1979, 20(4), 833–837.

[7] Ivanova AN, Furman GA, Bykov VI, Yablonskii GS. Catalytic mechanisms with reaction-rate self-oscillations. Dokl Akad Nauk SSSR 1978, 242(4), 872–875.

[8] Ivanova AN, Tarnopolskii BL. One approach to the determination of a number of qualitative features in the behavior of kinetic systems, and realization of this approach in a computer (critical conditions, autooscillations). Kinet Catal 1979, 20(6), 1271–1277.

[9] Ivanova AN, Tarnopolskii BL, Furman GA. Autooscillations of reaction-rate in heterogeneous oxidation of hydrogen in ideal-mixing reactor. Kinet Catal 1983, 24(1), 102–109.

[10] Kurkina ES, Semendyaeva NL. Mathematical modeling of spatial-temporal structures in a heterogeneous catalytic system. Comput Math Modeling 2012, 23(2), 133–157.

[11] Kurkina ES, Semendyaeva NL. Oscillatory dynamics of co oxidation on platinum-group metal catalyst. Kinet Catal 2005, 46(4), 453–463.

[12] Kurkina ES, Semendyaeva NL, Boronin AI. Mathematical modeling of nitrogen desorption from an iridium surface: a study of the effects of surface structure and subsurface oxygen 1. Kinet Catal 2001, 42(5), 703–717.

[13] Kurkina ES, Peskov NV, Slinko MM. Dynamics of catalytic oscillators locally coupled through the gas phase. Physica D: Nonlinear Phenomena 1998, 118(1), 103–122.

[14] Malinetskii GG, Shakaeva MS. Analysis of a cellular automaton serving as a model of oscillatory chemical reactions. Sov Phys Dokl 1991, 36(12), 826–829.

[15] Malinetskii GG, Shakaeva MS. Simulation of oscillatory reactions on a surface using cellular automata. Russ J Phys Chem 1995, 69(8), 1380–1384.

[16] Vanag VK. Probability cellular automaton-aided modeling of the stirring effect in the autocatalytic step of the Belousov–Zhabotinsky reaction. J Phys Chem 1996, 100(27), 11336–11345.

[17] Vanag VK. Study of spatially extended dynamical systems using probabilistic cellular automata. Phys Sci Progr 1999, 169(5), 481–505.

[18] Bykov VI. Modeling of critical phenomena in chemical kinetics. Moscow, URSS, 2014 (in Russian).

[19] Bykov VI, Tsybenova SB. Nonlinear models of chemical kinetics. Moscow, URSS, 2011 (in Russian).

[20] Bykov VI, Tsybenova SB, Merzhanov AG. Two-temperature models of combustion of heterogeneous systems. Dokl Phys Chem 2013, 450(2), 130–134.

[21] Aris R. Elementary chemical reactor analysis. Englewood Cliffs, NJ, USA, Prentice-Hall, 1969.

[22] Volter BV, Sal'nikov IE. Stability of work regimes of the chemical reactors. Moscow, Khimiya, 1982 (in Russian).

[23] Kholodniok M, Klich A, Kubichek M, Marek M. Methods of analysis of nonlinear dynamical models. Moscow, Mir, 1991 (in Russian).

[24] Perlmutter D. Stability of chemical reactors. Englewood Cliffs, NJ, USA, Prentice-Hall, 1972.

[25] Bykov VI, Tsybenova SB. Nonlinear base models of macrokinetics. Kinet Catal. 2012, 53(6), 737–741.

[26] Zhabotinskii AM. Concentration autooscillations. Moscow, Nauka, 1974.

[27] Turing AM. The chemical basis of morphogenesis. Phil. Trans. Roy. Soc. London B: Bio Sci 1952, 237(641), 37–72.

[28] Bykov VI, Kamenshchikov LP, Yablonskii GS. On a diffusion model of a catalytic reaction. React Kinet Catal Lett 1979, 12(4), 503–508.

[29] Bykov VI, Tsybenova S.B. The parametric analysis of Turing's two-dimensional model. XVI Int. Conf. on Chem. Reactors (CHEMREACTOR-16), Berlin, 2003, 286–289.

[30] Magnitskii NA, Sidorov SV. New methods for chaotic dynamics. Singapore, World Scientific, 2006.

[31] Bykov VI, Gilev SE, Gorban AN, Yablonskii GS. Imitation modeling of the diffusion on the surface of a catalyst. Dokl Akad Nauk SSSR 1985, 283(5), 1217–1220 (in Russian).

[32] Bykov VI, Tatarenko AA, Slinko MG. Structures in the adsorption layer on a catalyst surface and their macrokinetic description. Dokl Phys Chem 2003, 392(4), 264–267.

[33] Gilev SE, Gorban AN, Bykov VI, Yablonskii GS. Simulative modeling of processes on a catalyst surface. Dokl Akad Nauk SSSR 1982, 262(6), 1413–1416 (in Russian).

[34] Elenin GT, Krylov VV, Polezhaev AA, Chernavskii DT. Peculiarities of the formation of contrasting dissipative structures. Dokl Akad Nauk SSSR 1983, 271(1), 84–88 (in Russian).

[35] Elokhin VI, Matveev AV, Gorodetskii VV. Self-oscillations and chemical waves in CO oxidation on Pt and *Pd*: Kinetic Monte Carlo models. Kinet Catal 2009, 50(1), 40–47.

[36] Elokhin VI. Self oscillations and surface concentration waves in the CO oxidation reaction over Pt (100) and Pd (110): stochastic modelling. React Kinet Mech Catal 2016, 118(1), 87–97.

[37] Elokhin VI, Myshlyavtsev AV, Latkin EI, Resnyanskii ED, Sheinin DE, Bal'zhinimaev BS. Statistical lattice models of physicochemical processes in catalytic reactions: Autooscillations, adsorption on the rough surface, and crystallization. Kinet Catal 1998, 39(2), 246–267.

[38] Elokhin VI, Gorodetskii VV. Atomic scale imaging of oscillations and chemical waves at catalytic surface reactions: experimental and statistical lattice models. Surf Sci Ser 2006, 130, 159–189.

[39] Elokhin VI, Latkin EI, Matveev AV, Gorodetskii VV. Application of statistical lattice models to the analysis of oscillatory and autowave processes in the reaction of carbon monoxide oxidation over platinum and palladium surfaces. Kinet Catal 2003, 44(5), 692–700.

[40] Elokhin VI, Myshlyavtsev AV. Catalytic processes: nanoscale simulations. In: Schwarz JA, Contescu CI, Putyera K, eds. Dekker encyclopedia of nanoscience and nanotechnology. NY, New York, USA, Marsel Dekker Inc, 2009, 782–793.

[41] Zhdanov VP. Monte Carlo simulations of oscillations, chaos and pattern formation in heterogeneous catalytic reactions. Surf Sci Rep 2002, 45(7), 231–326.

[42] Bykov VI, Tsybenova SB, Slinko MG. Modeling of a reaction on a catalyst surface. Dokl Phys Chem 2003, 388(4), 67–70.

[43] Imbihl R, Ertl G. Oscillatory kinetics in heterogeneous catalysis. Chem Rev 1995, 95(3), 697–794.

[44] Slinko MM, Jaeger NI. Oscillating heterogeneous catalytic systems. Amsterdam, NL, Elsevier, 1994.

[45] Merzhanov AG. Combustion processes that synthesize materials. J Mater Process Technol 1996, 56(1), 222–241.

[46] Merzhanov AG. History and recent developments in SHS. Ceramics Int 1995, 21(5), 371–379.

[47] Merzhanov AG. The chemistry of self-propagating high-temperature synthesis. J Mater Chem 2004, 14(12), 1779–1786.

[48] Merzhanov AG, Filonenko AK, Borovinskaya IP. New phenomena during combustion of condensed systems. Dokl Akad Nauk SSSR 1973, 208(4), 892–894.

[49] Merzhanov AG. The chemistry of self-propagating high-temperature synthesis. J Mater Chem 2004, 14(12), 1779–1786.

[50] Gorban AN, Bykov VI, Yablonskii GS. Essays on chemical relaxation. Novosibirsk, Nauka, 1986 (in Russian).

[51] Bykov VI, Kiselev NV, Tsybenova SB. Macrokinetics of catalytic reactions on surfaces of various geometries. Dokl Chem 2008, 421(1), 161–164.

[52] Yablonskii GS, Bykov VI, Elokhin VI. The kinetics of model reactions of heterogeneous catalysis. Novosibirsk, Nauka,1984 (in Russian).

[53] Yablonskii GS, Bykov VI, Gorban AN. Kinetic models of catalytic reactions. Novosibirsk, Nauka, 1983 (in Russian).

[54] Yablonskii GS, Bykov VI, Gorban AN, Elokhin VI. Kinetic models of catalytic reactions. Amsterdam, Oxford, New York, Tokyo, Elsevier, 1991.

[55] Bykov VI, Gorban AN, Kamenshchikov LP, Yablonskii GS. Inhomogeneous stationary states in reaction of carbon-monoxide oxidation on platinum. Kinet Catal, 1983, 24(3), 520–524.

[56] Azatyan VV. New laws governing branched-chain processes and some new theoretical aspects. Chem Phys 1982, 1(4), 491–508 (in Russian).

[57] Azatyan VV, Andrianova ZS, Ivanova AN. The role played by chain avalanches in the developed burning of hydrogen mixtures with oxygen and air at atmospheric pressure. Russ J Phys Chem 2006, 80(7), 1044–1049.

[58] Azatyan VV, Ayvazyan RG, Kalkanov VA, et al. Kinetic features of silane oxidation. Chem Phys 1983, 2(8), 1056–1060 (in Russian).

[59] Azatyan VV, Soroka LB. Nonstationary nature of the state of the surface of a reaction volume and laws of the combustion of phosphorus vapor. Kinet Catal 1981, 22(2), 279–285.

[60] Azatyan VV, Vagner GG, Vedeshkin GK. Effect of reactive additives on detonation in hydrogen-air mixtures. J Phys Chem A 2004, 78(6), 1036–1044.

[61] Azatyan VV, Rubtsov NM, Tsvetkov GI, Chernysh VI. The participation of preliminarily adsorbed hydrogen atoms in reaction chain propagation in the combustion of deuterium. Russ J Phys Chem 2005, 79(3), 320–331.

[62] Mansurov ZA. Formation of soot from polycyclic aromatic hydrocarbons as well as fullerenes and carbon nanotubes in the combustion of hydrocarbon. J Eng Phys Thermophys 2011, 84(1), 125–159.

[63] Mansurov ZA. Soot formation in combustion processes (review). Combust Explos Shock Waves 2005, 41(6), 727–744.

[64] Mansurov ZA, Bobykov DU, Tashuta VN, Abil'gazinova SS. Oxidation of hexane in oscillatory conditions. Combust Explos Shock Waves 1991, 27(4), 421–424.

[65] Mansurov ZA, Matafonov AA, Nesterev VI. Oscillations in cold butane flames. Chem Phys 1988, 7(8), 1152–1154.

[66] Mantashyan AA. Kinetic manifestations of low-temperature combustion of hydrocarbons and hydrogen: Cool and intermittent flames. Combust Explos Shock Waves 2016, 52(2), 125–138.

[67] Bykov VI, Shkadinskii KG. Frontal phenomena in systems with hysteresis. Combust Explos Shock Waves 1985, 21(2), 193–197.

[68] Zhdanov VP. Periodic perturbation of the kinetics of heterogeneous catalytic reactions. Surf Sci Rep 2004, 55(1), 1–48.

[69] Boltenkov SA, Bykov VI, Tsybenova SB. Nonlinear interaction between the active surface and bulk of a solid. Dokl Phys Chem 2008, 421(1), 199–202.

[70] Bykov VI, Valiullin FH, Golovin YuM, et al. Modeling of coking processes of the flow channels of cooling systems of liquid rocket engines by thermal decomposition of hydrocarbon fuel. Cosmonautics and Rocket Eng 2010, 3(60), 100–109.

Part III: **Modelics everywhere**

This part of the book is devoted to several simple models of such complex systems as ecology (the interaction "prey–predator") and economics. Some processes of economics can also be referred to as the system "prey–predator". However, a significant difference of economic relations is the presence a man in them, who often thinks one thing, says another, and does a third. Nevertheless, simple models can be useful and here. They can highlight those main routes of the process dynamics to which phase trajectories of an economic system strive.

The purpose of the models considered here is to demonstrate that the experience of the mathematical modeling of physicochemical systems, their construction, and the parametric analysis of their simplified mathematical models can be useful in such related fields as ecology and economics. A detailed classification of these models and their parametric analysis remains to be done. A significant development has taken place in ecology (see, for example, a series of works on mathematical ecology, starting with Volterra's works [1–19]). Development in this area, and especially in mathematical economics, must rely on progresses in applied mathematics and computer science (see, for example, [20–28]).

6 Models of population dynamics: "prey–predator" models

6.1 "Prey–predator" model

In theoretical ecology a classical object of study is a system of two populations interacting with each other according to the "prey–predator" scheme [2, 3, 29]. We consider the simplest model of population dynamics of such populations, which is based on the following assumptions:

1. The population of prey multiplies exponentially with a small number of predators; an increase of population leads to competition; the rate of population reproduction decreases linearly with an increase in the number of populations (this assumption gives the logistic model of the population dynamics of isolated populations of prey).
2. The amount of food produced and consumed by a predator per unit of time depends on the number of prey; with a small number of prey, this dependence is directly proportional, and a large one leads to saturation; the consumed food is processed with some coefficient into the biomass of the predator; this process includes both a rise and reproduction of the predator.
3. In the absence of prey, the predator population dies out at a rate determined by mortality and competition.

The assumptions formulated assumptions (here we do not discuss their validity) lead to the following mathematical model [3]:

$$\frac{\partial x}{\partial t} = a(x)x - b(x)y ,$$

$$\frac{\partial y}{\partial t} = -c(y)y + db(x)y , \tag{6.1}$$

where x is the number of prey, y is the number of predators, and the function $a(x)$ is the specific growth rate of isolated population of prey. It is set in the form

$$a(x) = r\left(1 - \frac{1}{k}x\right) ;$$

the trophic function of predator is written down as

$$b(x) = \frac{bx}{1 + Ax} ;$$

and the specific rate of extinction of isolated predator population is

$$c(y) = c + gy .$$

The parameters r, k, A, c, g, d can be varied within given limits.

https://doi.org/10.1515/9783110464948-006

The system of two differential equations (6.1) can be considered as a generalization of the Lotka–Volterra classical model, in which the functions $a(x)$, $b(x)$, $c(y)$ are constants [1].

Mathematical model (6.1) is a special case of a more general model of the "prey–predator"

$$\frac{\partial x}{\partial t} = A(x) - B(x, y) \,,$$

$$\frac{\partial y}{\partial t} = -C(y) + D(x, y) \,. \tag{6.2}$$

Here, the functions of extinction–reproduction of predator and prey are defined in accordance with a given structure of their interaction [3]. We mention only some of the elementary factors determining the specific form of the model (6.2):

1. nonlinearity (often quadratic) of the dependence of the reproduction rate of the prey population from density at small values of the density;
2. competition in the prey population;
3. mortality of prey (with accounting for the nonlinear nature of reproduction at small densities of population);
4. saturation of the predator;
5. nonlinear (quadratic) dependence of the rate of predators eating the victims on the density of victim populations at low density values;
6. competition of predators for a prey;
7. competition of predators for resources other than a prey;
8. nonlinear dependence of the reproduction rate of the predator from the population density at small values of density.

All functions $A(x)$, $B(x)$, $C(y)$, $D(x, y)$ of model (6.2) contain parameters. The task of a parametric analysis of a system of two differential equations (6.2) fit into the program of parametric analysis of chemical kinetics models shown above. It includes the analysis of the number and stability of steady states, plotting of parametric and phase portraits of considered dynamic model, calculation of characteristic time dependencies. If the model is adequate for the system "prey–predator", then the obtained dependencies can serve as a basis for predicting system dynamics in a specific area of parameters and ultimately for solution of the optimal control by a system.

6.1.1 Nonlinearity of reproduction

Taking into account a factor of nonlinearity of reproduction in the prey population gives a model that can be written in the form:

$$\frac{\partial x}{\partial t} = \frac{ax^2}{N + x} - bxy \,,$$

$$\frac{\partial y}{\partial t} = -cy + dxy \,, \tag{6.3}$$

where N is the density of prey population, at which the rate of reproduction is a half the maximum possible. The replacement of variables

$$t = \tau/a, \quad x = Nu, \quad y = av/b$$

leads system (6.3) to the form

$$\frac{\partial u}{\partial \tau} = \frac{u^2}{1 + u} - uv ,$$

$$\frac{\partial v}{\partial \tau} = -yu + kuv , \tag{6.4}$$

where $y = c/a$ and $k = dN/a$. There is only one unstable equilibrium on the phase portrait of system (6.4) for any values of parameters y, k. All phase trajectories represent unwinding spirals, which strive for an infinitely remote limit cycle [3].

6.1.2 Competition in the prey population

Taking into account a factor of competition in the prey population leads to the mathematical model

$$\frac{\partial x}{\partial t} = ax(K - x)/K - bxy ,$$

$$\frac{\partial y}{\partial t} = -cy + dxy . \tag{6.5}$$

After the replacement of variables: $t = \tau/a$, $x = Ku$, $y = av/b$, this model is rewritten in a dimensionless form

$$\frac{\partial u}{\partial \tau} = u(1 - u) - uv ,$$

$$\frac{\partial v}{\partial \tau} = -yu + kuv . \tag{6.6}$$

For model (6.6) two options are possible [3]. When $y/k < 1$ there are a stable equilibrium and a saddle on the x-axis. When increasing the parameter y/k, these two equilibriums come closer and merge at $y/k = 1$. When increasing the parameter one equilibrium goes to negative area and the other becomes a stable node. This means that at strong intraspecific competition in the prey population, i.e., when its resources are very limited, the prey population is not able to feed predators. So, its population is doomed to extinction for any initial state of the system.

6.1.3 Saturation of the predator

Taking this factor into consideration leads to the model

$$\frac{\partial x}{\partial t} = ax - \frac{bxy}{1 + Ax} ,$$

$$\frac{\partial y}{\partial t} = -cy + \frac{dxy}{1 + Ax} . \tag{6.7}$$

After replacement of variables: $t = \tau/a$, $x = au/d$, $y = av/b$, this model can be written in the form

$$\frac{\partial u}{\partial \tau} = u - \frac{uv}{1 + \alpha u},$$

$$\frac{\partial v}{\partial \tau} = -yu + \frac{uv}{1 + \alpha u}, \qquad (6.8)$$

where y and α are dimensionless parameters. It is easy to show that the equilibrium exists only at $\alpha y < 1$, but it is always unstable. This means that oscillations of the density of the prey populations and the predator occur with indefinitely increasing amplitude. When increasing the parameter α, the system behavior is changed. So, at $\alpha > 1/(1 + y)$, and large densities of the prey population, the predator is unable to catch and return to the region of small densities. As a result, the density of the prey population grows unlimitedly. The density of the predator population also increases. In this case, we talk about the prey "eluding" the predator [3].

6.1.4 Competition for the predator

Competition for the predator in terms of prey is described a system.

$$\frac{\partial x}{\partial t} = ax - \frac{bxy}{1 + By},$$

$$\frac{\partial y}{\partial t} = -cy + \frac{dxy}{1 + By}. \qquad (6.9)$$

After the replacement of variables: $t = \tau/a$, $x = au/d$, $y = av/b$, the system has the form

$$\frac{\partial u}{\partial \tau} = u - \frac{uv}{1 + \beta v},$$

$$\frac{\partial v}{\partial \tau} = -yu + \frac{uv}{1 + \beta v}, \qquad (6.10)$$

where $\beta = aB/b$. At $\beta < 1$, equilibrium exists and is stable. With the growth of the parameter β, the equilibrium populations of predator and prey rise. At $\beta > 1$, the growth rate of the prey is always greater than the rate of grazing for any large density of predator population, i.e., the prey population increases indefinitely.

6.1.5 Competition of the prey and saturation of the predator

In the models considered above, (6.3) – (6.10) the determining factor was only one parameter. Now we give an example that takes into account two factors: the competition of the prey and saturation of the predator. For this case, a model is written as

$$\frac{\partial x}{\partial t} = ax\frac{K - x}{K} - \frac{bxy}{1 + Ax},$$

$$\frac{\partial y}{\partial t} = -cy + \frac{dxy}{1 + Ax}. \qquad (6.11)$$

After replacement of variables: $t = \tau/a$, $x = cu/d$, $y = av/b$, this model has the form

$$\frac{\partial u}{\partial \tau} = u - \frac{uv}{1 + \alpha u} - \varepsilon u^2,$$

$$\frac{\partial v}{\partial \tau} = -\gamma v\left(1 - \frac{u}{1 + \alpha u}\right), \qquad (6.12)$$

where $\alpha = Ac/d$, $\varepsilon = c/Kd$ and $\gamma = c/a$. A nontrivial equilibrium exists in system (6.12), which can be lost with the birth a stable limit cycle around it [3]. The increase of a value of the parameter a, characterizing the intensity of predator saturation, can lead to both the loss of equilibrium stability and its appearance.

6.1.6 Nonlinearity of eating of the prey by the predator and saturation of the predator

For the third type of trophic function a mathematical model takes the form

$$\frac{\partial x}{\partial t} = ax - \frac{bx^2 y}{1 + Ax^2},$$

$$\frac{\partial y}{\partial t} = -cy + \frac{dx^2 y}{1 + Ax^2}. \qquad (6.13)$$

The replacement of variables: $t = \tau/a$, $x = bu/d$, $y = adv/b^2$, leads the model to the form

$$\frac{\partial u}{\partial \tau} = u - \frac{u^2 v}{1 + \alpha u^2},$$

$$\frac{\partial v}{\partial \tau} = -\gamma v + \frac{u^2 v}{1 + \alpha u^2}, \qquad (6.14)$$

where $\gamma = c/a$, $\alpha = Ab/d^2$. In the study (6.14) it is convenient to use the parameter $u_0 = \sqrt{\gamma/(1 - \alpha\gamma)}$. A parametric analysis of model (6.14) shows [3] that at $u_0 > 1/\sqrt{a}$ equilibrium in the system is globally unstable. All trajectories go to infinity. The meaningful result of the study of model (6.14) is that the dominant role of one of the two factors considered in the system of prey–predator leads to global stabilization or destabilization of the equilibrium.

6.1.7 Competition of the predator for the prey and saturation of the predator

Accounting of the destabilizing factor of saturation of the predator and the stabilizing factor of competition for the prey leads to the following model [3, 29]:

$$\frac{\partial x}{\partial t} = ax - \frac{bxy}{(1 + Ax)(1 + By)},$$

$$\frac{\partial y}{\partial t} = -cy + \frac{dxy}{(1 + Ax)(1 + By)}. \tag{6.15}$$

By the replacement of variables: $t = \tau/a$, $x = au/d$, $y = av/b$, system (6.15) can be rewritten in the dimensionless form

$$\frac{\partial u}{\partial \tau} = u - \frac{uv}{(1 + \alpha u)(1 + \beta v)},$$

$$\frac{\partial v}{\partial \tau} = -\gamma v + \frac{uv}{(1 + \alpha u)(1 + \beta v)}, \tag{6.16}$$

where $\gamma = c/a$, $\alpha = Aa/d$, and $\beta = aB/b$.

The detailed qualitative analysis of conditions of bifurcations in the space of three parameters (α, β, γ) shows that a set of plots of the three surfaces – a saddle node, neutrality of the equilibrium and a loop of a separatrix of a saddle are the typical parametric boundary of system (6.16). The general structure of the boundaries on the parametric portrait is determined by its local structure. The presence of three parameters leads to a large variety of possible dynamic behavior of model (6.16).

6.1.8 Nonlinearity of reproduction of predator and competition of prey

The combination of these two factors leads to the model [3]:

$$\frac{\partial x}{\partial t} = ax\frac{K - x}{K} - bxy,$$

$$\frac{\partial y}{\partial t} = -cy + dx\frac{y^2}{N + y}. \tag{6.17}$$

Here a, K, b, c, d, and N are the parameters. The replacement of variables: $(t = \tau/a, x = au/d, y = av/b)$ leads the system to the following form

$$\frac{\partial u}{\partial \tau} = u(1 - v) - \varepsilon u^2,$$

$$\frac{\partial v}{\partial \tau} = -\gamma v + \frac{uv^2}{n + v}, \tag{6.18}$$

where $\gamma = c/a$, $\varepsilon = a/dK$, $n = bK/a$. The system (6.18) also contains three parameters. The origin of the phase portrait of system (6.18) for all parameter values is the equilibrium of saddle type. In addition, there is always an equilibrium ($u = 1/\varepsilon$, $v = 0$) on the x-axis (the unstable node). Classification of nonlocal bifurcations is a time-consuming problem [3]. In the space of three parameters (n, ε, γ) the dangerous borders of an ecosystem can be built by a combination of qualitative and numerical methods of studying the dynamic systems on the plane.

Above, we gave several examples of two-factor models of the prey–predator type, in which various options taking into account two types of interaction are considered. An even greater variety of models occurs when accounting for three factors. Two of them are given below.

6.1.9 Saturation of the predator, nonlinearity of eating of prey by the predator, and competition of the prey

Additional accounting for the competition of prey in model (6.13) leads to the system

$$\frac{\partial x}{\partial t} = ax - \frac{bx^2 y}{1 + Ax^2 y} - \varepsilon x^2 \,,$$

$$\frac{\partial y}{\partial t} = -cy + \frac{dx^2 y}{1 + Ax^2 y} \,. \tag{6.19}$$

We replace the variables: $(t = \tau/a, x = bu/d, y = (ad/b^2)v)$ and rewrite system (6.19) in the form

$$\frac{\partial u}{\partial \tau} = u - \frac{u^2 v}{1 + \alpha u^2} - \varepsilon u^2 \,,$$

$$\frac{\partial v}{\partial \tau} = -\gamma v + \frac{u^2 v}{1 + \alpha u^2} \,, \tag{6.20}$$

where $\gamma = c/a$, $\alpha = Ab^2/d^2$. Phase portraits of model (6.20) qualitatively coincide with the portraits of system (6.13). Varying the parameters only leads to their qualitative change.

6.1.10 Saturation of the predator, competition of the predator for the prey, and competition of prey

The set of these three factors leads to the model [3]:

$$\frac{\partial x}{\partial t} = ax - \frac{bxy}{(1 + Ax)(1 + By)} - \varepsilon x^2 \,,$$

$$\frac{\partial y}{\partial t} = -cy + \frac{dxy}{(1 + Ax)(1 + By)} \,, \tag{6.21}$$

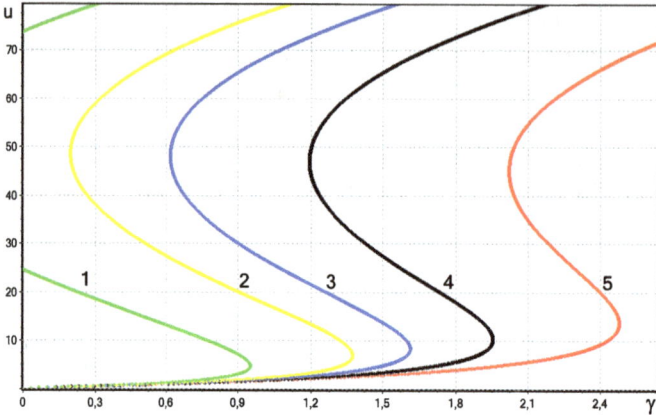

Fig. 6.1: Parametric dependencies $\gamma(u)$ at $\beta = 0.1, \varepsilon = 0.01, 1)\, \alpha = 0.5; 2)\, 0.35; 3)\, 0.3; 4)\, 0.25;$
5) 0.2

which by the replacement of variables: $(t = \tau/a, x = (a/d)u, y = (a/b)v)$ leads to the
form

$$\frac{\partial u}{\partial \tau} = u - \frac{uv}{(1 + \alpha u)(1 + \beta v)} - \varepsilon u^2 ,$$

$$\frac{\partial v}{\partial \tau} = -\gamma v + \frac{uv}{(1 + \alpha u)(1 + \beta v)} . \tag{6.22}$$

There are four parameters in model (6.22) – α, β, γ and ε. From the stationarity conditions for (6.22) we obtain

$$v = \frac{u}{\beta\gamma(1 + \alpha u)} - \frac{1}{\beta} ,$$

and

$$\varepsilon\alpha\beta u^3 + \beta u^2(\varepsilon - \alpha) + (1 - \alpha\gamma - \beta)u - \gamma = 0 . \tag{6.23}$$

Parametric dependencies are expressed from the stationary equation (6.23)

$$\alpha(u) = \frac{\gamma - u(1 - \beta) - \varepsilon\beta u^2}{\beta u^2(\varepsilon u - 1) - \gamma u}$$

$$\beta(u) = \frac{\gamma - u(1 + \alpha\gamma)}{u(1 + \alpha u)(\varepsilon u - 1)}$$

$$\gamma(u) = \frac{\varepsilon\alpha\beta u^2 + \beta u^2(\varepsilon - \alpha) + (1 - \beta)u}{1 + \alpha u}$$

$$\varepsilon(u) = \frac{\gamma + \alpha\beta u^2 - u(1 - \alpha\gamma - \beta)}{\beta u^2(1 + \alpha u)} . \tag{6.24}$$

An example of parametric dependencies (6.24) is shown in Fig. 6.1 with a change of
the second parameter α.

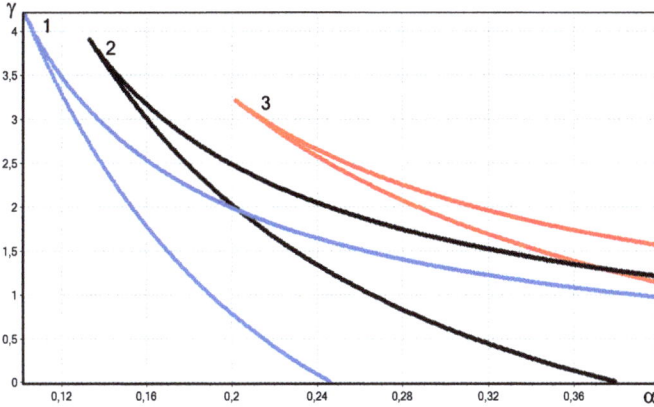

Fig. 6.2: Curves of multiplicity in (α, γ) plain at $\varepsilon = 0.01$, 1) $\beta = 0.15$; 2) 0.1; 3) 0.05

The elements of the Jacobian matrix for (6.22) take the form:

$$a_{11} = 1 - 2\varepsilon u - \frac{v}{(1 + \alpha u)^2 (1 + \beta v)}, \quad a_{12} = \frac{u}{(1 + \alpha u)(1 + \beta v)^2},$$

$$a_{21} = \frac{v}{(1 + \alpha u)^2 (1 + \beta v)}, \quad a_{22} = -\gamma + \frac{u}{(1 + \alpha u)(1 + \beta v)^2}.$$

Bifurcation curves

It is difficult to obtain expressions for plotting of curves of multiplicity for system (6.22) in any of the planes of parameters in an explicit form. So, we used the procedure of graphical plotting L_Δ, proposed above. An example of curves of multiplicity is given in Fig. 6.2 with varying of the β parameter.

The L_σ expression is obtained from the elements of Jacobian matrix,

$$L_\sigma = a_{11} + a_{22} = 1 - \gamma - 2\varepsilon u - \frac{u - \gamma(1 + \alpha u)}{\beta u(1 + \alpha u)^2} + \frac{\gamma^2(1 + \alpha u)}{u}.$$

To plot the curves of neutrality ($L_\sigma = 0$) in any parameter plane, the proposed graphical procedure can be used.

Parametric portraits

The mutual arrangement of curves of multiplicity (L_Δ) and neutrality (L_σ) specifies the parametric portrait of a system shown in Fig. 6.3 in the (β, α) plane. There are six regions on the parametric portrait, which differ by the number and the type of stability of the steady states. A unique unstable steady state exists in region 1, which provides auto-oscillations in the system. Regions 5 and 6 correspond to one stable steady state of a system. There are three steady states in regions 2, 3, and 4, where two can be stable and one steady state is unstable.

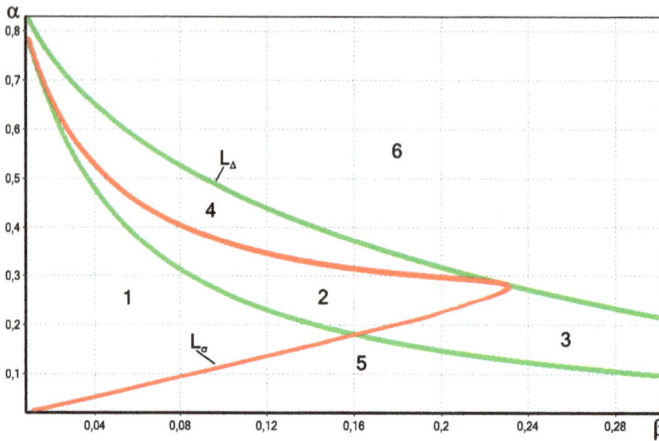

Fig. 6.3: Parametric portrait in (β, α) plain at $\gamma = 1$ and $\varepsilon = 0.01$

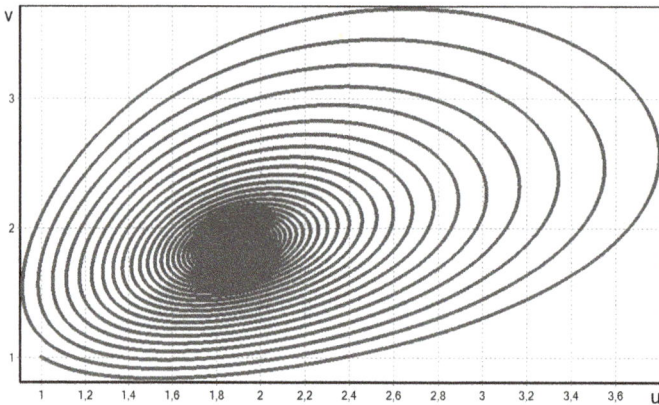

Fig. 6.4: Phase portrait in the (u, v) plane for region 3 at $\alpha = \beta = 0.2, \gamma = 1$, and $\varepsilon = 0.01$

Phase portraits

In accordance with the parametric portrait of our system the phase portraits for regions 2 and 3 are presented in Figs. 6.4–6.5. It is convenient to investigate the dynamic behavior of a system on the phase portraits. Approaching the stable steady state has the character of slowly damped oscillations (see Figs. 6.4 and 6.6). If the initial conditions are set in the neighborhood of the unstable steady state, the system goes to infinity after a few oscillations (see Figs. 6.5 and 6.7)

Time dependencies

Examples of the numerical integration of the original dynamical system (6.22) at different initial conditions and parameters are given in Figs. 6.6–6.8. Figure 6.8 shows

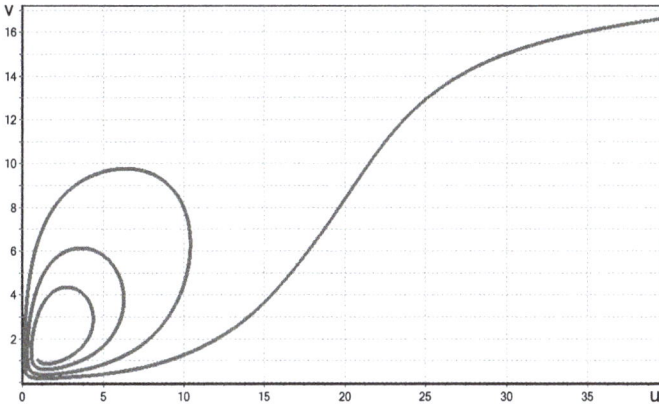

Fig. 6.5: Phase portrait in the (u, v) plane for region 2 at $\alpha = \beta = 0.2, \gamma = 1$, and $\varepsilon = 0.01$

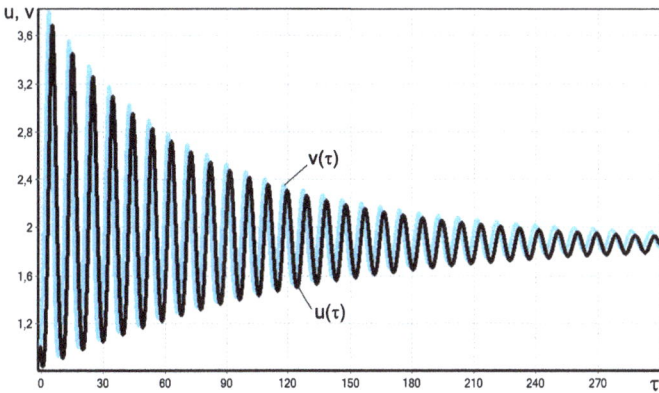

Fig. 6.6: Time dependencies $u(\tau)$ and $v\tau)$ for region 3 at $\alpha = \beta = 0.2, \gamma = 1$, and $\varepsilon = 0.01$

auto-oscillations, which exist in region 1. In region 3, we have damped oscillations (see Fig. 6.6).

Different variants of the system dynamics occur at variation of the parameter γ, which is defined for the simplified model (6.16). Depending on the values of the parameters in system (6.22), there are one or three nontrivial equilibriums or an equilibrium is absent $(\alpha + \varepsilon > 1/\gamma)$.

Along with (6.21), we can consider a few models, corresponding to different combinations of the prey–predator interaction. For example, competition of preys and competition of predators for resources other than a prey; the low critical density of the prey and the competition of the prey; the competition and symbiosis, etc. [3]. Sequential qualitative and numerical analysis of simplified models allows us to hope that the task of construction and analysis dynamics of a real model of the "resource-consumer" type is solvable. The study of a complex system in parts, analysis, and

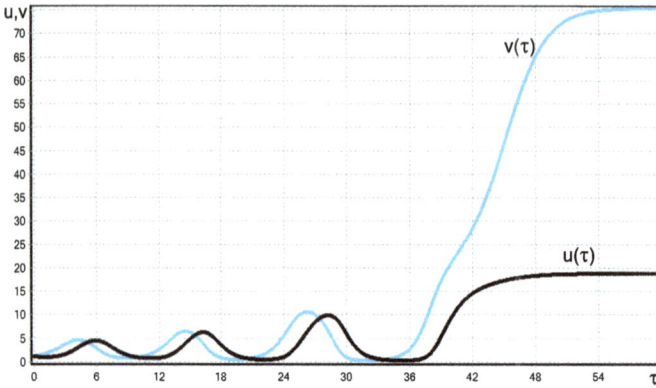

Fig. 6.7: Time dependencies $u(\tau)$ and $v\tau$) for region 2 at $\alpha = 0.25$, $\beta = 0.15$, $\gamma = 1$, and $\varepsilon = 0.01$

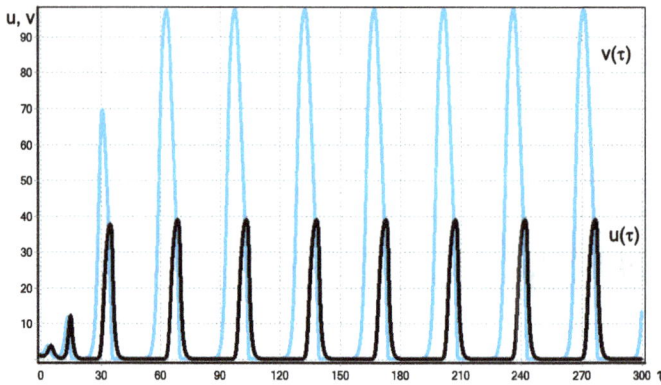

Fig. 6.8: Time dependencies $u(\tau)$ and $v\tau$) for region 1 at $\alpha = 0.3$, $\beta = 0.05$, $\gamma = 1$, and $\varepsilon = 0.01$

synthesis of its components may lead to an understanding of it, the allocation of a relatively small number of key parameters, and ultimately to rational control over it, or prediction of its development for quite a long period of time.

6.1.11 Three populations

In the dynamics of systems, a transition from two-dimensional to three-dimensional phase space is essential. Diversity of movement in a three-dimensional space, in which we live, is much more than on a plane. On the plane, two lines typically cross, but in three-dimensional space, they usually do not cross. Therefore, freedom of movement in space is significantly greater than on the plane. The requirement of the noncrossing of phase trajectories is much weaker than on the plane. Nevertheless, the experience of parametric analysis of system dynamics allows us to hope that an acceptable classification of possible phase portraits on varying the model parameters is possible [2–4, 10].

6.1.12 One-predator–two-prey and one-prey–two-predator systems

We give two models describing the three populations. The system of differential equations corresponding to the interaction of one-predator–two-prey type has the form:

$$\frac{\partial x_1}{\partial t} = a_1 x_1 - b_1 x_1 y \,,$$

$$\frac{\partial x_2}{\partial t} = a_2 x_2 - b_2 x_2 y \,,$$

$$\frac{\partial y}{\partial t} = -cy + d_1 x_1 y + d_2 x_2 y \,. \tag{6.25}$$

After the replacement of variables: $t = \tau/a_1$, $x_1 = a_1 u_1/d_1$, $x_2 = a_2 u_2/d_2$, $y = a_1 v/b_1$, system (6.25) can be rewritten as

$$\frac{\partial u_1}{\partial \tau} = u_1(1 - v) \,,$$

$$\frac{\partial u_2}{\partial \tau} = \gamma_1 u_2(n - v) \,,$$

$$\frac{\partial v}{\partial \tau} = -\gamma_2 v(1 - u_1 - u_2) \,, \tag{6.26}$$

where $\gamma_1 = b_2/b_1$, $\gamma_2 = c/a_1$, and $n = a_2 b_1/a_1 b_2$.

When $u \neq 1$, the system (6.26) has trivial (not lying on the coordinate planes) equilibriums. The equilibrium points of the system, except the origin, are the following:

$$A_1(u_1 = 1, \ u_2 = 0, \ v = 1), \quad A_2(u_1 = 0, \ u_2 = n, \ v = 1) \,.$$

On the coordinate planes, the points A_1, $A2$ are centers, and phase trajectories form a one-parameter family of closed trajectories [3]. Without loss of generality we can assume $n > 1$. Then, from the point A_1, the trajectory goes to inside the first phase octant. For $n < 1$ the situation is reversed. When $n = 1$, there is a whole straight line of nonisolated singular points: $v = 1$, $u_1 + u_2 = 1$ on the phase portrait.

The one-prey–two-predators system can be described by three differential equations:

$$\frac{\partial x}{\partial t} = ax - b_1 x y_1 - b_2 x y_2 \,,$$

$$\frac{\partial y_1}{\partial t} = -c_1 y_1 + d_1 x y_1 \,,$$

$$\frac{\partial y_2}{\partial t} = -c_2 y_2 + d_2 x y_2 \,. \tag{6.27}$$

The model (6.27) behaves similarly to (6.25). One predator population displaces the other. In this case, the predator, providing a minimum stationary density of the prey population, survives. Also, mixing of trajectories occurs [3].

6.1.13 Community: two-prey–one-predator

Interspecific competition among prey leads to effects that we consider here. The population dynamics of these three populations is described by the following system of differential equations [3]:

$$\frac{\partial x_1}{\partial t} = a_1 x_1 - b_1 x_1 y - e_{11} x_1^2 - e_{12} x_1 x_2 ,$$

$$\frac{\partial x_2}{\partial t} = a_2 x_2 - b_2 x_2 y - e_{21} x_1 x_2 - e_{22} x_2^2 ,$$

$$\frac{\partial y}{\partial t} = -cy + d_1 x_1 y + d_2 x_2 y . \tag{6.28}$$

By the replacement of variables: $t = \tau/c$, $x_1 = cu_3/e_{11}$, $x_2 = cu_2/e_{22}$, $y = v$, system (6.28) result in the form

$$\frac{\partial u_1}{\partial \tau} = u_1(\alpha_1 - \beta_1 v - u_1 - \varepsilon_1 u_2) ,$$

$$\frac{\partial u_2}{\partial \tau} = u_2(\alpha_2 - \beta_2 v - u_2 - \varepsilon_2 u_1) ,$$

$$\frac{\partial v}{\partial \tau} = v(1 - \delta_1 u_1 - \delta_2 u_2) , \tag{6.29}$$

where $\alpha_{1,2} = a_{1,2}/c$, $\beta_{1,2} = b_{1,2}/c$, $\varepsilon_1 = e_{12}c/e_{22}$, $\varepsilon_2 = e_{21}c/e_{11}$, $\delta_1 = d_1/e_{11}$, and $\delta_2 = d_2/e_{22}$.

It is possible to show [3] that for some values of parameters, the presence of a predator in the community can provide the coexistence of competing populations of prey, which is impossible in the absence of a predator. In addition, the possibility of the existence and uniqueness of a stable limit cycle (self-oscillations) is established in this system in some range of parameter values. Moreover, at fixed values of the parameters in the numerical study, (6.29), the quasistochastic mode of behavior of solutions (6.29) was discovered. As in other similar mathematical models with three-dimensional phase space, chaotization of solutions is observed in rather narrow ranges of parameter values. However, the existence of these modes indicates a significant three-dimensional structure of the phase trajectories of a system. Dynamical chaos is characteristic for systems of dimension three and higher.

Briefly we describe the results obtained from an environmental point of view. We restrict ourselves to the case $\varepsilon_1 \varepsilon_2 > 1$, when in the absence of a predator coexistence of two competing populations of the prey is impossible. On the introduction of a predator into community, the following regimes can be realized [3]:

1. globally stable regimes (one population of prey without a predator; one population of prey with a predator; stationary coexistence of all three populations);
2. trigger regimes (in the absence of a predator either one or the other population of prey; with a predator there coexists either one or the other prey population; stable coexistence of all three populations);

3. self-oscillatory regimes (coexistence of all three species is only possible in the self-oscillatory regime).

On the basis of the numerical and qualitative results obtained during the study of models, we use two deliberate schematic examples to explain. At first, we consider the system of "grass–deers–wolves". The extermination of wolves allows the deers to reproduce so much that they eat up all the grass, undermining the basis of their own existence, and then die off. In this case, the grass is replaced by, for example, bushes, and a completely different ecosystem occurs.

Another schematic example of a community "forest–pest (insects)–predators (birds)". The sequence of events may be as follows. A significant single artificial reduction the number of the insects leads to a reduction of the density of predator populations (insectivorous birds). The decrease in predators leads to increased reproduction of insects, eluding from control by the predator. This increase of insects causes increased consumption of their resources (needles or leaves) to below a critical level, after which the forest dies and is replaced by a qualitatively different ecosystem for a long time.

The examples of ecosystems considered and their models show the reversibility of the process of cognition of nature by means of mathematical modeling. On the one hand, a model should adequately reflect the described reality. On the other hand, reality must correspond to the features of the behavior of its mirror – an adequate model [30].

6.2 A mathematical model of immunology

Mathematical models of immunology can be attributed to typical models such as the "prey–predator" one, given above. In this section, we consider only one model, which describes the dynamics of interaction of "tumor–organism". In this part, the references [29–35] are intended as examples, and in no case do we claim them to be a complete review on the subject.

For a description of the interaction between specific immune forces and malignant formation, the simplest model describing the dynamics of the growth of active tumor cells x and lymphocyte killers y, can be used [23]:

$$\frac{\partial x}{\partial t} = \mu_x x - \gamma_x xy ,$$

$$\frac{\partial y}{\partial t} = \mu_y (x - \beta x^2) y - \chi_y y + \nu_y . \tag{6.30}$$

It assumes that tumor cells reproduce at a constant specific rate. Constraint of the growth rate of the tumor is due to destruction of malignant cells by the lymphocyte killers. The dynamics of lymphocytes is determined primarily by the rate of their reproduction. At small x, the tumor stimulates the proliferation of lymphocytes, and for

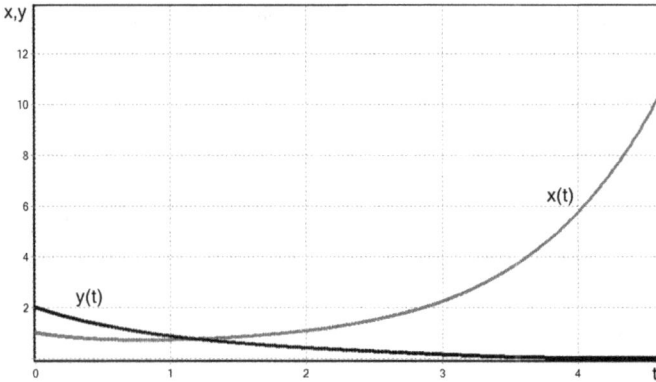

Fig. 6.9: Time dependencies $x(t)$ and $yt)$ at $\chi = 2$, $v = 0.2$, and $\mu = 0.5$

large x, it suppresses it. Also, the natural death of lymphocytes and their continuous influx from stem cells are taken into account. The death of lymphocytes at interaction with tumor cells is clearly not taken into account.

After the introduction of dimensionless variables and parameters ((6.30)), the model can be rewritten in the following form [22]:

$$\frac{\partial x}{\partial t} = (1 - y)x ,$$

$$\frac{\partial y}{\partial t} = \mu((x - x^2 - \chi)y + v) , \tag{6.31}$$

where μ, χ, and v are dimensionless parameters; x and y are dimensionless variables.

The system (6.31) has three steady states

$$x_1 = 0, \quad y_1 = v/\chi , \tag{6.32}$$

$$x_{2,3} = 0.5 \pm \sqrt{0.25 - (\chi - v)}, \quad y_{2,3} = 1 . \tag{6.33}$$

Depending on the ratio of the parameters χ and v, the steady states (6.33) can be complex, or two real numbers of opposite sign, or both are positive. When $x_1 = 0$, the absolute failure of the system of immune protection takes place: all phase trajectories of system (6.31) tend to a point (∞, 0), i.e., with any initial conditions, the tumor grows indefinitely, and the population of lymphocytes is depleted (see Fig. 6.9). This case corresponds to a very large ratio of coefficients of the natural death of lymphocytes $\chi > 1$, so that the influx and reproduction cannot compensate for their death.

At sufficiently small values of χ the phase portrait is as shown in Fig. 6.10 Here, the final result also depends on the initial data, but the positive outcome is reduced to the complete disappearance of the tumor. The three steady states have the following coordinates: (0; 2), (1.092; 1), and (∞; 0).

In natural conditions (without prior stimulation of the immune system) it is not necessary to expect a significant influx of immune lymphocytes. The of stable steady

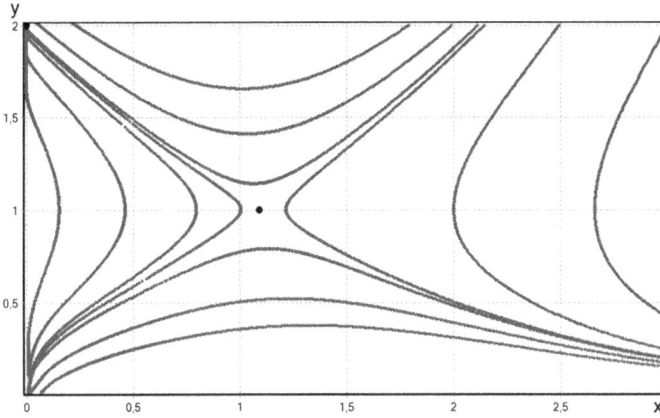

Fig. 6.10: Phase portrait in (x, y) at $\chi = 0.1$, $v = 0.2$, and $\mu = 0.5$

state position can be close enough to the y-axis, i.e., the tumor is resorbed to such an extent that it can no longer be detected. The existence of this equilibrium can be identified by the significant tension in the immune system.

Thus, model (6.30) allows us to describe a variety of outcomes of the fight of the immune forces with the tumor. Like any model, it has its advantages and disadvantages. For example, it does not contain the concrete assumptions of the reasons of oppression of immunity in an organism that has a tumor. However, accounting for additional factors and research of their influence on system dynamics is another story.

6.3 One model of economic dynamics

Economic systems, no doubt, are an example of complex systems. Nonlinearity, dynamics, the hierarchical structure, and the presence of a large number of control parameters exist in these models [12, 13, 26]. As mentioned above, here we cannot give a complete review of the literature on modeling economic dynamics. Our goal in this section is only a demonstration of one mathematical model with an economic meaning. Like any model, it contains the parameters. By varying them it is possible to study the transformation of its phase and parametric portraits. If it is generally admitted that a model is adequate, i.e., it correctly reflects certain selected characteristics of a real system, the results of the parametric analysis of the model may be useful in predicting the development of the real economic system. For the economy, the ability to predict the development of events, perhaps, is the most valuable quality of the corresponding model.

In this section, we consider a model that describes the economic dynamics and the effectiveness of investments in a company, industry, or system. We start with the

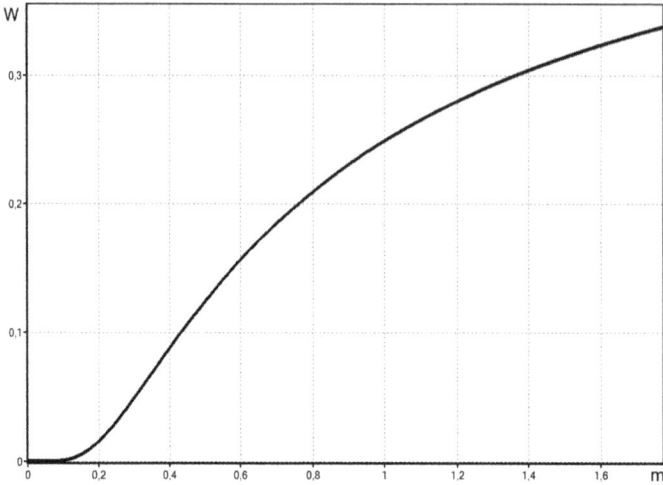

Fig. 6.11: View of function $W(m)$ at $w_0 = 0.5$ and $\alpha = 0.7$

classical scheme:

$$P \xrightarrow{m} P' ,$$ (6.34)

where P is goods, m is money, and P' is manufactured goods. Setting the intensity dependence of the manufacturing of goods P' from raw materials P in the following form:

$$W(m) = W_0 \exp(-\alpha/m) ,$$ (6.35)

where W_0 and α are the parameters, and m is the amount of financing for this enterprise. The function view $W(m)$ is shown in Fig. 6.11; it has the nonlinear character of a curve with saturation.

The derivative of the function $W(m)$ is easily determined as

$$\frac{dW}{dm} = \frac{\alpha}{m^2} W > 0 .$$ (6.36)

Write down the second derivative from (6.36)

$$\frac{d^2 W}{dm^2} = -\frac{2\alpha}{m^3} W + \frac{\alpha^2}{m^4} W ,$$

The maximum derivative of the function $W(m)$ is reached at the point

$$m^* = \frac{\alpha}{2} .$$

The value of m^* is an inflection point of the $W(m)$ function. In dependence on (6.35), the value of m^* separates the areas of convexity and concavity. So, at oscillation of values of m at $m < m^*$ the average value of $\bar{W}$ will be more than W, and at $m > m^*$ it is the opposite.

Let us enter a response function of the intensity of production from the volume of investments $W(m)$ and write down the main balance ratio in the form

$$\frac{dm}{dt} = \beta(m_0 - m) + \gamma W(m) , \qquad (6.37)$$

where β, γ, and m_0 are the parameters. The model (6.37) describes dynamics of the process of financial provision of production $m(t)$ depending on volumes of its financing W. Summand $\beta(m_0 - m)$ reflects the initial and current costs of finance on the production organization. Equilibrium in (6.37) is determined from the equality

$$\gamma W(m) = \beta(m - m_0) . \qquad (6.38)$$

It is obvious that equation (6.38) can have either only one or three solutions. In the case of a multiplicity of steady states their extremes are stable, and the average state is unstable. The only steady state is stable.

At a variation of parameters the dependencies of steady states can be characterized by hysteresis. If in some area of the change of parameters, a multiplicity of steady states is observed, then with the change of some parameters the system can sharply pass from one steady state to another. Such sharp reorganizations of phase portraits are characteristic for nonlinear dynamic systems. The hysteresis phenomenon most strongly testifies to the nonlinearity of a system.

The dynamic model (6.37) has the dimension of the phase space of 1, i.e., it cannot describe the oscillations. At the same time, real economic systems are often characterized by oscillatory modes. For their description, the simplest model (6.37) should be added with at least one more equation.

The presence of hysteresis, which is determined from the conditions (6.38), and sufficient degrees of freedom in the phase space create the conditions for possible existence of auto-oscillatory regimes of economic dynamics.

By analogy with (6.37) we will write down a mathematical model of the economic system that shows auto-oscillations. Suppose that in system (6.34) (system goods–money–goods) both p and m (goods, money) are variables. In this case, by analogy with (6.37) a basic two-component model of economic dynamics is written down as:

$$\frac{\partial p}{\partial t} = v(p_0 - p) + \beta W(m, \, p) ,$$
$$\frac{\partial m}{\partial t} = \mu(m_0 - m) + \gamma W(m, \, p) , \qquad (6.39)$$

where v, p_0, β, μ, m_0, and γ are the parameters. The function W has the form

$$W(m, \, p) = W_0 p \exp(-\alpha/m) . \qquad (6.40)$$

The mathematical model (6.39)–(6.40) is nonlinear and has a phase space of dimension two. The right parts of system (6.39) are equated to zero and transformed. So we

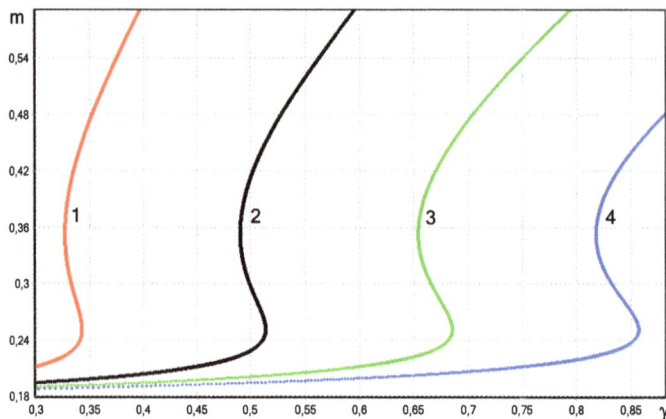

Fig. 6.12: Parametric dependencies $\gamma(m)$ at 1) $\mu = 0.2$; 2) 0.3; 3) 0.4; 4) 0.5

obtain the equation of steady states depending on one variable m:

$$p = \frac{v p_0}{v - \beta w_0 \exp(-\alpha/m)} \; ;$$

$$\mu(m_0 - m) + \frac{v \gamma p_0 w_0 \exp(-\alpha/m)}{v - \beta w_0 \exp(-\alpha/m)} = 0 \qquad (6.41)$$

We write down parametric dependencies from (6.41):

$$\mu(m) = \frac{v \gamma p_0 w_0 \exp(-\alpha/m)}{(\beta w_0 \exp(-\alpha/m) - v)(m_0 - m)} \; ,$$

$$\gamma(m) = \frac{\mu(m_0 - m)(\beta w_0 \exp(-\alpha/m) - v)}{v p_0 w_0 \exp(-\alpha/m)} \; ,$$

$$p_0(m) = \frac{\mu(m_0 - m)(\beta w_0 \exp(-\alpha/m) - v)}{v \gamma w_0 \exp(-\alpha/m)} \; . \qquad (6.42)$$

The example of parametric dependencies (6.42) is given in Fig. 6.12 for the following set of parameters: $\alpha = w_0 = 1$, $\beta = -1$, $p_0 = 2.5$, $m_0 = 0.18$, and $v = 0.15$. Parameter μ is varied.

The elements of the Jacobian matrix for the right sides of (6.39) have the form

$$a_{11} = -v + \beta w_0 \exp(-\alpha/m) \; ,$$

$$a_{12} = \frac{\alpha \beta v p_0 w_0 \exp(-\alpha/m)}{v - \beta w_0 \exp(-\alpha/m)} \; ,$$

$$a_{21} = \gamma w_0 \exp(-\alpha/m) \; ,$$

$$a_{22} = -\mu + \frac{\alpha \gamma v p_0 w_0 \exp(-\alpha/m)}{m^2(v - \beta w_0 \exp(-\alpha/m))} \; .$$

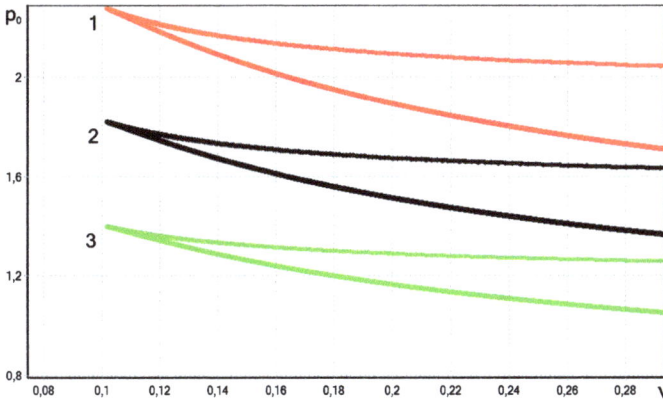

Fig. 6.13: Curves of multiplicity $L_\Delta(v, p_0)$ at 1) $\gamma = 0.8$; 2) 1; 3) 1.3

The curves of multiplicity of steady states are obtained from the stationarity equation (6.41) and the condition $\Delta = 0$. For example,

$$L_\Delta(v, p_0): \quad v(m) = \frac{\beta m^2 w_0 \exp(-\alpha/m)}{m^2 + \alpha(m_0 - m)},$$

$$p_0 = p_0(m, v(m)).$$

$$(6.43)$$

From (6.43) and the condition $\sigma = 0$, we have the curves of neutrality, for example, in the (v, p_0) plane,

$$L_\sigma v, p_0): \quad v(m) = \beta w_0 \exp(-\alpha/m) - \mu - \alpha\mu \frac{m_0 - m}{m^2},$$

$$p_0 = p_0(m, v(m)).$$

$$(6.44)$$

where the function $p_0(m)$ is taken from (6.42). The curves of multiplicity and neutrality according to the formulas (6.43) and (6.44) are plotted in Figs. 6.13 and 6.14.

The parametric portrait of the system is represented in Fig. 6.15. There is the only unstable steady state in region 2, which corresponds to the existence of auto-oscillations in system (6.39). Inside the curve L_Δ (regions 3 and 4) we have three steady states, which can be stable or unstable. Region 1 corresponds to one stable steady state.

Time dependencies of system (6.39) for the region 2 are presented in Fig. 6.16.

By analogy with the models of combustion theory considered above, this simple model of economic dynamics has auto-oscillations. For example, on the condition of the only and unstable steady state in the phase plane (p, m) a stable limit cycle exists. From any initial data the phase trajectories in the plane (p, m) tend to the limit cycle. In the dynamic mode the trajectories $p(t)$ and $m(t)$ make undamped oscillations (Fig. 6.16). Such behavior is observed in a certain area of parameters of model (6.39) (region 2). The corresponding bifurcation analysis of this model was carried out on the

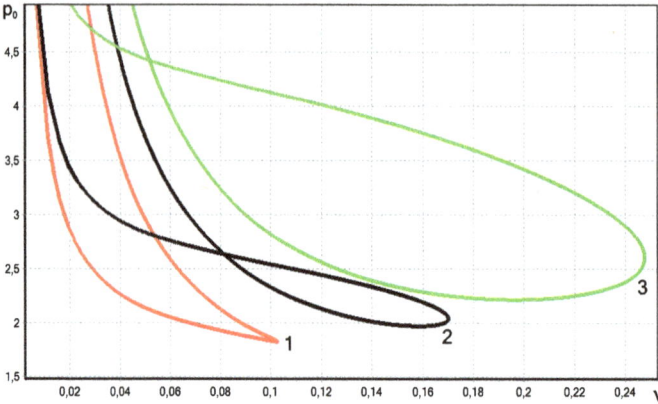

Fig. 6.14: Curves of neutrality $L_\sigma(v, p_0)$ at 1) $\alpha = 1$; 2) 1.1; 3) 1.22

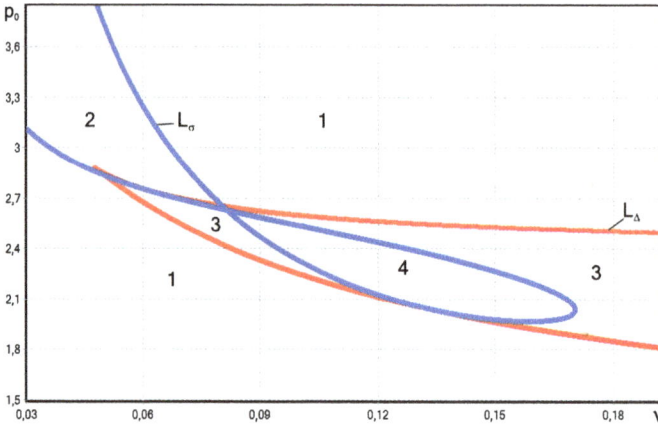

Fig. 6.15: Parametric portrait in (v, p_0) at $\alpha = 1.1$

basis of the scheme, which was repeatedly demonstrated on a series of basic models of chemical kinetics and combustion theory.

The method of analogy of various types of models shown in this book is productive, from our point of view, and in the field of economic dynamics. It is necessary, on the one hand, to clearly see that the basic models are simpler than real systems. On the other hand, if these models are adequate, then they reflect real features of interacting subsystems. The value of the basic models has a gnoseological character. Their construction and use (at regular correlation with real systems) allow us to develop an understanding of a concrete situation. Typically, a real system is characterized by a huge number of various factors that determine its behavior.

However, the number of parameters (or rather the ratio of some complexes of parameters) is not so large. Conditions for uniqueness and stability of steady states, the

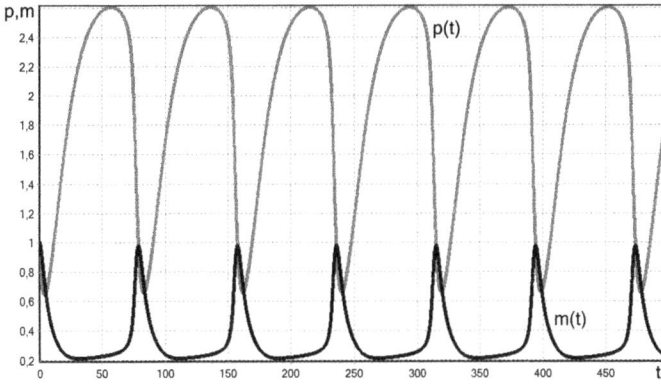

Fig. 6.16: Time dependencies $p(t)$ and $m(t)$ at $v = 0.06$, $p_0 = 3$ and $\alpha = 1.1$

existence, or the absence of oscillations is defined by one or two conditions. The ability to identify several determining parameters in the specific process also corresponds to understanding the possible linear and nonstationary properties of the process studied. The system built from basic models is an important stage of such an understanding.

6.4 Environmental management model

In the simplest option, the mathematical models of environmental management can be carried to models of prey–predator type, some of which are considered above. However, experience of the last years, and furthermore, plans for the near future, show that problems of environmental management are coming to the fore. They will become critical if they are not resolved. Humanity will face a problem of existence, if a harmonious balance with nature is not found.

Consider a model of resource–consumer type. Wood, fish, oil, gas, fertile soils, and other "gifts" of nature can act as a resource, for example. These resources are renewable. Each resource has its own characteristic time and conditions of restoration. Moreover, the reasonable user of nature must know these properties of a resource and the optimum way to consider them. Examples of the predatory use of natural resources are enough [7–9, 18].

Let us write down the simplest of possible models of environmental management in the form of system of two nonlinear differential equations with parameters:

$$\frac{\partial r}{\partial t} = f_r(r, p) - g_r(r, p) \,,$$

$$\frac{\partial p}{\partial t} = f_p(r, p) - g_p(r, p) \,, \tag{6.45}$$

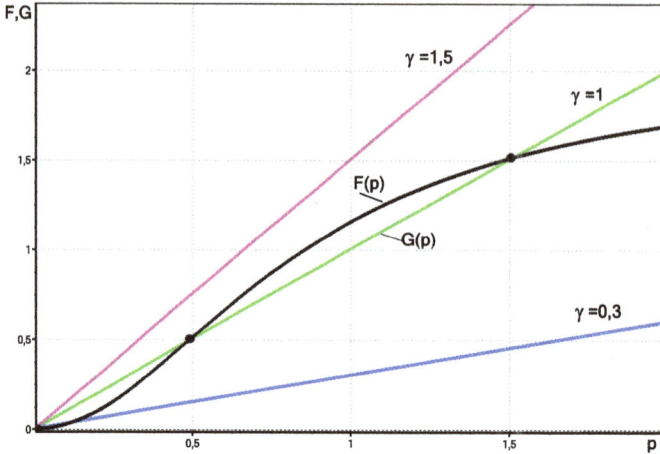

Fig. 6.17: Diagram of steady states for model (6.46) at $v = 0.7$, $\mu = 0.6$, $\alpha = 2$, $\beta = 1$, and 1) $\gamma = 0.3$; 2) 1; 3) 1.5

where r is the variable characterizing the quantity of the resource, p is the variable characterizing the user activity (for example, the number of fish produced for a certain period). The functions f and g characterize, similarly to the prey–predator models, the intensity of recovery and the death of resources. For example, one of the specific models (6.45) can be considered as a system of the form

$$\frac{\partial r}{\partial t} = \frac{\alpha r^2}{\beta + r^2} - \gamma p ,$$

$$\frac{\partial p}{\partial t} = -v p^2 + \mu r p , \tag{6.46}$$

where r is the number of resources, p the number (intensity) of users, and α, β, γ, v, and μ are the parameters. The equilibrium states of system (6.46) are determined from the equality to zero of the right sides of the equations (6.46). In particular, they can be determined from the conditions of the crossing of the curves of consumption $G(p) = \gamma p$ and formation $F(p) = \alpha(v/\mu)^2 p^2 / (\beta + (v/\mu)^2 p^2)$ resources, which are shown in Fig. 6.17. There are one or three such points of intersection, depending on the ratios of the model parameters.

A human, as a reasonable "predator", has to vary the parameters of interaction with nature so that the sheep are intact and the wolves are sated. In each case, knowing (or selecting) the functions F and G and the corresponding parameters it is possible to estimate on a model those critical conditions, at which a significant return from the resource can be obtained and this resource cannot be destroyed. Criteria that will characterize the necessary balance can have an ecological or an economic sense. A detailed parametric analysis of adequate mathematical models of the type (6.46) will help the reasonable user to control the intensity of use of this or that natural resource.

We quite understand that the simplified dynamic models of (6.45) type cannot describe all interactions of humans with nature. For each specific situation of modeling of the relations "resource–consumer", perhaps, at the first stage of research we have to complicate the model significantly, for example, increase the number of dynamic variables and use more complex functions f and g. However, in the process of research of the model and a comparison of its decisions with specific data and experience, the reverse movement is also possible: from a complex to a more simple model. It often happens that the simplicity of a model corresponds to the degree of understanding of a specific ecological situation. In a simplified basic model there are factors that define the behavior of a system in general. As the poet said: As the poet Alexander Blok wrote: "Erase random features and you will see that the world is beautiful".

The pendulum "from simple to difficult and back" can also be detected in many sciences. It especially deserves attention on creating a scientifically-based interaction "human–nature".

Bibliography

[1] Volterra V. Lessons on the mathematical theory of the struggle for survival. (reprint of the original 1931 Gauthier-Villars edition), Editions Jacques Gabay, Sceaux, 1990.

[2] Bazykin AD, Kuznetsov YA, Khibnik AI. Portraits of bifurcations: bifurcation diagrams of planar dynamical systems. Znanie, Moscow, 1989.

[3] Bazykin AD. Mathematical biophysics of interacting populations. Nauka, Moscow, 1985. (in Russian).

[4] Bazykin AD. Nonlinear dynamics of interacting populations. World Scientific Publishing, Danvers, USA, 1998.

[5] Svirezhev YM. Nonlinear waves, dissipative structures and catastrophes in ecology. Nauka, Moscow, 1978.

[6] Svirezhev YM, Logofet DO. Stability of biological communities. Nauka, Moscow, 1978. [English version: Stability of biological communities, Mir, Moscow, 1983].

[7] Forrester JW. World dynamics. Wright-Allen Press, Cambridge, MA, 1971.

[8] Forrester JW. Industrial dynamics. J. of the Operational Research Society. 1997, 48(10), 1037–1041.

[9] Jorgensen SE, Svirezhev YM. Towards a thermodynamic theory for ecological systems. Elsevier, Amsterdam, 2004.

[10] Dynamics of chemical and biological systems. Eds. Bykov VI, Nauka, Novosibirsk, 1989.

[11] Fomin SV, Berkinblit MB. Mathematical problems in biology. Nauka, Moscow, 1973. (In Russian)

[12] Goodwin RM. Chaotic economic dynamics. Clarendon Press, Oxford, 1990.

[13] Blatt JM. Dynamic economic systems: a post-Keynesian approach. ME Sharpe Inc, NY, 1983.

[14] Gorban AN, Bykov VI, Yablonskii GS. Essays on chemical relaxation. Nauka, Novosibirsk, 1986.

[15] Malinetskii GG. Mathematical foundations of synergetics: Chaos, structures, computational experiment. URSS, Moscow, 2015.

[16] Marry J. Nonlinear differential equations in biology (lectures on models). Clarendon Press, Oxford, 1977. (Russian translation, Mir, Moscow, 1983)

[17] Merrill SJ. A mathematical model of tumor growth and cytotoxic blocking activity. Math. Biosciences. 1979, 47(1–2), 79–89.

[18] Marchuk GI. Mathematical models in environmental problems. Vol. 16. Elsevier, North Holland, 2011.

[19] Murray JD. Lectures on nonlinear differential equation models in biology. Clarendon Press, Oxford, 1977.

[20] Poluektov RA. Dynamical theory of biological populations. Nauka, Moscow, 1974.

[21] Razzhevaikin VN. Selection functionals in autonomous models of biological systems with a continuous age and spatial structure. Computational Mathematics and Mathematical Physics. 2010, 50(2), 322–329.

[22] Romanovskii YM, Stepanova NV, Chernavskii DS. Mathematical biophysics. Nauka, Moscow, 1984.

[23] Romanovskii YM, Stepanova NV, Chernavskii DS. Mathematical modeling in biophysics. Nauka, Moscow, 1975.

[24] Rubin AB. Thermodynamics of biological processes. Moscow St. Univ. Press, Moscow, 1976.

[25] Rubin AB, Pyt'eva NF, Riznichenko GY. Kinetics of biological processes. Moscow St. Univ. Press, Moscow, 1977.

[26] Strogatz SH. Nonlinear dynamics and chaos: with applications to physics, biology, chemistry, and engineering. Westview press, USA, 2014.

[27] Malinetskii GG. The future of applied mathematics. Lectures for young researchers. URSS, Moscow, 2005 (in Russian).

[28] Ivanitskii GR, Krinskii VI, Sel'kov EE. Mathematical biophysics of the cell. Nauka, Moscow, 1978.

[29] Bazykin AD, Berezovskaya FS, Isaev AS, Khlebopros RG. Dynamics of forest insect density: bifurcation approach. J. of Theoretical Biology. 1997, 186(3), 267–278.

[30] Bell GI. Mathematical model of clonal selection and antibody production. J. of Theoretical Biology. 1970, 29(2), 191–232.

[31] Bellman RE. Mathematical methods in medicine. World Scientific Publishing Co., NJ, USA, 1983.

[32] Weidlich W. Sociodynamics: A systematic approach to mathematical modeling in the social sciences. Nonlinear Phenomena in Complex Systems. 2002, 5(4), 479–487.

[33] Khlebopros RG, Okhonin VA, Fet AI. Catastrophes in nature and society. Mathematical modeling of complex systems. World Scientific Publishing Co., NJ, USA, 2007.

[34] Gorban AN, Khlebopros RG. Demon of Darwin: Idea of optimality and natural selection. Nauka, Moscow, 1988.

[35] Marchuk GI, Petrov RV, Romanyukha AA, Bocharov GA. Mathematical model of antiviral immune response. I. Data analysis, generalized picture construction and parameters evaluation for hepatitis B. J. of Theoretical Biology. 1991, 151(1), 1–40.

Conclusion

To sum up the results. In our book, a gallery of models of chemical kinetics and thermokinetics, mathematical biology (prey–predator model), and economics is presented. The unifying idea for us was the conception of the basic models and their detailed parametric analysis. The history of our study is that the main attention was paid to the nonstationary and nonlinear models of chemical kinetics.

We hope that our experience in building and analyzing basic models can be useful not only for specialists in the field of chemistry and chemical technology, but also in the related fields of biology, ecology, and economics. The formal kinetic approach is rather universal. It consists in the choice of leading factors (they are usually a few – one, two, or three) in each of the studied complex systems. Success can be provided if it is possible to select a system of basic models describing some nonlinear or nonstationary behavior of the studied system. Studying critical phenomena of various types is of great importance. They determine the restructuring of a system at varying parameters, which is characterized by the appearance of fundamentally new properties. For example, at changing environmental conditions the deformation and restructuring of the main dependencies, fields, and structures occur. Exactly such bifurcation points separate some properties of a system from its other properties.

Outside of our consideration the problem of solving the inverse tasks remains. In our opinion, this is an important and technically difficult task of building adequate mathematical models. It fully can be solved based on the experience of the study of basic models, the properties of which are known a priori.

A few words about the tasks to the future. The question as to why at first glance a simple dynamic model (i.e., the Lorentz model, in which there are two nonlinearities of multiplication type) shows a complex behavior down to deterministic chaos, still remains unresolved for us. Where is the measure of complexity, which determines the existence of strange attractors? Are there constructive algorithms for building of dynamic models with predetermined properties? These questions for dynamical systems with polynomial right parts are especially actual. Further, is it possible to formulate conditions in the form of inequalities on model parameters, at which a given type of dynamics is implemented, while remaining within polynomial models with uncertain coefficients? Because it is often so that a system with large right parts has a trivial behavior, but a simple model shows a greater variety of dynamics.

Distributed models of the reaction–diffusion type as a rule correspond to the ideal geometry. However, the real surface is far from smooth; it is most often rough. In our opinion, in this case, integral equations are adequate models. Parametric analysis of such nonlinear integral models and formation within them of a system of basic models are problems for the future.

Solution of optimal control problems is the next step that is logically derived from the parametric analysis. Optimization, optimal control, and design of optimal con-

https://doi.org/10.1515/9783110464948-007

trollers – these are fundamentally important problems, whose solution is impossible without a preliminary parametric analysis of the studied mathematical models of real processes.

The pendulum "from simple to difficult and back" can also be detected in many sciences, and, generally, in all interactions between the human beings and nature.

"Simple", "complex" and "simple" again.

In our opinion, the ideal strategy of modeling should include elements of both the *bottom-up* and *top-down* approaches. An ideal approach should be the dynamic one formulated as follows:

1. Start with the simple model, making it more complex "at the state of desperate need" or rather if it is very necessary.
2. Make the model simpler at the earliest convenience.

Both "necessity" and "convenience" are determined by the goal of modeling and the available information as well. To sum up, the art of modeling is very much related to *modelics*, i.e., the ability to construct the simple and understandable models.

In conclusion, we note that the construction of a system of basic models is aimed at solving the main task, i.e., the task of understanding the specifics of studied system, highlighting the key factors that determine model dynamics at varying the parameters in a wide range. Quoting one of the main characters of a famous film, we say: "Happiness is when you are understood."

Index

https://doi.org/10.1515/9783110464948-008

www.ingramcontent.com/pod-product-compliance
Lightning Source LLC
Chambersburg PA
CBHW080713220326
41598CB00033B/5404